CW00956987

Explorations in QUANTUM COMPUTING

Colin P. Williams
Scott H. Clearwater

Explorations in QUANTUM COMPUTING

With 78 Illustrations

Colin P. Williams
Mail Stop 525-3660
Jet Propulsion Laboratory
California Instiute of Technology
Pasadena, CA 91109-8099
USA

Library of Congress Cataloging-in-Publication Data
Williams, Colin P.
Explorations in quantum computing / by Colin P. Williams and Scott H. Clearwater.
p. cm.
ISBN 0-387-94768-X (alk. paper)
1. Electronic digital computers—Design and construction. 2. Quantum Physics. I. Clearwater, Scott H. II. Title.
TK788.3.W46 1997
004—dc21 97-2159

Printed on acid-free paper.

Camera-ready copy prepared using the author's Microsoft Word files.
Printed and bound by Hamilton Printing Co., Rensselaer, NY.
Printed in the United States of America.

9 8 7 6 5 4 3

ISBN 0-387-94768-X Springer-Verlag New York Berlin Heidelberg SPIN 10756653

TELOS, The Electronic Library of Science, is an imprint of Springer-Verlag New York. Its publishing program encompasses the natural and physical sciences, computer science, mathematics, economics, and engineering. All TELOS publications have a computational orientation to them, as TELOS' primary publishing strategy is to wed the traditional print medium with the emerging new electronic media in order to provide the reader with a truly interactive multimedia information environment. To achieve this, every TELOS publication delivered on paper has an associated electronic component. This can take the form of book/diskette combinations, book/CD-ROM packages, books delivered via networks, electronic journals, newsletters, plus a multitude of other exciting possibilities. Since TELOS is not committed to any one technology, any delivery medium can be considered. We also do not foresee the imminent demise of the paper book, or journal, as we know them. Instead we believe paper and electronic media can coexist side-by-side, since both offer valuable means by which to convey information to consumers.

The range of TELOS publications extends from research level reference works to textbook materials for the higher education audience, practical handbooks for working professionals, and broadly accessible science, computer science, and high technology general interest publications. Many TELOS publications are interdisciplinary in nature, and most are targeted for the individual buyer, which dictates that TELOS publications be affordably priced.

Of the numerous definitions of the Greek word "telos," the one most representative of our publishing philosophy is "to turn," or "turning point." We perceive the establishment of the TELOS publishing program to be a significant step forward towards attaining a new plateau of high quality information packaging and dissemination in the interactive learning environment of the future. TELOS welcomes you to join us in the exploration and development of this exciting frontier as a reader and user, an author, editor, consultant, strategic partner, or in whatever other capacity one might imagine.

TELOS, The Electronic Library of Science
Springer-Verlag New York, Inc.

TELOS Diskettes

Unless otherwise designated, computer diskettes packaged with TELOS publications are 3.5" high-density DOS-formatted diskettes. They may be read by any IBM-compatible computer running DOS or Windows. They may also be read by computers running NEXTSTEP, by most UNIX machines, and by Macintosh computers using a file exchange utility.

In those cases where the diskettes require the availability of specific software programs in order to run them, or to take full advantage of their capabilities, then the specific requirements regarding these software packages will be indicated.

TELOS CD-ROM Discs

For buyers of TELOS publications containing CD-ROM discs, or in those cases where the product is a stand-alone CD-ROM, it is always indicated on which specific platform, or platforms, the disc is designed to run. For example, Macintosh only; Windows only; cross-platform, and so forth.

TELOSpub.com (Online)

Interact with TELOS online via the Internet by setting your World-Wide-Web browser to the URL: http://www.telospub.com.

The TELOS Web site features new product information and updates, an online catalog and ordering, samples from our publications, information about TELOS, data-files related to and enhancements of our products, and a broad selection of other unique features. Presented in hypertext format with rich graphics, it's your best way to discover what's new at TELOS.

TELOS also maintains these additional Internet resources:

gopher://gopher.telospub.com
ftp://ftp.telospub.com

For up-to-date information regarding TELOS online services, send the one-line e-mail message:

send info to: *info@TELOSpub.com.*

I dedicate this book to my parents, Leslie and Cynthia Williams, for the exciting, diverse, and inquisitive life they gave me.

Colin P. Williams

Dedicated to people like Santideva who realized that "For one who desires the impossible, mental afflictions and disappointment arise; but for one who is free of expectations, there is unblemished prosperity."

Scott H. Clearwater

Acknowledgments

We would like to thank Chris Adami, Richard Doyle, Paul Fahn, Raymond Laflamme, Seth Lloyd and Peter Shor for their careful reading of this manuscript and for suggesting improvements and corrections. Their comments were truly insightful and were invaluable in helping us to appreciate the subtleties of quantum computing. Any remaining errors are, of course, entirely ours.

We would also like to thank Keisha Sherbecoe for creating our web site and for fielding the numerous copies of the manuscript between authors, reviewers and Springer-Verlag, New York. In addition, we'd like to thank Ralph Abrahams for assembling the components of our multi-platform CD-ROM and Anthony Guardiola for dealing with the final typesetting of the manuscript. We also appreciate Karen Phillips assistance with the cover design.

But most of all we thank our editor, Allan Wylde, for his advice and unflagging support throughout this project. Allan signed this book at a time when quantum computing was relatively unknown. It took a publisher with vision to see the potential for this field.

Last but not least Colin thanks his wife Patricia for her patience and encouragement throughout the writing process that steals so much time from family life. Patricia, guess what? We've finished!

Preface

Over the past 40 years there has been a dramatic miniaturization in computer technology. So far, however, our basic understanding of what a computer can do has not changed. This is partly due to the fact that the tiny components inside a computer still behave, in all important respects, in accordance with the laws of classical physics. As the trend in miniaturization continues, however, there will come a point at which we will be forced to use quantum physics to describe the elementary operations of our computers. Remarkably, the quantum laws are often very different from the classical laws. Consequently, at such microscopic scales, the very theory describing what computers can do must be revised. The field of "quantum computing" has arisen in anticipation of this inevitable step.

Initially, researchers in quantum computing tried to understand how the basic operations of a conventional computer might be accomplished using quantum mechanical interactions. However, they soon realized that quantum physics offered something genuinely new. By exploiting delicate quantum phenomena that have no classical analogues, it is possible to do certain computational tasks much more efficiently than can be done by any classical computer, even a supercomputer. Moreover, these same quantum phenomena allow unprecedented tasks to be performed such as teleporting information, breaking supposedly unbreakable codes, generating true random numbers, and communicating with messages that betray the presence of eavesdropping. Already, a fully quantum scheme for sending and receiving ultra-secure messages has been implemented over a distance of 30 km — far enough to wire the financial district of any major city. Moreover, the United States government is quietly funding research in code-breaking, using quantum computers.

Explorations in Quantum Computing is the first book on quantum computing. It is completely self-contained and, in particular, does not presume any prior knowledge of quantum physics or theoretical computer science. It covers the foundations of computer science, quantum physics, quantum cryptography, quantum teleportation, and cutting-edge quantum technology. The book is based on over 300 peer-reviewed research papers, dozens of Web sites, numerous technical conferences, and countless private conversations. It also describes the key technological hurdles that must be overcome in order to make quantum computers a reality.

In addition, we believe that it is much better to play with computer simulations of quantum devices than it is to merely study equations. So most of the chapters come with electronic supplements that contain simulators, software tools, and animations that bring the ideas of quantum computing to life in fresh and exciting ways. The software is written in the Mathematica programming language which makes it easy to mix numerical, symbolic, and graphical outputs and allows users to perform their own experiments in real-time. The code will run on just about any computer platform including PCs, Macintoshes, NeXT machines, and UNIX boxes.

Quantum computing is currently one of the hottest topics in computer science, physics, and engineering. It has already sent tremors through the theoretical bedrock established over the past 50 years. It is changing the way scientists think about the fundamental operations, capabilities, and ultimate limits of computers and will surely lead to a revolution in the computer industry itself. Colleges are beginning to offer courses in quantum computing and new post-doctoral positions are becoming available. It has ignited a truly interdisciplinary effort with researchers from computer science, physics, and nuclear engineering pooling their talents to study these remarkable devices. Consequently, we hope that *Explorations in Quantum Computing* will appeal to students, researchers, teachers, and professionals from a wide variety of disciplines, as well as those who are merely curious to learn more about this burgeoning field.

Last but not least we hope, sincerely, that *Explorations in Quantum Computing* will encourage young scientists to choose quantum computing for their career paths. There is far too much of a schism today between the training of computer scientists and that of physicists. Only when students have been trained, from the start, to wed physics and computation can the prejudices of a bygone era be eliminated.

Colin P. Williams
Scott H. Clearwater
August 1997

Contents

CD-ROM Contents

Readme file

Legal notice

Mathreaders (for Mathematica Version 2.2)

Mathreaders (for Mathematica Version 3.0)

Mathematica Notebooks for Version 2.2

Anims2	Animations of quantum systems
BraKet.ma	Basic tools for Dirac notation
ErrorCo.ma	Simulation of quantum error correction
Feynman.ma	Simulation of a quantum computer
Interfer.ma	Analysis of interference effects
OTPExamp.ma	One time pad cryptosystem
Qbugs.ma	Ssimulating bugs in quantum computers
Qcdata.txt	A database of quantum computing papers
Qcrypt.ma	Simulation of quantum cryptography
RSAExamp.ma	RSA pubic-key cryptosystem
SearchEn.ma	Search engine for the database

ShorFact.ma	Simulation of Shor's algorithm
Teleport.ma	Simulation of quantum teleportation
TimingFa.ma	Illustration of the difficulty of factoring

Mathematica Notebooks for Version 3.0

Anims2	Animations of quantum systems
BraKet.nb	Basic tools for Dirac notation
ErrorCo.nb	Simulation of quantum error correction
Feynman.nb	Simulation of a quantum computer
Interfer.nb	Analysis of interference effects
OTPExamp.nb	One time pad cryptosystem
Qbugs.nb	Simulating bugs in quantum computers
Qcdata.txt	A database of quantum computing papers
Qcrypt.nb	Simulation of quantum cryptography
RSAExamp.nb	RSA pubic-key cryptosystem
SearchEn.nb	Search engine for the database
ShorFact.nb	Simulation of Shor's algorithm
Teleport.nb	Simulation of quantum teleportation
TimingFa.nb	Illustration of the difficulty of factoring

CHAPTER 1

Computer Technology Meets Quantum Reality

"There's plenty of room at the bottom."
— Richard Feynman

Over the past 40 years there has been a dramatic miniaturization in computer technology. If current trends continue, by the year 2020 the basic memory components of a computer will be the size of individual atoms. At such scales, the current model of computation, based on a mathematical idealization known as the Universal Turing Machine, is simply invalid. A new field, called "quantum computing" is emerging that is re-inventing the foundations of computer science and information theory in a way that is consistent with quantum physics — the most accurate model of reality that is currently known.

Remarkably, this new theory predicts that quantum computers can perform certain computational tasks exponentially faster than any conventional computer. Moreover, quantum effects allow unprecedented tasks to be performed such as teleporting information, breaking supposedly "unbreakable" codes, generating true random numbers, and communicating with messages that betray the presence of eavesdropping. These capabilities are of significant practical importance to banks and government agencies. Indeed, a quantum scheme for sending and receiving ultra-secure messages has already been implemented over a distance of 30 km — far enough to wire the financial district of any major city[Marand95].

Although modern computers already exploit some quantum phenomena they do not make use of the full repertoire of quantum phenomena that Nature provides. Harnessing these phenomena will take computing technology to the very brink of what is possible in this Universe.

In this chapter we demonstrate the necessity of treating computers as physical systems rather than merely mathematical idealizations. We review the trends that have taken place in computer miniaturization and show how they lead inexorably towards quantum limits.

1.1 Computers as Physical Systems

The first computers were physical systems designed for particular computational purposes. Ancient humans built megalithic structures such as Stonehenge in England to predict astronomically significant events such as a shift in the seasons. These devices could perform at best a handful of different computational tasks and reprogramming them to do different tasks would have required a Herculean effort!

A step towards a more flexible computer appeared when the need for trade promoted the use of pebbles or beads to represent numbers. By maneuvering pebbles in the sand, more complicated calculations could be done than were possible in people's heads. Moreover, the steps in the calculation were in clear view and could be scrutinized by both the buyer and the seller. The habit of using tokens to represent numbers was developed to a fine art in the Middle Ages when the abacus gained widespread use. Typically, this instrument consisted of a wooden frame supporting a set of parallel dowels on which movable beads were strung. The position of a bead on a dowel represented a particular number. By moving the beads in accordance with certain rules, simple arithmetic calculations could be performed.

The problem with the abacus was that it did not actually *do* the calculation. It merely provided a means to represent the intermediate steps. By the 17th century some people began to see the possibility of having machines carry out the intermediate steps mechanically, thereby avoiding the possibility of human error. In particular, Blaise Pascal, Gottfried Wilhelm Leibniz, and Wilhelm Schickard each developed their own mechanical calculators[Pratt87]. Leibniz's was notable because it was able to multiply numbers together directly instead of achieving multiplication through the naive technique of repeated addition. However, none of these machines were terribly reliable and the practical needs of the day did not yet require them and so they went largely unnoticed for over 100 years.

By the mid-19th century, mathematical calculations had become a routine necessity in many fields. Navigators and astronomers, for example, demanded precise latitude and longitude calculations, bankers needed interest rate calculations, and merchants required exchange rate calculations. Although these calculations were not difficult, they were extremely laborious. In an effort to streamline the process, many tables of multiplications, divisions, and logarithms were published. People, called "computers," were hired to draw up the required volumes. Unfortunately, despite their best efforts, the human computers occasionally made arithmetic errors in figuring an answer and transcription errors in copying an answer into a manuscript. Even more errors were introduced by printers when they typeset the manuscripts. In fact, Charles Babbage, an English mathematician, became so frustrated with the errors he kept encountering in such tables that he one day exclaimed, "I wish to God these calculations had been executed by steam!"[Palfreman91]. And the idea of a mechanical computer was born.

Babbage began work on his mechanical computer, called Difference Engine No.1, in 1821[Morrison61]. The name came from the use of the "method of differences" to evaluate the formulae Babbage sought to tabulate. The method of differences works for any expression of the form $a_0 + a_1 x + a_2 x^2 + \mathrm{L} + a_n x^n$ and only requires numbers to be added together once a certain constant, "the constant difference," is known. As such expressions arise in the Taylor series expansion of other functions, such as sine and cosine, the Difference Engine could calculate just about any function that was of practical significance in the 19th century.

Babbage's design called for many precisely engineered parts that would have taxed the limits of the manufacturing capabilities of the day. He hired Joseph Clement, a renowned engineer, to build the first Difference Engine. The machine consisted of a series of columns of interlocking gears. The method of differences required the operator to set the "constant difference" on the column farthest to the left and various "orders of difference" on columns farther to the right. The calculation would be effected by turning a crank twice clockwise and twice counterclockwise. This caused the number on one column to be added to the number on the adjacent column. An auxiliary mechanism was to impress the final answer into a sheet of copper plate thus eliminating the need for a compositor and hence the possibility of typesetting errors.

Although Babbage never completed Difference Engine No.1 he did succeed in building part of a working prototype that is now on display at the Science Museum in London. However, since the downfall of the project, speculation raged as to whether the design was ever feasible. If not, the Difference Engine hardly deserved the reputation of being the world's first computer. In 1991, in an effort to lay the matter to rest, officials at the Museum of Science in London commissioned the building of a full-size Difference Engine[Swade93]. This machine, called appropriately Difference Engine No.2, was built out of bronze, cast iron, and steel in strict accordance with Babbage's design and weighed in at an imposing three tons. Despite its mammoth proportions, it performed flawlessly, as Babbage had predicted, on its first significant calculation, the tabulation of the first 100 values of the powers of 7.

Even earlier, in 1853, a crude version of a Difference Engine based on the Babbage design had been built by a Swedish printer George Scheutz and his son. The Scheutz machine was engineered to less exacting tolerances than those demanded by Babbage but worked well enough nonetheless. In fact, one of the Scheutz machines was bought by the Dudley Observatory in Albany, New York and used to produce astronomical tables. Thus the Difference Engine was certainly a viable design for a computer, albeit one that was specialized to do a particular kind of computation. By 1837, however, Babbage was anticipating the next step in computer development: the programmable computer. In fact it might have been this vision, rather than a dispute with Joseph Clement over where the Difference Engine was to be assembled, that caused Babbage to abandon his first project in favor of developing a more sophisticated contraption, which he called the "Analytical Engine".

Whereas a given Difference Engine could perform only one type of calculation, the Analytical Engine was to have been capable of being instructed to perform many different types of calculations. The instructions needed to configure the Analytical Engine for a particular computational task were stored in a pattern of holes punched through a series of linked cards. Babbage had borrowed this idea from the Jacquard loom which had used the same technique to configure the loom to weave different patterns in textiles. Had it ever been built, the Analytical Engine would have been even more of a monster than Difference Engine No.2. Doron Swade, curator of the Science Museum, suggests the Analytical Engine would have been "the size of a small locomotive"[Palfreman91]. It would also have had remarkably advanced capabilities including the ability to solve simultaneous algebraic equations and mimic logical reasoning.

Table 1.1 The first function tabulated by Difference Engine No.2, a full-size replica of Difference Engine No.1, in 1991. This machine is based on a design by Charles Babbage in 1823.

i	7^i
1	7
2	49
3	343
4	2401
5	16807
6	117649
7	823543
8	5764801
9	40353607
10	282475249
11	1977326743
12	13841287201
13	96889010407
14	678223072849
15	4747561509943
16	33232930569601
17	232630513987207
18	1628413597910449
19	11398895185373143
20	79792266297612001
⋮	⋮

Ironically, mechanical designs for computers are back in vogue today, although on a vastly smaller scale. Eric Drexler, research fellow at the Institute for Molecular Manufacturing in Palo Alto, California, has proposed a model computer that uses interconnected "sliding rods" made up of roughly a hundred thousand atoms each, to accomplish the basic logical operations of a computer[Drexler92]. This may sound like a lot of atoms but, in comparison to ordinary-sized objects, is very few indeed. These machines will not be true quantum computers, however, but rather ultra-small versions of conventional computers.

Thus the idea of a computer as a mechanical device has a pedigree that stretches back many hundreds of years. In the early part of this century, largely under pressure from the military for code-breaking and ballistic calculations, computer technologies were developed that were first electromechanical (based on telephone relays) and then electronic (based on vacuum tubes). Each stage of progress succeeded in mapping the basic operations of a computer onto smaller and faster physical systems. Nevertheless, the computer was always a tangible object: an engineered device whose design seemed to continually change with the tasks required of it.

However, one of the greatest advances in computer science came when Alan Turing, Alonso Church, Kurt Gödel, and Emil Post each independently created *mathematical* models of the computing process[Turing37, Church41, Gödel31, Post44][1]. These models, which incidentally all turned out to be equivalent to one another, sought to characterize the essence of the computing process in an effort to circumscribe the limits of actual computing machinery. By eliminating what they assumed were extraneous details, such as how the basic operations of a computer were implemented, these men succeeded in elevating computer science to the status of a branch of pure mathematics. Hence, the field of theoretical computer science was born.

Many remarkable insights followed from their efforts. In particular, Alan Turing showed, using his Turing machine model, that there was no mechanical procedure that could determine the truth or falsity of all mathematical conjectures. As the mathematical model of the computing process was so all-encompassing, it was natural to believe that it must be applicable to any device that was capable of computing, including human beings. As other mathematicians had distilled logical reasoning down to a simple set of rules, the early computer scientists therefore felt justified in believing that computers would one day "think." Indeed, in specialized areas, progress has been quite impressive. In 1997 a computer called Deep Blue was the first machine to defeat Garry Kasparov, chess champion of the World, in a match. Whether this proves Deep Blue can think is highly questionable.

With the benefit of 20-20 hindsight we now know that it was a mistake to have eliminated from the theory all trace of the way the basic operations of a computer were implemented. This is because the theory implicitly made assumptions about how information could be stored and manipulated inside a computer. These assumptions, however, are based on common sense intuitions. Unfortunately, these intuitions cease to be valid on a sufficiently small scale. The new theory of quantum computation allows for the possibility of using the complete gamut of physical phenomena to accomplish the storage and manipulation of information. This turns out to be extremely liberating and opens up a completely new vista for computer technology. But when might these devices be upon us?

---

[1] For an excellent review of these ideas see [Penrose89].

1.2 Technological Issues

The fact that a real computer is a physical system prompts us to think about the space, time, and energy implications of trying to make computers faster. To make computers faster, their components have to be squeezed closer together because the signals that need to be passed around inside the computer cannot travel faster than the speed of light, roughly 186,000 miles per second. In addition, we also need to drive the components at a higher clock speed. In order to pack more components into a given volume the components need to be made smaller. In other words, we have to find smaller physical systems that can achieve the same functionality as those they replace. This is what happened when vacuum tubes were replaced by transistors.

However, size is not the only issue. The components inside conventional computers give off a certain amount of heat as a side effect of their operation. If the components are simply packed closer together without also improving energy efficiency the computer will melt itself as it computes! Already supercomputers have to use a liquid gas refrigerant to cool them down. Consequently, not only do the basic operations need to be mapped onto smaller physical processes but they also have to be made more energy efficient. Solving either of these problems helps; solving both will be essential to achieve further miniaturization and speedup.

Let us take a look at the space and energy trends that have taken place, and still persist, in the computer industry today. In the 1970s Gordon Moore, the cofounder of Intel, noticed that the memory capacity of a chip approximately doubled every one and a half years even though a chip remained, physically, about the same size[Malone95]. This meant that every 18 months only half as many atoms were needed to store a bit of information. Robert Keyes of IBM has analyzed the number of atoms needed to represent one bit of information as a function of time. His findings are plotted in Figure 1.1.

From the figure we can see why, in 1959, Nobel Laureate Richard Feynman was inspired to entitle an address to the American Physical Society as "There's Plenty of Room at the Bottom"[Feynman60]. Indeed there are potentially many orders of magnitude available for further size reduction. However, we can see that the exponential trend, which has held since 1950, reaches a limit of one atom per bit around the year 2020. At the one-atom-per-bit level and, realistically, even a little before this, it will be *necessary* to use quantum

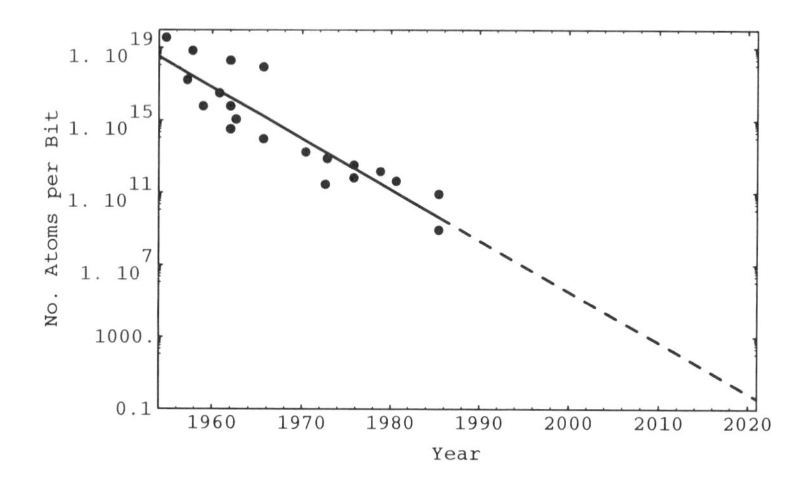

Fig. 1.1 The number of atoms needed to represent one bit of information as a function of calendar year. As the vertical axis is on a logarithmic scale, the straight line fit suggests the trend is exponential. Extrapolation of the trend suggests that the one-atom-per-bit level is reached in about the year 2020. Adapted from [Keyes88].

effects to read bits from and write bits to the memory register of an ultrasmall computer. So even on memory grounds alone, there is a strong reason to investigate the operating principles and feasibility of quantum devices.

However, this is not the complete story. Over the same period the number of transistors per chip has risen exponentially. Simultaneously, the speed of the chips has been increasing exponentially. Extrapolating this trend suggests the computers available in 2020 will operate at about 40 GHz (40,000 MHz). By comparison, in 1996 Intel released the Pentium-Pro processor that operated at a speed of 200 MHz. Clearly, if transistor density and clock speed continue to follow their exponential trends without an accompanying increase in energy efficiency, then the chips of the next millennium will simply melt themselves as they compute.

Fortunately, in addition to the exponential improvement in transistor density and clock speed, there has been a concomitant improvement in energy efficiency. The early computers generated relatively large amounts of heat per logical operation. As computer components have become smaller, and transistor density per chip has increased, the components have had to be made more energy efficient per logical operation to avoid thermal damage to the semiconductors during normal operation. Extrapolating the energy dissipation trend suggests that computers will reach the 1 kT level in about 2020 (where k is Boltzmann's constant $k = 1.3805\, 10^{-23}\, J\, K^{-1}$. and $T \approx 300\, K$ at room temperature). 1 kT is the typical amount of

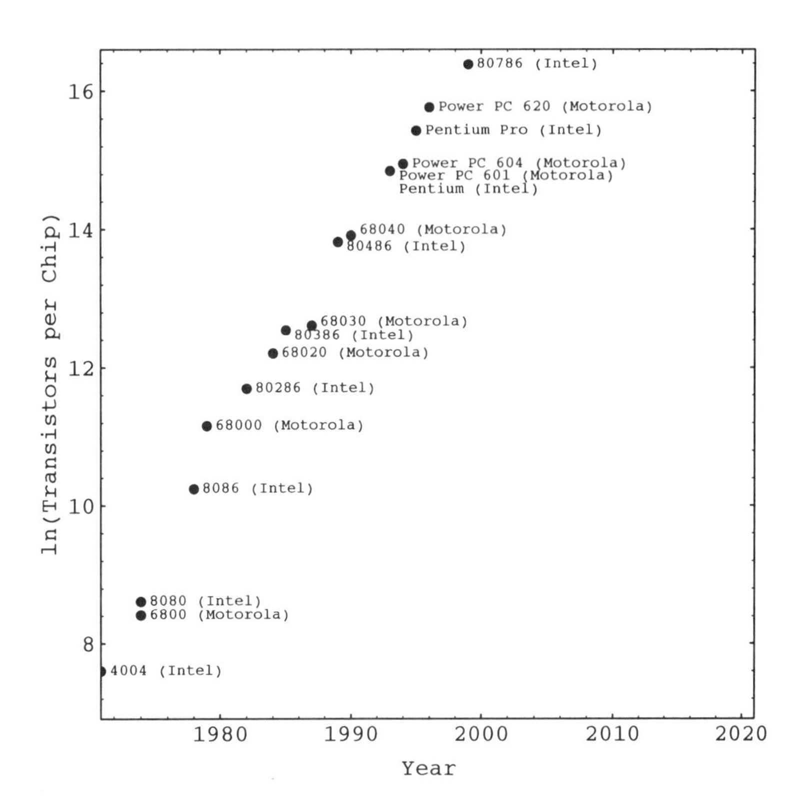

Fig. 1.2 Logarithm of the number of transistors per chip as a function of calendar year and the processors that achieved these transistor densities. Notice the near straight line indicating that the number of transistors per chip has risen exponentially over the past 25 years. Adapted from [Hutcheson96].

energy in thermal noise at the atomic level. Consequently, 1 kT marks a practical order of magnitude threshold for controllable quantum devices.

Simply put, if current trends in miniaturization and energy efficiency continue as they have for the past 25 to 30 years, computing technology should arrive at the quantum level at or before the year 2020. At this point every aspect of computer operation, from loading programs, running such programs, and reading the answers will be dominated by quantum effects. Even the design of algorithms will have to be rethought to make the best possible use of the quantum possibilities.

How can we be sure that these trends will be extrapolated? Some have argued that the exponential rate of miniaturization must stop long before the quantum limit is reached. They base these arguments on what has happened to the automobile and airplane. Although innovation was rampant at the outset of these industries, eventually

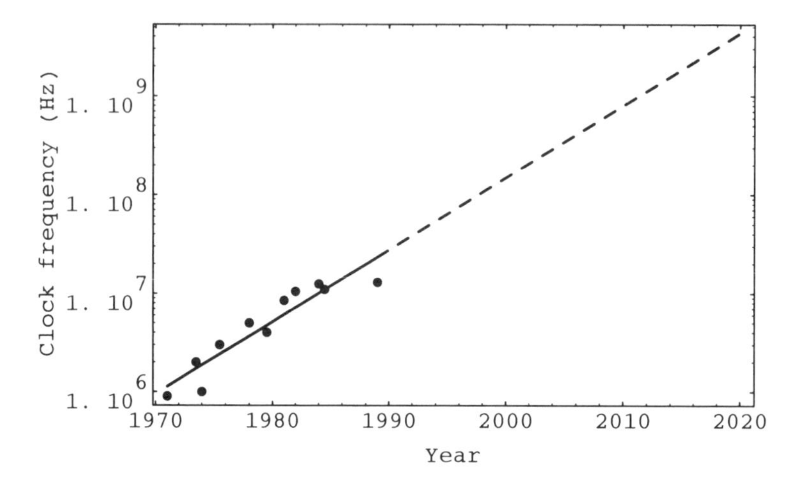

Fig. 1.3 Clock speed (Hz) versus calendar year. Note the exponential increase of clock speed with time. Adapted from [Malone95].

there came a slowdown in the breakthroughs and what we are left with are minor variations of a theme. Nowadays, all cars are basically the same and commercial aircraft are even more so.

However, car sizes and performance (speed and maneuvering) are limited not by technology but by human physiology. People can react only so fast. Race cars are purposely slowed down so drivers have time to react. Aircraft are not built to withstand excessive G-forces because pilots black out in abrupt, tight turns. There are psychological limits as well. What would public reaction be to a plane crash with 1,000 people aboard? Even in the absence of a catastrophe, would you want to be the first person to board or last person to deplane a 1000-seat aircraft?

The difference between vehicular technology and computing technology is that vehicles are constrained by the physiological and psychological limitations of the humans who use them but computers are not. People will not want to drive faster than a certain speed because of the danger involved, but who can imagine being afraid to compute too fast?

There are some caveats to this simplified analysis, of course. Even if quantum computers are possible theoretically, this does not necessarily mean they will be technologically feasible. For one thing, quantum-scale computing devices will be extremely vulnerable to stray interactions with the environment, cosmic rays, and errors in microfabrication. Such considerations will necessitate using error-correction techniques and perhaps massive redundancy. However, there is no reason, in principle, why quantum computers cannot be

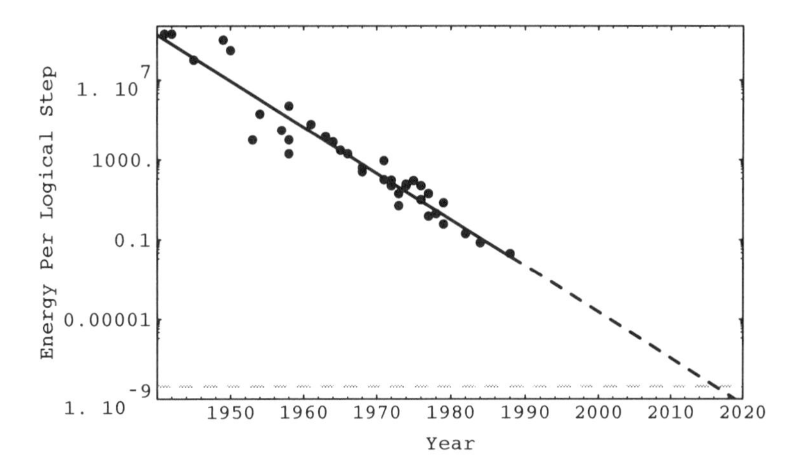

Fig. 1.4 Energy (pico-Joules) dissipated per logical operation as a function of calendar year. The 1 kT level is indicated by a dashed line. Adapted from [Keyes88].

built and it remains for the next generation to solve the technological problems that might arise before quantum computers can be brought to fruition. We come back to the feasibility of quantum computers in Chapter 10.

1.3 Economic and Environmental Issues

It takes more than mere scientific possibility to ignite a technological revolution, of course. There is an altogether different reason why a move *away* from conventional computer technology seems inevitable. Money.

In spite of competitive business pressures to be efficient, the cost of a new semiconductor plant has been doubling every three years[Hutcheson96]. In 1994, Motorola proposed building a $2.4 billion plant. If the tri-annual doubling trend continues, by 2020 the cost of building a semiconductor plant will be $1 trillion, which would be over 5% of the projected Gross Domestic Product of the entire United States! Clearly, this economic trend is not sustainable. To be profitable, Motorola would need sales in 2020 to be $10 trillion (over half the predicted GDP). So either computer performance will plateau or some new technology must be found that can deliver superior performance more efficiently.

As we show, quantum computers are predicted to be significantly more efficient at certain computational tasks than conventional computers. Moreover, quantum phenomena allow entirely new kinds of tasks to be performed, including faithful simulations of molecular processes, superdense information coding, and ultra-secure communication. It is too soon to tell whether our computing needs in 2020 will be dominated by the kinds of tasks at which quantum computers excel or indeed whether they can be manufactured cheaply. Nevertheless, they offer a distinct and relatively unexplored alternative to conventional computer technology that may have unforeseen financial payoffs.

Apart from the raw economics of building computers, there is also the cost of running them. Currently, 5% of the total power generation of the USA is consumed by computers[Malone95]. Assuming that the total number of computers remains constant and extrapolating the clock speed, memory capacity, and energy efficiency trends, we estimate that the computers of 2020 will be operating at 40 GHz, have 160 Gb RAM, and dissipate about 40 W of power. This is comparable to the power requirements of current machines and would therefore still consume roughly 5% of the current total power generation capacity. Unfortunately, it is hard to believe that the number of computers will not grow. At 40 W power consumption, with the cost of building a new power plant so high, fossil fuels continuing to dwindle, fission power in disfavor with the public, and fusion power still many decades away, the drain computers impose on our power supply could become significant.

In this regard, quantum computers might again offer an advantage. Theoretically speaking, quantum computers are examples of what are called reversible computers. In principle, any energy you need to feed into them to do a particular computation should be redeemable at the end of the computation, resulting in no net energy consumption. In practice, it is most unlikely that such a perfect energy balance will be achievable, certainly if you want the computer to work at a reasonable speed. A more realistic advantage of a quantum computer over any classical computer (no matter how advanced) is that certain types of computation can be done using far fewer logical operations on a quantum computer than a classical computer. If you can reduce the number of logical operations to achieve a given result, you might be able to save on the energy required for the computation. This is because quantum computers are able to work on all possible inputs to a program simultaneously and then extract partial information about all the different answers at the end. In contrast a classical computer has to work on each input separately. If there are very many inputs all of which need to be tested, the energy savings could be appreciable. Whether practical problems can be couched in a way that makes them amenable to this kind of quantum parallelism, as it is called, is under active inquiry.

The energy argument, incidentally, also suggests that there is a notion of complexity missing from current computer science. In addition to characterizing how the memory requirements should scale up with increasingly larger problems (i.e., "space complexity") and how the number of computational steps should scale up with increasingly larger problems (i.e., "time complexity"), there should also be an "energy complexity"; how the energy required to perform a computation scales with the size of the problem. This would not only involve the energy incurred in doing the computation but also the energy needed to pose the problem to the computer and to read out the final answer.

You can see that thinking about computers as physical systems forces a quite different perspective on the nature of computation.

1.4 Social and Political Pressures

We have seen how the technological trends suggest a movement towards quantum computing devices and how the economic trend suggests a movement away from conventional computing devices. But who is going to figure out how quantum computer's work? For quantum computing to take off, there has to be a pool of quantum physicists and computer scientists who want to work on the problem. A knowledge of theoretical computer science and quantum physics is an unusual combination of skills; so what factors will motivate a cross-fertilization of ideas? From where do these physicists and computer scientists come?

The demise of the Superconducting Super Collider, a ten billion dollar particle accelerator project that was to have been built in Texas, has put a large number of quantum physicists on the job market. With particle physics closed off as a career option and other areas of physics suffering from reduced funding, these researchers and students are looking for a promising new research vector. Moreover, with the changing economic climate, the US government has shifted its research funding away from the pure sciences, like physics, towards projects that seem more commercially relevant in the near term.

A similar situation exists in theoretical computer science. Viewed in a positive way, the foundational work has already been done and theoretical computer science has, by and large, moved on to higher-level questions such as the nature of computer networks, the theory of programming languages, computer security, and artificial intelligence. Indeed, funding for basic computer science research is now quite scarce with most of the research money being directed towards applications-oriented projects such as the information superhighway. Quantum computing not only offers the potential for a new wave of foundational research, but it could also galvanize a brand new computer industry. Consequently, the field will attract many students who want to make basic contributions as well as those who want to build quantum computing devices.

Most important, however, there is now a perceived need within government agencies to understand the capabilities and feasibility of quantum computers. The United States Central Intelligence Agency is already funding efforts to build a prototypical quantum computer that is capable of factoring large integers. This task is important to the CIA as the security of many top-secret codes rests on the presumption that factoring is hard. For a quantum computer, however, factoring would be an easy problem. Thus if anyone could build a quantum computer, the security of sensitive encrypted information would be compromised.

Other agencies are interested in quantum computing from a more scientific perspective. For example, quantum computers might be able to perform certain simulations of physical systems much more efficiently than can any classical computer. In addition, quantum computers will open the way to performing experiments that test the more esoteric predictions of quantum theory. Such studies might eventually shed light on the fundamental nature of reality.

The preceding state of affairs is just what is needed for quantum computing research to take off with a full complement of talent and enthusiasm.

1.5 The 2020 Vision

As any actual computer must, first and foremost, be a physical device, the correct theory of computation ought to be a branch of physics rather than a branch of mathematics (as is currently the case). We have shown how extrapolation of the trend in miniaturization of computer components suggests that they will reach the quantum scale by about the year 2020. Moreover, we have argued that there are economic, political, and sociological reasons that will continue to drive this trend.

Having faster computers always seems desirable. What we *want* to do with our computers has always co-evolved with what we *can* do with our computers. The computer industry has sustained a phenomenal growth rate because it continually creates novel functionality and makes us feel we simply have to have it. For example, a decade ago the average person either did not use a computer at all or did so in the office for word processing or spreadsheets. Today, home computers abound, fax/modems are commonplace, multimedia books and games have never been more popular, and the Internet craze is positively booming. What is next?

Within a decade we expect to see talking interfaces, software agents, translating telephones, household robots, and so on. This is an industry fueled by human imagination and a thirst for speed. There seems to be no end to how fast and how powerful we can make our computers. Or is there? Just what are the capabilities of computing machinery and how might these capabilities change as we move to the quantum level? Are there any tasks that computers cannot perform no matter how powerful they become? We take up these questions in the next chapter.

CHAPTER 2

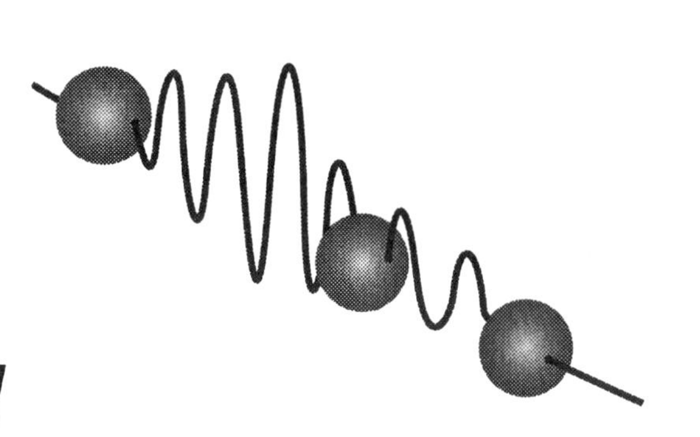

The Capabilities of Computing Machinery

"It is unworthy of excellent men to waste hours like slaves in the labor of calculation."
— Baron Gottfried von Leibnitz

What can computers do? Ask a young child and you might be told, "They let me draw pictures and play games." Ask a teenager and you might hear, "They let me surf the Web and go online with my friends." Ask an adult and you might discover, "They're great for word processing and keeping track of my finances." What is remarkable is that the toddler, the teenager, and the parent might all be talking about exactly the same machine! By loading the appropriate software it seems as if we can make the computer do just about anything.

But what exactly are the capabilities of computers? In particular, what problems can computers solve efficiently and what problems can they solve at all? Are there any problems that will never be solved no matter how powerful our computers become? Does it matter whether computers are implemented as gears, vacuum tubes, transistors, or integrated circuits? These questions, and many others besides, can be answered by developing a mathematical model, or theory, of computation and exploring its ramifications.

It might seem preposterous to believe that there *is* a theory circumscribing the capabilities of computers. Each year new computers appear that are significantly faster than their predecessors and thousands more programs become available. Many people upgrade their computers precisely because they believe their new machines can do more than the ones they replace. If the theory of computation is based on particular machines and particular software, then it will quickly become outdated. This point was understood by the early theorists Alan Turing, Alonso Church, Kurt Gödel, and Emil Post. They independently developed *mathematical* modelsof the computing process that were intended to be free of any assumptions pertaining to how computers were actually implemented. Superficially, their models were quite distinct from one another. Gödel identified effective or algorithmic processes with the class of general recursive functions, Church identified them with the so-called λ-definable functions (which you will have encountered if you have ever used the LISP programming language), and Turing identified them with the class of functions computable by a hypothetical computing called a Turing machine.

It turns out, however, that these competing models of "computation" are all equivalent to one another. This was something of a surprise as there was no reason to expect their equivalence a priori. Moreover, any one of the models alone might be open to the criticism that it provided an incomplete account of computation. But the fact that three radically different views of computation all turned out to be equivalent was a clear indication that the most significant aspects of computation had been characterized correctly.

Unfortunately, we now know that although these models were intended to be mathematical abstractions of computation that were free of physical assumptions, they do, in fact, harbor rather subtle physical assumptions. These assumptions appear to be perfectly valid in the world we see around us but they cease to be valid on sufficiently small scales. To appreciate at what point the classical models break down, it is useful to take a look at the context in which the models arose.

2.1 How the Turing Machine Came About

The most influential model of computation was that invented by Alan Turing in 1936[Turing37]. A Turing machine is an idealized mathematical model of a computer that can be used to understand the limits of what computers can do[Hopcroft84]. It is not meant to be a practical design for any actual machine but rather a crude caricature which, nevertheless, captures the *essential* features of any real computer. A Turing machine's usefulness stems from being sufficiently simple to allow mathematicians to prove theorems about its computational capabilities and yet sufficiently complex to accommodate any actual classical digital computer, no matter how it is implemented.

The idea for a Turing machine grew out of an attempt to answer a question posed by David Hilbert, an eminent German mathematician, at the turn of the century. In 1900, Hilbert gave an address at the International Congress of Mathematics held in Paris concerning what he believed to be the 23 most challenging mathematical problems of his day. The last problem on his list, asked whether there was a mechanical procedure by which the truth or falsity of any mathematical conjecture could be decided. In German, the word for "decision" is "entscheidung," so Hilbert's 23rd problem became known as the *Entscheidungsproblem*.

Hilbert's motivation for asking this question stemmed from the trend towards abstraction in mathematics. Before Hilbert's time, mathematics was largely a practical matter, concerned with making statements about real-world objects. However, in the 1800s mathematicians began to invent, and then reason about, imaginary objects to which they ascribed properties that were not necessarily compatible with "common sense." The truth or falsity of statements made about such imaginary objects could not be determined by appeal to the real world. In an attempt to put mathematical reasoning on a sound footing, Hilbert advocated a "formalist" approach to proofs. To a formalist, symbols cease to have any meaning other than that implied by their relationships to one another. No inference is permitted unless there is an explicit rule that sanctions it, and no information about the meaning of any symbol enters into a proof from the outside. Thus the very philosophy of mathematics that Hilbert advocated lent itself to automation, and hence Hilbert proposed the *Entscheidungsproblem*.

Turing heard about Hilbert's *Entscheidungsproblem* during a course of lectures he attended at Cambridge University in England. He began thinking about how he could answer it. Although Hilbert probably meant "mechanical" figuratively, Turing interpreted it rather literally. Turing wondered whether a *machine* existed that could always decide the truth or falsity of any mathematical proposition.

In order to address the *Entscheidungsproblem*, Turing realized that he needed a model of the process that a mathematician goes through when attempting to prove some mathematical conjecture.

Mathematical reasoning is an enigmatic activity. We do not really know what goes on inside a mathematician's head, but we can examine the result of that process in the form of the notes the mathematician creates whilst developing a proof. Mathematical reasoning consists of combining axioms (facts taken to be true without proof) and prior conclusions, using explicit logical rules, to infer new conclusions. If you look over a mathematician's shoulder during a proof derivation you would see a record of his or her thought process.

Turing abstracted the process appearing in notes into four principal ingredients: a set of transformation rules that allowed one mathematical statement to be transformed into another, a method for recording each step in the proof, an ability to go back and forth over the proof to combine earlier inferences with later ones, and a mechanism for deciding which rule to apply at any given moment. This is the essence of the proof process — at least the visible part of the proof process.

Next Turing sought to simplify these steps in such a way that a machine could be made to imitate them. Mathematical statements are built up out of a mixture of ordinary alphanumeric characters, parentheses, and special mathematical symbols. Turing realized that the symbols themselves were of no particular significance. All that mattered was that they were used consistently and that their number was finite. Moreover, once you know you are dealing with a finite alphabet, you can place each symbol in one-to-one correspondence with a unique pattern of any two symbols (such as 0 and 1). Hence, rather than deal with a rich array of esoteric symbols, Turing realized that a machine only needed to be able to read and write two kinds of symbol, 0 and 1, say, with blank spaces or some other convention to identify the boundaries between distinct symbols.

Similarly, the fact that the scratch pad on which the mathematician writes intermediate results is two-dimensional is of no particular importance. You could imagine attaching the beginning of one line of a proof to end of the previous line, making one long continuous strip of paper. So, for simplicity, Turing assumed that the proof could be written out on a one-dimensional strip of paper or a "tape."

Moreover, rather than allowing freeform handwriting, it would clearly be easier for a machine to deal with a tape marked off into a sequence of identical cells and only permitting one symbol, or a blank, to be written inside each cell.

Finally, the process of a mathematician going back and forth over previous conclusions in order to draw new ones could be captured by imagining that there is a "read/write" head going back and forth along the tape.

When a mathematician views an earlier result it is usually in some context. A mathematician might read a set of symbols, do something, and come back to read those same symbols again later, and do something else. Thus the context in which a set of symbols is read can affect the subsequent actions. Turing captured this idea by defining the "head" of his Turing machine to be in certain "states," corresponding to particular contexts. The combination of the symbol being read under the head and the state of the machine determined what symbol to write on the tape, which direction to move the head, and which state to enter next.

This is clearly a crude model of the proof process. Nevertheless it turned out to be curiously powerful. No matter what embellishments people dreamt up, Turing could always argue that they merely were refinements to some existing part of the model rather than being fundamentally new features. Consequently the Turing machine model was indeed the essence of the proof process.

2.2 Deterministic Turing Machines

By putting the aforementioned mechanistic analogues of human behavior into a mathematical form, Turing was led to the idea of a "deterministic Turing machine" (see Fig. 2.1). The Turing machine model consists of an infinitely long tape that is marked off into a sequence of cells on which may be written a 0, a 1, or a blank, and a read/write head that can move back and forth along the tape scanning the contents of each cell. The head can exist in one of a finite set of internal "states" and contains a set of instructions (constituting the "program") that specify, given the current internal state, how the state must change given the bit (i.e., the binary digit 0 or 1) currently being read under the head, whether that bit should then be changed, and in which direction the head should then be advanced.

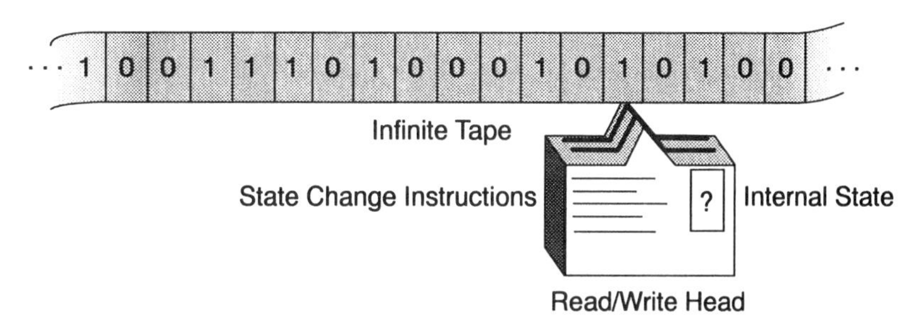

Fig. 2.1 A deterministic Turing machine.

Initially the tape is set up in some standardized state such as all cells containing 0 except for a few that hold the program and any initial data. Thereafter the tape serves as the scratch pad on which all intermediate results and the final answer (if any) are written.

Roman Maeder, a Swiss computer scientist, has written a particularly good example of a simulator for a Turing machine [Maeder93]. Maeder's simulator shows how recursive functions (functions that call themselves) can be implemented in Turing machines. Thus, the simulator demonstrates not only the basic operations of a Turing machine, but also its connection to the model of computation based on recursive function theory.

Despite its lack of adornments, the Turing Machine model has proven to be remarkably durable. In the 60-odd years since its inception, computer technology has advanced considerably. Nevertheless, the Turing machine model remains as applicable today as it was back in 1936. Although we are apt to think of multimillion dollar supercomputers as being more powerful than humble desktop machines, the Turing machine model proves otherwise. Given enough time and memory capacity there is not a single computation that a supercomputer can perform that a personal computer cannot also perform. In the strict theoretical sense, they are equivalent.

2.3 Probabilistic Turing Machines

An alternative model of classical computation is to imbue a deterministic Turing machine with the ability to make a random choice, such as flipping a coin. The result is a probabilistic Turing machine. Surprisingly, many problems that take a long time to solve on a deterministic Turing machine can often be solved very quickly on a probabilistic Turing machine (PTM). However, in the probabilistic model of computation there are often tradeoffs between the time it takes to return an answer to a computation and the probability that the answer returned is correct. Alternatively, if you *require* a correct answer there is then uncertainty in the length of time the probabilistic algorithm must run. Consequently, a new issue enters the computational theory, namely, the correctness of an answer and its relationship to the running time of an algorithm.

Whereas a deterministic Turing Machine, in a certain state, reading a certain symbol, has precisely one successor state available to it, the PTM has multiple legitimate successor states available. The choice of which state is the one ultimately pursued is determined by

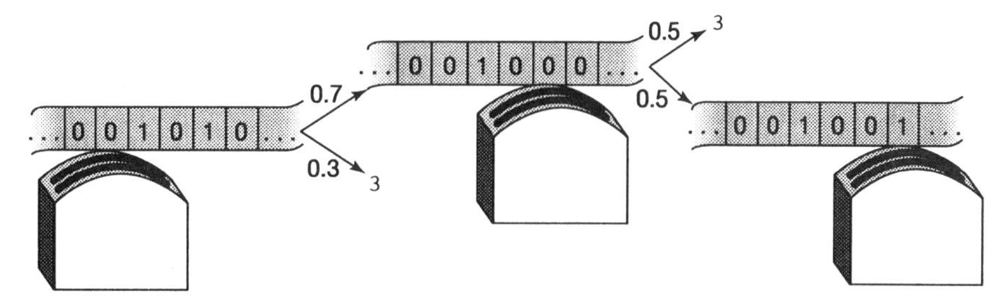

Fig. 2.2 In a probabilistic classical Turing machine there are multiple possible successor states, only one of which is actually selected. Unselected paths are terminated (×). The probabilities of transitioning between various states are shown. Notice that the sum of the probabilities on all the paths emanating from a state is 1.

the outcome of a random choice (possibly with a bias in favor of some states over others). In all other respects the PTM is just like a DTM.

Despite the superficial difference between PTMs and DTMs, computer scientists have proved that anything computable by a probabilistic Turing machine can also be computed by a deterministic Turing machine, although in such cases the probabilistic machine is often more efficient[Gill77]. However, in the early 1980s some scientists began to question the correctness of these models of computation. The models are certainly fine as *mathematical* abstractions but are they consistent with known *physics*?

We now know that the Turing Machine model is flawed. In spite of Turing's best efforts, some remnants of classical physics crept into the models. The evident advances in technology have been of a quantitative nature: more memory, more instructions per second, greater energy efficiency. Similarly, although certainly having a huge social impact, apparent revolutions, such as the explosion of the Internet, have merely provided new conduits for information to be exchanged. They have not altered the fundamental capabilities of computers in any way whatsoever.

However, as computers become smaller, eventually their behavior must be described using the physics appropriate for small scales, that is, quantum physics. The effects on the theory of computation of a reformulation of the basic operations of a computer in a way that is consistent with quantum physics is much more significant. It admits the possibility of harnessing entirely new kinds of physical phenomena, quantum coherent phenomena, in the service of computation. This is a *qualitative* change with profound ramifications on the capabilities of computing machinery, but it was not until the early 1980s that scientists began to speculate on what a quantum account of computation might be like. The ultimate product of their inquiries was the Quantum Turing Machine (QTM).

2.4 Quantum Turing Machines

The first quantum mechanical description of a Turing machine was given by Paul Benioff in 1980. Benioff was building on earlier work by Charles Bennett. Bennett had shown that a reversible Turing machine was a theoretical possibility[Bennett73]. If you thought of the reversible Turing machine as a dynamical system, then given knowledge of its state at any one moment would allow you to predict its state at all future and all past times. No information was ever lost and the entire computation could be run forwards or backwards.

This fact struck a chord with Benioff, for he realized that any isolated quantum system had a dynamical evolution that was reversible in exactly this sense. Thus it ought to be possible to devise a quantum system whose evolution over time mimicked the actions of a classical reversible Turing machine. This is exactly what Benioff did.

Benioff's machine was not quite a true quantum computer however, because although between computational steps the machine existed in an intrinsically quantum state (called a "superposition," explained in the next chapter), at the end of each step the "tape" of the machine was always back in one of its classical states: a sequence of bits. Thus, from a complexity perspective, Benioff's design could do no more or less than a classical reversible Turing machine.

The possibility that quantum mechanical effects might offer something genuinely new was first hinted at by Richard Feynman, of Caltech, in 1982, when he showed that no classical Turing machine could simulate certain quantum phenomena without incurring an exponential slowdown but that a "universal quantum simulator" could do so. Unfortunately, Feynman did not provide a design for such a simulator, so his idea had little immediate impact.

The key step to enabling the study of the computational power of quantum computers came in 1985, when David Deutsch, of Oxford University, described the first true quantum Turing machine (QTM). This was a Turing machine whose read, write, and shift operations were all accomplished by quantum mechanical interactions and whose "tape" could exist in states that were highly nonclassical. In particular, whereas a conventional classical Turing machine could only encode a 0, 1, or blank in each cell of the tape, the QTM could encode a blend, or "superposition," of 0 and 1 simultaneously. Thus the QTM had the potential for encoding many inputs to a problem simultaneously on the same tape, and performing a calculation on all the inputs in the time it took to do just one of the calculations classically. This technique was dubbed "quantum parallelism."

The idea that each bit in a QTM can be a blend of a 0 and a 1 can be illustrated by drawing each quantum bit, or "qubit," as a small vector contained in a sphere. "Straight up" represents the (classical) binary value 0 and "straight down" represents the (classical) binary value 1. When the vector is at any other orientation, the angle the vector makes with the vertical axis is a measure of the ratio of 0-ness to 1-ness in the qubit. Likewise, the angle through which the vector is rotated about the vertical axis is a measure of the "phase," which is a quantum mechanical notion explained in the next chapter. Thus, drawing qubits as vectors contained in spheres we can depict Deutsch's quantum Turing machine in Fig. 2.3.

Quantum Turing machines (QTMs) are best thought of as quantum mechanical generalizations of probabilistic Turing machines (PTMs). In a PTM, if you initialize the tape in some initial configuration and run the machine, without inspecting its state, for some time t, then the state of the machine is described by a probability distribution over all the possible states reachable, from the initial state, in time t.

Likewise, in a QTM if you start the machine off in some initial configuration, and allow it to evolve for a time t, then its state at time t is described by a superposition of all states reachable in time t. The key difference is that in a classical PTM only one particular computational trajectory is followed, but in the QTM all computational trajectories are followed and the resulting superposition is the result of summing over all possible trajectories achievable in time t.

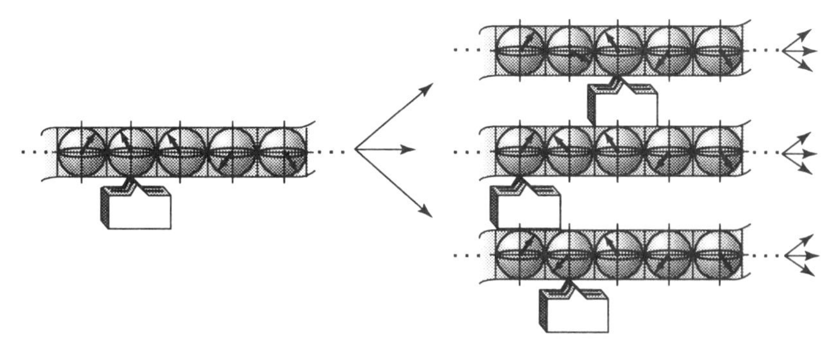

Fig. 2.3 In the quantum Turing machine each cell on the tape can hold a qubit whose state is represented as an arrow contained in a sphere. All paths are pursued simultaneously. Instead of probabilities on each path we now have amplitudes. Amplitudes are complex numbers whose square moduli are probabilities.

Armed with this model of an abstract quantum Turing machine, several researchers have been able to prove theorems about the capabilities of quantum computers. This effort has focused primarily on computability (what problems the machines can do), complexity (how the memory and time resources scale with problem size), and universality (whether one machine can simulate all others efficiently). Let us take a look at each of these concepts and compare the perspective given to us by classical computing and quantum computing.

2.5 Universality

The Turing machine model had something of a catalytic effect on computer science. In the 1930s computer science was a rather fledgling field. People dabbled with building computers but very few machines actually existed. Those that did had been tailor-made for specific applications. However, the concept of a Turing machine ushered in a new possibility. Turing realized that one could encode the transformation rules of any particular Turing machine, T say, as some pattern of 0s and 1s on the tape that is fed into some special Turing machine, called U. U had the effect of reading in the pattern specifying the transformation rules for T and thereafter treated any further input bits exactly as T would have done. Thus U was a universal mimic of T and hence was called the Universal Turing Machine. The possibility of one machine simulating another gave a theoretical justification for pursuing the idea of a *programmable* computer.

In fact the ability to prove that all the competing models of classical computation were equivalent led Church to propose the following principle, which has subsequently become known as the Church-Turing thesis[Shapiro90]: "Any process that is effective or algorithmic in nature defines a mathematical function belonging to a specific well-defined class, known variously as the recursive, the λ-definable, or the Turing computable functions." In Turing's words, "Every function which would naturally be regarded as computable can be computed by the universal Turing machine."

In 1982, Richard Feynman observed that it did not appear possible for a Turing machine to simulate certain quantum physical processes without incurring an exponential slowdown[Feynman82]. This led Feynman to ask whether all quantum systems can be simulated efficiently on a quantum computer and whether a quantum computer can be simulated efficiently on a classical probabilistic computer. Although Feynman gave a few examples of one quantum system simulating another he did not prove, conclusively, that a universal quantum simulator was possible. This last step was accomplished by Seth Lloyd in 1996[Lloyd96] and Christof Zalka [Zalka96].

However, the apparent discrepancy between Feynman's observations and the Church-Turing thesis led David Deutsch in 1985 to propose reformulating the Church-Turing thesis in physical terms. Thus Deutsch prefers: "Every finitely realizable physical system can be perfectly simulated by a universal model computing machine operating by finite means." This can only be made compatible with Feynman's observation on the efficiency of simulating quantum systems by basing the universal model computing machine on quantum mechanics itself.

2.6 Computability

Computability theory addresses the question of which problems can be solved (or which questions can be decided) in a finite amount of time on a computer. If there is no algorithm, with respect to a particular model of a computer, that can guarantee to find an answer to a given problem in a finite amount of time, that answer is said to be *uncomputable* with respect to that computer. One of the great breakthroughs in classical computer science was the recognition that all of the candidate models for computers, Turing machines, recursive functions, and λ-definable functions were equivalent in terms of what they could and could not compute.

It was, you will recall, a particular question regarding computability that was the impetus behind the Turing machine idea. Hilbert's *Entscheidungsproblem* had asked whether there was a mechanical procedure for deciding the truth or falsity of any mathematical conjecture, and the Turing machine model was invented to prove that there was no such procedure.

To construct this proof, Turing used a technique called *reductio ad absurdum*, in which you begin by assuming the truth of the opposite of what you want to prove and then derive a logical contradiction. The fact that your one assumption coupled with purely logical reasoning leads to a contradiction proves that the assumption must be faulty. In this case the assumption is that there *is* a procedure for deciding the truth or falsity of any mathematical proposition and so showing that this leads to contradictions allows you to infer that there is in fact no such procedure.

Turing reasoned that if there *were* such a procedure, and it were truly mechanical, it could be executed by one of his Turing machines with an appropriate table of instructions. But a "table of instructions" could always be converted into some finite sequence of 1s and 0s. Consequently, such tables can be placed in an order, which meant that the things these tables represented (i.e., the Turing machines) could also be placed in an order.

Similarly, the statement of any mathematical proposition could also be converted into a finite sequence of 1s and 0s; so they too could be placed in an order. Hence Turing conceived of building a table whose vertical axis enumerated every possible Turing machine and whose horizontal axis, every possible input to a Turing machine.

But how would a machine convey its decision on the veracity of a particular input, that is, a particular mathematical proposition? You could simply have the machine print out the result and halt. Hence the *Entscheidungsproblem* could be couched as the problem of deciding whether the *ith* Turing machine acting on the *jth* input would ever halt. Thus Hilbert's *Entscheidungsproblem* had been refashioned into Turing's *Halting Problem*.

Turing wanted to prove that there was no procedure by which the truth or falsity of a mathematical proposition could be decided; thus his proof begins by assuming the opposite, namely, that there *is* such a procedure. Under this assumption, Turing constructed a table whose *(i,j)th* entry was the output of the *ith* Turing machine on the *jth* input, if and only if the machine halted on that input, or else some special symbol, such as ⊗, signifying that the corresponding Turing machine did not halt on that input.

Next Turing replaced each symbol ⊗ with a bit, 0.

Now because the rows enumerate all possible Turing machines and the columns enumerate all possible inputs (or, equivalently, mathematical propositions) all possible sequences of outputs, that is, *all computable sequences of 1s and 0s, must be contained somewhere in this table.* However, as any particular output is merely some sequence of 1s and 0s it is possible to change each one in some systematic way, for example by flipping one of the bits in the sequence. Consider flipping each element on a diagonal slash through the table. The sequence of outputs along the diagonal differs in the *ith* position from the sequence generated by the *ith* Turing machine acting on the *ith* input. Hence this sequence cannot appear in *any* of the rows in the table. However, by construction, the infinite table is supposed to contain *all* computable sequences and yet here is a sequence that we can clearly compute and yet cannot appear in any one row. Hence Turing established a contradiction and the assumption underpinning the argument must be wrong. That assumption was "there exists a procedure that can decide whether a given Turing machine acting on a given input will halt." As Turing showed that the Halting problem was equivalent to the *Entscheidungsproblem*, the impossibility of determining whether a given Turing machine will halt before running it shows that the *Entscheidungsproblem* must be answered in the negative too. In other words, there is no procedure for deciding the truth or falsity of all mathematical conjectures.

This came as a bit of a shock to Hilbert and most other mathematicians but worse was to come! In 1936 Kurt Gödel proved two important theorems that illustrated the limitations of formal systems. A formal system L is called "consistent" if you can never prove both a proposition P and its negation NOT(P) within the system. Gödel showed that "Any sufficiently strong formal system of arithmetic is incomplete if it is consistent." In other words there are sentences P and NOT(P) such that neither P nor NOT(P) is provable using the rules of the formal system L. As P and NOT(P) express contradictory sentences, one of them must be true. So there must be true statements of the formal system L that can never be proved. Hence Gödel showed that truth and theoremhood (or provability) are distinct concepts.

In a second theorem, Gödel showed that the simple consistency of L cannot be proved in L. Thus a formal system might be harboring deep-seated contradictions.

The results of Turing and Gödel are quite surprising. Do similar problems arise in the quantum theory of computation?

Quantum Computability

In the 1980s some scientists began to think about the possible connections between physics and computability. The connections turn out to be rather deep[Lloyd93b]. For one thing, we must distinguish between Nature, which does what it does, and physics, which provides models of Nature expressed in mathematical form. The fact that physics is a mathematical science means that it is ultimately a formal system.

Physicist Christopher Moore has shown that a particle moving in a three-dimensional potential can be equivalent to a Turing machine, and hence is capable of universal computation[Moore90]. A corollary of this is that even if the initial conditions are known exactly, almost all questions concerning their long-term dynamical behavior are undecidable.

Asher Peres and Wojciech Zurek have articulated three reasonable desiderata of a physical theory[Peres82], namely, determinism, verifiability, and universality (i.e., the theory can describe anything). They conclude that:

> "Although quantum theory is universal, it is not closed. Anything can be described by it, but something must remain unanalyzed. This may not be a flaw of quantum theory: It is likely to emerge as a logical necessity in any theory which is self-referential, as it attempts to describe its own means of verification. In this sense it is analogous to Gödel's undecidability theorem of formal number theory: the consistency of the system of axioms cannot be verified because there are mathematical statements which can neither be proved nor disproved by the use of the formal rules of the theory, although their truth may be verified by metamathematical reasoning."

In a later paper Peres points out an amusing paradox[Peres85]. He shows that it is possible to set up three quantum observables such that two of the observables do not "commute" (and hence, as explained in the next chapter, cannot be measured simultaneously). Nevertheless, we can use the rules of quantum mechanics to predict, *with certainty*, the value of *both* these variables individually. Hence we arrive at an example system that we can say things about but which we can never determine experimentally — an analogue of Gödel's undecidability theorem.

These results are all consequences of treating physics as a formal system. However, is it possible to make more pointed statements about computability and quantum computers?

The first work in this area appeared in David Deutsch's original paper on quantum Turing machines[Deutsch85]. Deutsch argued that quantum computers could compute certain outputs, such as true random numbers, that are not computable by any deterministic Turing machine. Classical deterministic Turing machines can only compute functions, that is, mathematical procedures that return a single, reproducible, answer. However, there are certain computational tasks that cannot be performed by evaluating any function. For example, there is no *function* that generates a true random number. Consequently, a Turing machine can only feign the generation of random numbers.

In the same paper, Deutsch also introduced the idea of quantum parallelism. Quantum parallelism refers to the process of evaluating a function once on a blend or "superposition" of all possible inputs to the function to produce a superposition of outputs. Thus all the outputs are computed in the time taken to evaluate just one output classically. Unfortunately, you cannot obtain all of these outputs explicitly because a measurement of the final superposed state would yield only one output. Nevertheless, it is possible to obtain certain *joint* properties of all of the outputs.

In 1991 Richard Jozsa gave a mathematical characterization of the class of functions (i.e., joint properties) that were computable by quantum parallelism[Jozsa91]. He discovered that if f is some function that takes integer arguments in the range 1 to m and returns a binary value, and if the joint property function J that defines some collective attribute of all the outputs of f, takes m binary values and returns a single binary value, then only a fraction $\left(2^{2^m} - 2^{m+1}\right)/2^{2^m}$ of all possible joint property functions are computable by quantum parallelism. Thus quantum parallelism alone is not going to be sufficient to solve all the joint property questions we might wish to ask.

Of course, you could always make a QTM simulate a classical TM and compute a particular joint property in that way. Although this is feasible, it is not desirable, because the resulting computation would be no more efficient on the quantum computer than on the classical machine. However, the ability of a QTM to simulate a TM means that the class of functions computable on QTMs exactly matches the class of functions computable on classical TMs.

2.7 Proving Versus Providing Proof

Many decades have now passed since Turing first dreamt of his machine and in fact today there are a number of programs around that actually perform as artificial mathematicians in exactly the sense Turing anticipated. Current interest in them stems not only from a wish to build machines that can perform *mathematical* reasoning but also more general kinds of logical inference such as medical diagnosis, dialogue management, and even legal reasoning. Typically, these programs consist of three distinct components: a reservoir of knowledge about some topic (in the form of axioms and rules of inference), an inference engine (which provides instructions on how to pick which rule to apply next), and a specific conjecture to be proved.

In one of the earliest examples, SHRDLU, a one-armed robot, was given a command in English which was converted into its logical equivalent and then used to create a program to orchestrate the motion of the robot arm[Winograd72].

In a more contemporary example, the British Nationality Act was encoded in first-order logic and a theorem prover used to uncover logical inconsistencies in the legislation.

Similarly, the form of certain legal arguments can be represented in logic which can then be used to find precedents by revealing analogies between the current case and past examples.

So although most people would think themselves far removed from the issue of "theorem proving," they could be in for a surprise if the tax authorities decided to play these games with the tax laws!

Today's artificial mathematicians are far less ingenious than their human counterparts. On the other hand, they are infinitely more patient and diligent. These qualities can sometimes allow artificial mathematicians to churn through proofs upon which no human would have dared embark. Take, for example, the case of map coloring. Cartographers conjectured that they could color any planar map with just four different colors so that no two adjacent regions had the same color. However, this conjecture resisted all attempts to construct a proof for many years. In 1976 the problem was finally solved with the help of an artificial mathematician. The "proof," however, was somewhat unusual in that it ran to some 200 pages. For a human to even check it, let alone generate it, would be a mammoth undertaking. Table 2.1 shows a summary of some notable milestones in mathematical proof by humans and machines.

Despite differences in the "naturalness" of the proofs they find, artificial mathematicians are nevertheless similar to real mathematicians in one important respect: their output is an explicit sequence of reasoning steps (i.e., a proof) that, if followed meticulously, would convince a skeptic that the information in the premises combined with the rules of logical inference would be sufficient to deduce the conclusion. Once such a chain were found the theorem

Table 2.1 Some impressive mathematical proofs.

Mathematician	Proof Feat	Notable Features
Daniel Gorenstein (1980)	Classification of finite simple groups	Created by human. 15,000 pages long.
Kenneth Appel and Wolfgang Haken (1976)	Proved the Four-Color Theorem	Created by computer. Reduced all planar maps to combinations of 2,000 special cases and then exhaustively checked each case.
Andrew Wiles (1993)	Proved Fermat's Last Theorem	Created by human. 200 pages long. Only 0.1% of all mathematicians are competent to judge its veracity.
Laszlo Babai et al.	Invented probabilistic proof checking	Able to verify that a complex proof is "probably correct" by replicating any error in the proof in many places in the proof thereby amplifying the chances of the error being detected.

would have been proved. The important point is that the proof chain is a tangible object that can be inspected at leisure.

Surprisingly, this is not necessarily the case with a QTM. In principle, a QTM could be used to create some proof that relied upon quantum mechanical interference among all the computations going on in superposition. Upon interrogating the QTM for an answer you might be told, "Yes your conjecture is true," but there would be no way to exhibit all the computations that had gone on in order to arrive at the conclusion. Thus, for a QTM, the ability to prove something and the ability to provide the proof trace are quite distinct concepts.

Worse still, if you tried to peek inside the QTM as it was working, to glean some information about the state of the proof at that time, you would invariably disrupt the future course of the proof.

2.8 Complexity

Whereas "computability" concerns which computational problems computers can and cannot do, "complexity" concerns how efficiently they can solve the ones they can do. Efficiency is an important consideration for real-world computing. The fact that a computer can solve a particular kind of problem, in principle, does not guarantee that it can solve it in practice. If the running time of the computer is too long, or the memory requirements too great, then an apparently feasible computation can still lay beyond the reach of any practicable computer.

Computer scientists have developed a rich classification scheme for describing the complexity of various algorithms running on different kinds of computers. The most common measures of efficiency employ the rate of growth of the time or memory needed to solve a problem as the size of the problem increases. Of course "size" is an ambiguous term. Loosely speaking, the "size" of a problem is taken to be the number of bits needed to state the problem to the computer. For example, if an algorithm is being used to factor a large integer N, the "size" of the integer being factored would be roughly $\log_2 N$.

The reason complexity classifications are based on the rates of growth of running times and memory requirements, rather than absolute running times and memory requirements, is to factor out the variations in performance experienced by different makes of computers with different amounts of RAM, swap space, and processor speeds. Using a growth rate-based classification, the complexity of a particular algorithm becomes an intrinsic measure of the difficulty of the problem the algorithm addresses.

Although complexity measures are independent of the precise make and configuration of computer, they are related to a particular mathematical *model* of the computer such as a deterministic Turing machine or a probabilistic Turing machine. It is now known, for example, that many problems that are intractable with respect to a deterministic Turing machine can be solved efficiently, with high probability, on a probabilistic Turing machine.

There are many criteria by which you could assess how efficiently a given algorithm solves a given type of problem. For the better part of the century, computer scientists focused on worst-case complexity analyses. These have the advantage that, if you can find an efficient algorithm for solving some problem, *in the worst case*, then you can be sure that you have an efficient algorithm for any instance of such a type of problem.

This estimation can be somewhat misleading however. Recently some computer scientists have developed average case complexity analyses. Moreover, it is possible to understand the finer grain structure of complexity classes and locate regions of especially hard and especially easy problems within a supposedly "hard" class[Williams94]. Nevertheless, one of the key questions is whether some algorithm runs in polynomial time or exponential time.

Polynomial Versus Exponential Growth

Computer scientists have developed a rigorous way of quantifying the difficulty of a given type of problem. The classification is based on the mathematical form of the function that describes how the computational cost incurred in solving the problem scales up as larger problems are considered. The most important quantitative distinction is between polynomially growing costs (which are deemed tractable) and exponentially growing costs (which are deemed intractable).

Exponential growth will always exceed polynomial growth eventually, regardless of the order of the polynomial. For example, Fig. 2.4 shows a comparison of the growth of the exponential function $exp(L)$ with the growth of the polynomials L^2, L^3 and, L^4. As you can see, eventually, whatever the degree of the polynomial in L, the exponential becomes larger. A good pair of example problems that illustrate the radical difference between polynomial and exponential growth are multiplication and factoring.

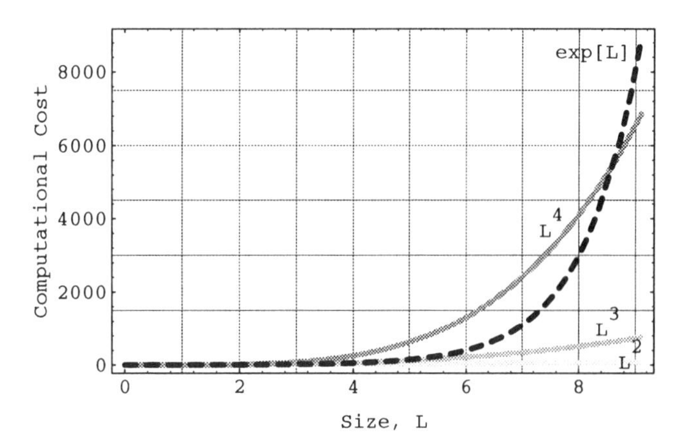

Fig. 2.4 Comparison of exponential versus polynomial growth rates. An exponential always beats a polynomial eventually.

It is relatively easy to multiply two large numbers together to obtain their product, but it is extremely difficult to do the opposite; namely, to find the factors of a composite number:

$$10433 \times 16453 = ? \quad \text{(is easy)}$$

$$? \times ? = 200949083 \quad \text{(is hard)}$$

If, in binary notation, the numbers being multiplied have L bits, then multiplication can be done in a time proportional to L^2, a polynomial in L.

For factoring, the best known classical algorithms are the Multiple Polynomial Quadratic Sieve[Silverman87] for numbers involving roughly 100 to 150 decimal digits, and the Number Field Sieve[Lenstra90] for numbers involving more than roughly 110 decimal digits. The running time of these algorithms grows subexponentially (but superpolynomially) in L, the number of bits needed to specify the number to be factored N (so $L \approx \log_2 N$). The best factoring algorithms require a time of the order $\exp\left(L^{1/3} \log(L)^{2/3}\right)$. Richard Crandall has charted the progress in feats of factoring over the past three decades (see Table 2.2)[Crandall96]. In the early 1960s computers and algorithms were only good enough to factor numbers with 20 decimal digits, but by 1994 that number had risen to a 129 digit numbers, but only after a Herculean effort. As we show later in the book, the presumed difficulty of factoring large integers is the basis for the security of so-called public key cryptosystems that are in widespread use today. When one of these

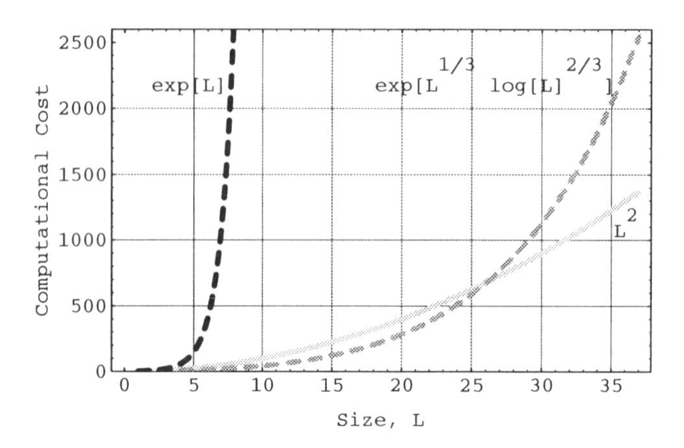

Fig. 2.5 The best factoring algorithms grow subexponentially (but super-polynomially) in *L*, the number of bits needed to specify the number being factored.

systems was invented the authors laid down a challenge prize for anyone who could factor the following 129 digit number (called RSA129).

RSA129 =

1143816257578888676692357799761466120102182967212423625
6256184293570693524573389783059712356395870505898907514
7599290026879543541

In 1994 a team of computer scientists using a network of work-stations succeeding in factoring RSA129. The resulting factors are:

3490529510847650949147849619903898133417764638493387843
990820577

and

3276913299326670954996198819083446141317764296799294253
9798288533.

Table 2.2 Progress in factoring large composite integers. 1 MIP-Year is the computational effort of a machine running at 1 million instructions per second for 1 year.

Year	Factorizable	Effort (MIP-Years)
1964	20 decimal digits	0.000009
1974	45 decimal digits	0.001
1984	71 decimal digits	0.1
1994	129 decimal digits	5000

Extrapolating the observed trend in factoring suggests that it would take about 2.9 billion MIP-Years to factor a 200-digit number.

Classical Complexity Classes

Knowing the exact functional forms for the rates of growth of the number of computational steps for various algorithms allows computer scientists to classify computational problems based on difficulty. The most useful distinctions are based on classes of problems that either can or cannot be solved in polynomial time, in the worst case. Problems that can be solved in polynomial time are usually deemed "tractable" and are lumped together into the class "P." Problems that cannot be solved in polynomial time are usually deemed "intractable" and may be in one of several classes. Of course it is possible that the order of the polynomial is large, such as 12, making a supposedly "tractable" problem rather difficult in practice. Fortunately, such large polynomial growth rates do not arise that often, and the polynomial/exponential distinction is a pretty good indicator of difficulty.

The next most interesting class after P is the class NP. Many algorithms for tackling a variety of practical problems involve trying to extend partial solutions into bigger partial solutions that eventually lead to a full solution. If a partial solution is found that cannot be extended further, the algorithm has to back up and try another alternative. Unfortunately, the number of computational steps required by a deterministic algorithm in order to guarantee that it will find a solution if often exponential in the size of the problem. If so, that type of problem is effectively intractable. However, many such problems also have the feature that once a candidate solution has been found its correctness can be tested efficiently, that is, in polynomial time. Thus if you could magically "guess" the right solution, the problem *could* be solved efficiently. Such a possibility means that there is, in principle, an efficient *nondeterministic* algorithm for solving the problem. Computer scientists therefore lump all such problems into a complexity class called NP, which stands for "nondeterministic polynomial time."

The question of great interest to computer scientists is whether the class P is the same as the class NP, that is, whether P = NP. So far, nobody knows.

Within the class NP there is a special subclass of problems that share a peculiar property: a problem in the subclass can be converted into another problem in the class by the action of some polynomial time algorithm. Consequently, the class of problems is called

"NP-complete" because they are in NP and consist of the "complete" set of problems mappable into one another in polynomial time. Thus the fate of one NP-complete problem is intimately bound to its kin. Either all NP-complete problems are tractable or none of them are! They stand or fall together. Currently, around 1000 distinct NP-complete problems are known. They crop up in many real-world problems such as scheduling, route planning, and matching.

Table 2.3 Some classical complexity classes.

Classical Complexity Class	Intuitive Meaning	Examples
P (or PTIME)	Polynomial-time: the running time of the algorithm is, in the worst case, a polynomial in the size of the input. All problems in P are tractable.	Multiplication
NP	Nondeterministic polynomial time: a candidate answer can be checked for correctness in polynomial time.	Factoring composite integers
NP-complete	A subset of problems in NP that can be mapped into one another in polynomial time. If just one of the problems in this class is shown to be tractable, then they must all be tractable.	Scheduling Satisfiability Traveling salesman problem
ZPP	Can be solved, with certainty, by PTMs in average case polynomial time.	
BPP	Can be solved in polynomial time by PTMs with probability > 2/3. Probability of success can be made arbitrarily close to 1 by iterating the algorithm a certain number of times.	Problems in BPP are tractable

Are NP-complete problems the hardest problems of all? No. Not by a long way in fact. There is a towering hierarchy of monstrous problems whose running times are doubly, triply, quadruply exponential in the size of the input; that is, their running times grow as 2^{2^N} or $2^{2^{2^N}}$ and so on. One example is Presburger arithmetic, a superficially innocuous looking logic that allows mathematicians to prove properties about formulae built out of positive integers and variables whose values can be positive integers. Presburger arithmetic is, at best, doubly exponential.

Even though NP-complete problems are not the hardest problems of all, they crop up in practical applications sufficiently often that there is a desperate need for efficient algorithms to tackle them. Consequently, it would be a great achievement if someone could show how to make a quantum computer solve an NP-complete problem more efficiently than a classical (deterministic and probabilistic) computer. This goal is being actively pursued by several researchers.

By way of setting things up for the future, we should point out the classical complexity behavior of two important problems: primality testing and factoring, which arise in public key cryptography. Primality testing (via probabilistic algorithms) is fast. In a popular cryptosystem, RSA, the recipient must choose two large primes. If proving that a number is prime was slow, then RSA would not be feasible. Factoring, however, is not fast on any type of classical computer; that is, there is no polynomial-time solution, deterministic or probabilistic. In RSA-cryptography an eavesdropper must factor a publicly known number in order to obtain the private key. If factoring were easy RSA would be insecure. We have more to say about factoring on a quantum computer later.

Quantum Complexity Classes

Just as there are classical complexity classes, so too are there quantum complexity classes. As quantum Turing machines are quantum mechanical generalizations of probabilistic Turing machines, the quantum complexity classes resemble the probabilistic complexity classes. There is a tradeoff between the certainty of your answer being correct versus the certainty of the answer being available within a certain time bound. In particular, the classical classes P, ZPP, and BPP become the quantum classes QP, ZQP, and BQP. These mean, respectively, that a problem can be solved with certainty in worst-case polynomial time, with certainty in average-case polynomial time, and with probability greater than 2/3 in worst-case polynomial time, by a quantum Turing machine.

Table 2.4 Some new, quantum, complexity classes and their known relationships to classical complexity classes.

Quantum Class	Class of Computational Problems That Can ...	Relationships to Classical Complexity Classes (if known)
QP	... be solved, with certainty, in worst case polynomial time by a quantum computer.	$P \subset QP$ (the quantum computer can solve more problems in worst-case polynomial time than can the classical computer).
BQP	... be solved in worst case polynomial time by a quantum computer with probability > 2/3 (i.e., the probability of error is bounded, hence the "B" in "BQP").	$BPP \subseteq BQP \subseteq PSPACE$ (i.e., it is not known whether QTMs are more powerful than PTMs). Shor proved factoring is in BQP.
ZQP	... be solved with zero error probability in expected polynomial time by a quantum computer.	$ZPP \subset ZQP$.

Statements about the relative power of one type of computer over another can be couched in the form of subset relationships among complexity classes. Thus QP is the class of problems that can be solved, with certainty, in polynomial time, on a quantum computer, and P is the set of problems that can be solved, with certainty, in polynomial time on a classical computer. As the class QP contains the class P (see Table 2.4) this means that there are more problems that can be solved efficiently by a quantum computer than by any classical computer. Similar relationships are now know for some of the other complexity classes too, but there are still many open questions remaining.

The study of quantum complexity classes began with David Deutsch in his original paper on quantum Turing machines (QTMs). The development of the field is summarized in Table 2.5.

In Deutsch's original paper he presented the idea of quantum parallelism. Quantum parallelism allows you to compute an exponential number of function evaluations in the time it takes to do just one function evaluation classically. Unfortunately, the laws of quantum mechanics make it impossible to extract more than one of these answers explicitly. The problem is that although you can indeed calculate all the function values for all possible inputs at once, when you read off the final answer from the tape, you will only obtain one of the many outputs. Worse still, in the process, the information about all the other outputs is lost irretrievably. So the net effect is that you are no better off than had you used a classical Turing machine. So, as far as function evaluation goes, the quantum computer is no better than a classical computer.

Deutsch realized that you could calculate certain *joint properties* of all of the answers without having to reveal any one answer explicitly. The example he gave concerned computing the XOR (exclusive-or) of two outputs. Suppose there is a function f that can receive one of two inputs, 0 and 1, and that we are interested in computing the XOR of both function values, that is, $f(0) \oplus f(1)$. The result could, for example, be a decision as to whether to make some particular stock investment tomorrow based on today's closing prices. Now suppose that, classically, it takes 24 hours to evaluate each f. Thus if we are stuck with a single classical computer, we would never be able to compute the XOR operation in time to make the investment the next day. On the other hand, using quantum parallelism, Deutsch showed that half the time we would get no answer at all, and half the time we would get the guaranteed correct value of $f(0) \oplus f(1)$. Thus the quantum computer would give useful advice half the time and never give wrong advice.

Richard Jozsa refined Deutsch's ideas about quantum parallelism by showing that many functions, for example, SAT (the propositional satisfiability problem) cannot be computed by quantum parallelism at all[Jozsa91]. Nevertheless, the question about the utility of quantum parallelism for tackling computational tasks that were *not* function calculations remained open.

In 1992 Deutsch and Jozsa exhibited a problem, that was not equivalent to a function evaluation, for which a QTM was exponentially faster than a classical deterministic Turing Machine (DTM). The problem was rather contrived, and consisted of finding a true statement in a list of two statements. It was possible that both statements were true, in which case either statement would be acceptable as the answer. This potential multiplicity of solutions meant that the problem could not be reformulated as a function evaluation. The upshot was that the QTM could solve the problem in a time that was a polynomial in the logarithm of the problem size (poly-log time), but that the DTM required linear time. Thus the QTM was exponentially faster than the DTM. The result was only a partial success, however, as a probabilistic Turing machine (PTM) could solve it as efficiently as could the QTM. But this did show that a quantum computer at least could beat a deterministic classical computer.

So now the race was on to find a problem for which the QTM beat a DTM *and* a PTM. Ethan Bernstein and Umesh Vazirani

[Bernstein93] analyzed the computational power of a QTM and found a problem that did beat both a DTM and a PTM. Given any Boolean function on n-bits Bernstein and Vazirani showed how to sample from the Fourier spectrum of the function in polynomial time on a QTM. It was not known if this were possible on a PTM. This was the first result that hinted that QTMs might be more powerful than PTMs.

The superiority of the QTM was finally clinched by André Berthiaume and Gilles Brassard who constructed an "oracle" relative to which there was a decision problem that could be solved with certainty in worst-case polynomial time on the quantum computer, yet cannot be solved classically in probabilistic expected polynomial time (if errors are not tolerated). Moreover, they also showed that there is a decision problem that can be solved in exponential time on the quantum computer, that requires double exponential time on all but finitely many instances on any classical deterministic computer. This result was proof that a quantum computer could beat both a deterministic and probabilistic classical computer but it was still not headline news because the problems for which the quantum computer were better were all rather contrived.

The situation changed when, in 1994, Peter Shor, building on work by Dan Simon, devised polynomial-time algorithms for factoring and discrete logarithms. The latter two problems are believed to be intractable for any classical computer, deterministic or probabilistic. But more important, the factoring problem is intimately connected with the ability to break the RSA cryptosystem which is in widespread use today. Thus if a quantum computer could break RSA, then a great deal of sensitive information suddenly becomes vulnerable, at least in principle. Whether it is vulnerable in practice depends, of course, on the feasibility of designs for actual quantum computers.

Of course, computer scientists would like to develop a repertoire of quantum algorithms that can, in principle, solve significant computational problems faster than any classical algorithm. Unfortunately, the discovery of Shor's algorithm for factoring large composite integers, was not followed by a wave of new quantum algorithms for lots of other problems. To date, there are only about seven quantum algorithms known. These are the algorithms of Deutsch/Jozsa[Deutsch92] (true statement problem), Simon [Simon94], Shor[Shor94] (factoring), Kitaev[Kitaev95] (factoring), Grover (database search[Grover96] and estimating the median [Grover96a]), and Durr/Hoyer[Durr96] (estimating the mean). Many of these quantum algorithms rely upon a quantum version of the Fourier transform.

Table 2.5 Historical development of quantum complexity theory.

Year	Advance in Quantum Complexity Theory
Benioff (1980)[Benioff80]	Shows how to use quantum mechanics to implement a Turing Machine (TM).
Feynman (1982)[Feynman82]	Shows TMs cannot simulate quantum mechanics without exponential slowdown.
Deutsch (1985)[Deutsch85]	Proposes first universal QTM and the method of quantum parallelism. Quantum parallelism returns a correct answer or no answer, but never an incorrect answer.
	Proves QTMs have the same complexity class with respect to function calculation as TMs. Remarks that certain computational tasks do not require function calculation. Exhibits such a task that is solved faster on a QTM than on a TM.
Jozsa (1991)[Jozsa91]	Describes classes of functions that can and cannot be computed by quantum parallelism.
Deutsch and Jozsa (1992)[Deutsch92]	Exhibit a contrived problem that the QTM solves with certainty in poly-log time but which requires linear time of a deterministic TM (DTM). Thus the QTM is exponentially faster than the DTM. Unfortunately, this problem is also easy for a probabilistic TM (i.e. , a PTM), so it is not a complete victory over classical machines.
Berthiaume & Brassard (1992)[Berthiaume92]	Proved $P \subset QP$ (strict inclusion). The first definitive complexity separation between classical and quantum computers.
Bernstein and Vazirani (1993)[Bernstein93]	Describe a universal QTM that can simulate any other QTM efficiently (Deutsch's QTM could simulate other QTMs but only with an exponential slowdown).
	Show how to sample from the Fourier spectrum of a Boolean function on *n* bits in polynomial time on a QTM (set up for Simon's paper and Shor's paper).
Yao (1993)[Yao93]	Shows complexity theory for quantum circuits matches that of QTMs. This legitimizes the study of quantum circuits (which are simpler to design and analyze than QTMs).
Berthiaume and Brassard (1994)[Berthiaume94]	Prove randomness alone is not what gives QTMs the edge over TMs.
	Prove there is a decision problem that is solved in polynomial time by a QTM but which requires exponential time, in the worst case, on a DTM or PTM. First time anyone showed a QTM beat a PTM.
	Prove there is a decision problem that is solved in exponential time on a QTM but which requires *double* exponential times on a DTM on all but a few instances.
Simon (1994)[Simon94]	Lays foundational work for Shor's algorithm.
Shor (1994)[Shor94]	Discovers a polynomial time quantum algorithm for factoring large integers. This is the first *significant* problem for which a quantum computer is shown to outperform any type of classical computer. Factoring is related to breaking codes in widespread use today.
Grover (1996)[Grover96]	Discovers a quantum algorithm for finding a single item in an unsorted database in square root of the time it would take on a classical computer

The Equivalence of Quantum Turing Machines and Quantum Circuits

Turing machines are fine as abstract models of computers but they are a far cry from how real computers are implemented. A better model from a practical standpoint is the quantum circuit model.

A quantum circuit consists of a set of quantum gates with "wires" indicating how the outputs from one gate become the inputs to other gates. Each quantum gate is reversible in the sense that its outputs can be inferred from its inputs and vice versa. Mathematically, the transformation between inputs and outputs is described by a unitary operator. As we show in the next chapter, this is very important from the point of view of describing the overall operation of the circuit in quantum mechanical terms.

Of greater concern here is whether the quantum circuit model admits the same complexity classes as the quantum Turing machine model. Certainly in the classical context Turing machines and circuits are indeed equivalent. Fortunately, the equivalence between quantum circuits and quantum Turing machines was proved by Andrew Yao in 1993[Yao93]. Yao showed that any function computable in polynomial time on a QTM had a polynomial-sized quantum circuit. Although this is a theoretical result, it legitimizes much of the work on quantum algorithms as, ultimately, those algorithms are likely to be implemented as specialized quantum circuits rather than quantum "programs" running on a universal quantum Turing machine.

CHAPTER 3

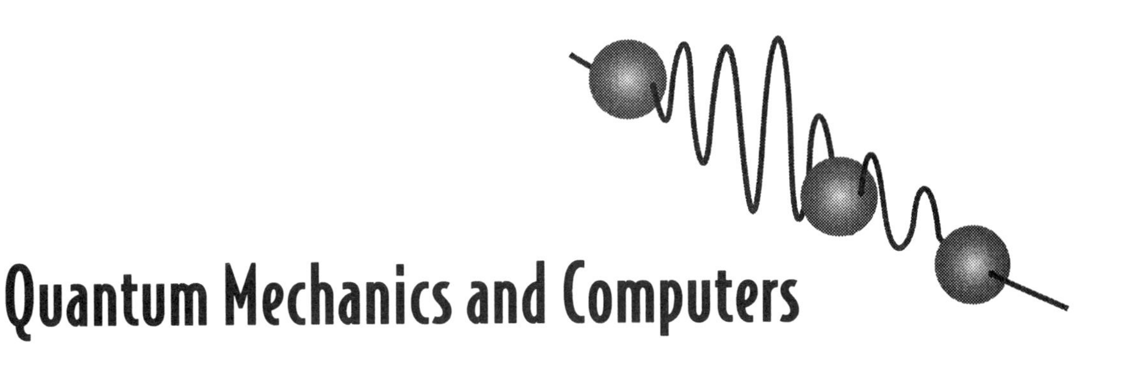

Quantum Mechanics and Computers

"I've got some grad student. He's thinking about the meaning of quantum mechanics. He's doomed!"
— John McCarthy

The theory of computation, developed by Alan Turing, Alonso Church, Emil Post, and Kurt Gödel, was thought to be free of any assumptions regarding how a computer was implemented. By making theoretical computer science a branch of pure mathematics, it became possible to prove all sorts of theorems regarding what was and was not computable, and how efficiently certain computations could be done, and it also revealed hidden equivalences between superficially different problems. This work culminated in some of the intellectual highlights of the 20th century including Gödel's result that in any formal system rich enough to describe arithmetic there were theorems that could never be proved, and the Church-Turing thesis, which asserts that any computable function can be computed on a Turing machine.

3.1 Physics and Computers

Around the mid to late 1970s, however, some scientists began to question the possible connections between physics and computation. Initially, these efforts focused on understanding the thermodynamics of classical computation, asking questions such as how little energy could be expended to accomplish a particular computation, how much heat is dissipated when a bit is erased from memory, are there fundamental limits on the rate at which information can be processed[Deutsch82], and whether there is an upper bound on the maximum speed at which a machine can compute[Margolus96].

The resulting hypothetical computers served as ideal thought experiments that revealed some intriguing connections between logical reversibility and thermodynamic reversibility. Certain logical operations, such as the NOT operation, are logically reversible because, given knowledge of the output bit, you can infer the input bit unambiguously and vice versa. Conversely, other operations, such as the bitwise AND of two bits, are logically irreversible because it is not always possible to infer the input bits given only knowledge of the output bit. For example, if the output bit is a 0, the input bits could be 00, 10, or 01. Unfortunately, IBM physicist Charles Bennett [Bennett87], building on the work of Szilard[Szilard29], showed that any irreversible logical operation had to dissipate a certain minimum amount of energy. In fact, this is true whenever a computation throws information away such as taking the bitwise AND of the bits 0 and 1 or erasing a bit from a memory register.

However, a logically reversible operation need not dissipate any energy. In fact, Bennett was able to show that a complete logically reversible Turing machine was a theoretical possibility[Bennett73].

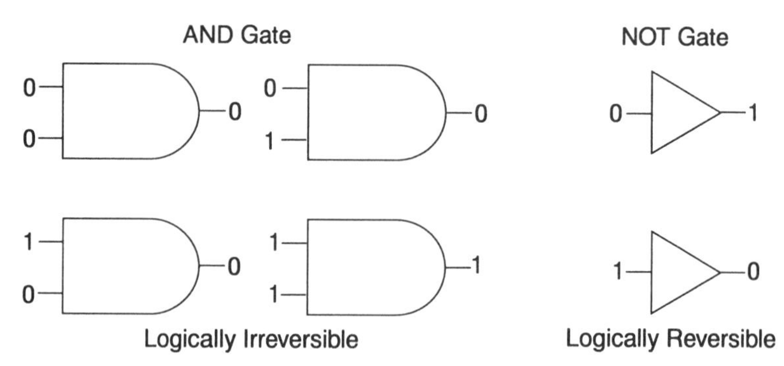

Fig. 3.1 Shows how the outputs of an AND gate and a NOT gate are related to their possible inputs. The NOT gate is logically reversible because you can determine the input from the output and vice versa. However, the AND gate is logically irreversible because it is impossible to tell exactly what the inputs must have been if all you are told is the output was a 0.

However, to actually build a reversible computer, one would need reversible logic gates. So it became an interesting question as to whether there existed a classical logic gate that was both reversible (in the sense that its inputs could be used to infer its outputs and vice versa) and universal (in the sense that any computable function could be computed using a circuit built out of such gates).

A priori it was not at all clear that such a gate was possible. However, in 1982 Ed Fredkin and Tommaso Toffoli of M.I.T. proved that the laws of classical physics permitted a reversible, universal, classical logic gate[Fredkin82]. The actual proof is rather peculiar in that it envisions a logic gate that resembles a pinball machine, with bits being encoded as hard billiard balls shooting into one side of the gate, bouncing off one another and internal reflecting planes, and finally emerging from the other side of the gate. In such a "ballistic" computer, computation occurs at finite speed as a result of the dynamical evolution of perfect machinery started off in a precise initial condition. As each interaction inside the computer is reversible, the computer must function reversibly overall. Recently, the Fredkin gate was redesigned in a more practicable manner using photons to encode bits and nonlinear optics to control the output of an interferometer[Milburn89].

Such dissipationless computing schemes require perfect apparatus, which is, strictly speaking, impossible to obtain. The convex shape of the balls used to encode bits means that a small initial error would grow by about a factor of two with each interaction. Hence a slightly misaligned ball or reflecting plane would send a computation awry after just a few dozen interactions. The model can be redesigned in a manner less susceptible to errors by using "square" balls, but "square" balls are arguably harder to come by in Nature!

An alternative approach is to conceive of computation taking place in a dissipative environment, such as molecular dynamics. In this model, computation occurs as a net drift on top of an underlying "Brownian motion." Treating DNA replication as a computation problem for such a machine, Charles Bennett showed that the energy dissipated per logic step is only about 20 to 200 kT, which is far less than the energy dissipation per logical step in any conventional computer. Although dissipation is present, it is proportional to the computational velocity. Hence the Brownian motion computer can be made arbitrarily close to being reversible by driving the computation forward sufficiently slowly, becoming truly reversible in the limit of infinitesimal net computational velocity.

Neither the billiard ball scheme nor the Brownian motion scheme are satisfactory from a practical perspective as the billiard ball model requires an unattainable precision and the Brownian motion model is only reversible when the computation advances infinitesimally slowly. Nevertheless, the models demonstrated that the laws of classical physics allowed for the existence of a reversible computer, in principle. This point came to be crucial to the development of models of quantum computers, as we demonstrate shortly.

Although, initially, most theoretical computer scientists paid scant attention to the thermodynamic investigations, from an engineering perspective they demonstrated the potential for reversible logic devices and paved the way to technological advances such as reversible CMOS devices that dissipate almost no energy.

In addition to thermodynamic considerations, other physicists investigated the cosmological limits of computation[Tipler86]. If the Universe is finite, then presumably there is a limit to what can be computed due to the finiteness of the number of particles available for encoding bits! Such considerations are of little significance for day to day computing, but illustrate that there are limits to what is computable given that any real computer has to be a physically realizable device, built out of materials available in our Universe. Indeed, such considerations have led some to question the ultimate limits of what is knowable.

However, neither the thermodynamic nor cosmological limits revealed anything fundamentally new about the mathematical properties of computation. The thermodynamic and cosmological limits augmented the classical theory rather than superseded it.

The situation began to change in 1980, when Paul Benioff, a physicist at Argonne National Laboratory, formulated a model of a reversible Turing machine whose read, write, and shift operations were accomplished using *quantum mechanical* interactions [Benioff80]. Although the resulting machine still behaved, in computational respects, as a classical computer, this marked the first attempt to bring quantum mechanics into the heart of the mathematical model of a computer. Benioff's machine was also important in that, theoretically at least, it could be designed so that it dissipated no energy[Benioff82]. Nevertheless, whether they would ever be feasible to build was an open question. As IBM physicist Rolf Landauer put it [Landauer91] "They are not patent disclosures."

That a physically valid model of computation might confer something new was first hinted at by Richard Feynman who showed that no classical computer could simulate an arbitrary quantum system efficiently[Feynman82]. He speculated that a true quantum computer could do so. Hence this was the first hint that the computational efficiency of a quantum device, specifically a quantum simulator, might exceed the capabilities of a classical machine. This conjecture was finally proven to be correct by Christof Zalka and Seth Lloyd in 1996[Zalka96, Lloyd96].

At the very least, basing the model of computation on what is physically realizable as opposed to what is mathematically feasible obliges us to reformulate the celebrated theorems of computer science in physical terms. In particular, as physicist David Deutsch of Oxford University has pointed out, theorems such as the Church-Turing thesis ought really to be couched in terms of physical realizability. Deutsch proposes that the original Church-Turing thesis be replaced with[Deutsch85]:

> "Every finitely realizable physical system can be perfectly simulated by a universal model computing machine operating by finite means."

In order to appreciate how model computing devices would work, we now introduce the basic mathematical machinery of quantum mechanics and show how it can be applied to describe the operation of a quantum computer.

3.2 Taking the Quantum Leap

Quantum physics arose at the turn of the century in response to the failure of classical physics to provide the right predictions concerning the outcomes of experiments on light and elementary particles. Since then it has been subjected to the most intense scrutiny by the scientific establishment. Yet experiment after experiment has confirmed the predictions of quantum physics no matter how preposterous they might have seemed.

Despite such impressive successes quantum physics remains an enigma. It asks us to abandon many of our most deeply held assumptions about the nature of reality. It is often at odds with what most people regard as common sense. Yet it is currently the best tool for understanding the physics of the microworld and hence the operations of a quantum scale computer.

Quantum physics provides answers to three basic questions:

- How do you describe the state of a physical system?
- How does the state change if the system is not observed?
- How do you describe observations and their effects?

3.3 Quantization: From Bits to Qubits

There are a number of properties that quantum systems possess that lend themselves to computational applications. For example, at the quantum level, the values of certain observable quantities are restricted to a finite set of possibilities. The significance of this is that, in any computer, each bit must be stored in the state of some physical system. The various states must be sufficiently stable to ensure that the system will not flip spontaneously from a state representing a 0 to a state representing a 1. Contemporary machines use voltage levels. Babbage's Difference Engine used the position of gear teeth.

In a quantum computer, each bit could be represented by the state of a simple 2-state quantum system such as the spin state of a spin-$\frac{1}{2}$ particle. The spin of such a particle, when measured, is always found to exist in one of two possible states, represented as $|+\frac{1}{2}\rangle$ (spin-up) or $|-\frac{1}{2}\rangle$ (spin-down). This intrinsic "discreteness" is called quantization. As the spin of a particle is quantized we can use one spin state to represent the binary value 0, and the other spin state to represent the binary value 1. In fact, there is nothing special about spin systems. Any 2-state quantum system, such as the direction of polarization of a photon, or the discrete energy levels in an excited atom, would work equally well.

Once you have a way of encoding the binary values 0 and 1 in the states of a physical system, you can envisage making a complete memory register out of a chain of such systems.

3.4 State Vectors and Dirac Notation

How do we describe the state of quantum systems, such as spin-$\frac{1}{2}$ particles mathematically?

In quantum physics, the state of a quantum system is described by a vector in a Hilbert space. A mathematical "space" is just a fancy way of saying that something, in this case the representation of physical state, depends upon many independent coordinates that can be pictured as defining a set of perpendicular axes. The axes correspond to the possible states in which the physical system can be found. These possible states are called "eigenstates." The projection of the vector onto the axes shows the relative contributions of each eigenstate to the whole state, rather like the components of a classical vector in ordinary Euclidean space. We can

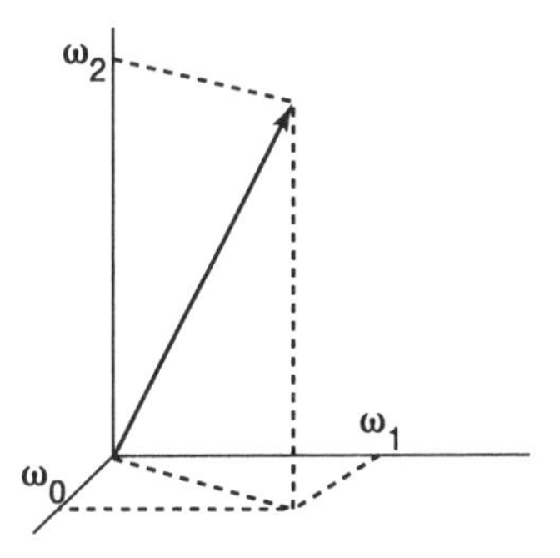

Fig. 3.2 The state of a quantum system can be described by a vector in a Hilbert space. The "lengths" of the axes are marked off in complex numbers. As the state evolves, the vector rotates in this space.

draw a picture such as Fig. 3.2 as an aid to visualizing what a state vector might look like.

You should not interpret such pictures too literally, however, because the complete mathematical pedigree of the Hilbert space asserts that it is a "complex linear vector space." The label "complex" signifies that the components of the state vector have lengths that are complex numbers, that is, numbers of the form $a+ib$ where a and b are real numbers and $i=\sqrt{-1}$. Unfortunately, it is impossible to draw a line on a piece of paper that truly has a complex-valued length, so the axes drawn in the figure can only hint at the structure of an authentic Hilbert space.

The label "linear vector space" signifies that if you can add and multiply vectors by numbers you end up with other vectors that lie in the same Hilbert space. In fact, as a physical system evolves, its state vector can be thought of as rotating with its base anchored to the axes' origin.

State vectors are usually written using a special angular bracket notation called a "ket vector" $|\psi\rangle$. The word "ket" was coined by Paul Dirac, a famous British physicist, who wanted a shorthand notation for writing the formulae that arise in quantum mechanics. Many of these formulae turn out to be products of a row vector with a related column vector (a state vector). Dirac wrote such products using a bracket notation, for example, $\langle\psi|\psi\rangle$. Thus the row vectors, such as $\langle\psi|$, became known as "bra" vectors and the column vectors, such as $|\psi\rangle$, became known as "ket" vectors so that when you put them together, $\langle\psi|\psi\rangle$, became a **bracket**.

You can think of the ket vector notation as being analogous to ordinary vector notation that writes a vector as a symbol in a boldfaced font or as a symbol with a line beneath it. A state vector is just a particular instance of a ket vector. Like ordinary vectors, state vectors are specified by a particular choice of basis vectors (the eigenstates) and a particular set of complex numbers, corresponding to the amplitudes with which each eigenstate contributes to the complete state vector. A simple 2-state system (the basic building block of a quantum memory register) can, by definition, be in one of two possible states. Consequently, such a system has two eigenstates, and hence its state vector has exactly two components. Thus you can write the state as

$$|\psi\rangle = \omega_0|\psi_0\rangle + \omega_1|\psi_1\rangle \equiv \begin{pmatrix} \omega_0 \\ \omega_1 \end{pmatrix}$$

where the weights ω_i are complex numbers and the eigenstates $|\psi_i\rangle$ form a complete orthogonal basis for the state vector $|\psi\rangle$. By "complete" we mean that *any* state vector in the Hilbert space can be represented as a weighted sum of just the $|\psi_i\rangle$. By "orthogonal" we mean that the eigenstates are all "perpendicular" to one another. The eigenstates therefore define a system of axes in the Hilbert space similar to the way unit vectors define a system of axes for regular vectors in Euclidean space.

Once the state vector $|\psi\rangle$ is known, the expected value of any observable attribute of the system can be calculated. The state vector $|\psi\rangle$ contains complete information about the associated system. This is similar to classical physics in which the complete state is known once the time-dependent functions for position and momentum are determined. Some physicists, such as Stephen Hawking, have even attempted to unravel the properties of the state vector that describes the entire Universe![Hartle83]

3.5 Superposition

This ability for quantum systems to exist in a blend of all their allowed states simultaneously, rather than existing in just one allowed state at a time, is known as the Principle of Superposition. Nothing like this is possible classically.

To get a more intuitive feel for superpositions, it is useful to picture the state of a 2-state quantum system, such as that used to encode a quantum bit (or qubit), as a vector contained in a sphere (see Fig. 3.3).

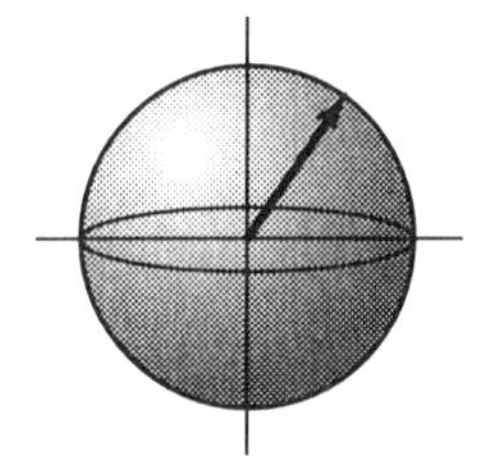

Fig. 3.3 A single qubit can be visualized as a vector contained in a sphere.

The angle this vector makes with the vertical axis is related to the relative contributions of the $|\psi_0\rangle$ and $|\psi_1\rangle$ eigenstates to the whole state. The angle through which the vector is rotated about the vertical axis corresponds to the "phase." The phase factors do not affect the relative contributions of the eigenstates to the whole state but they are crucially important in so-called quantum interference effects that we come to shortly. Thus a state can have the same proportions of 0-ness and 1-ness but actually have different amplitudes due to different phase factors. Figure 3.4 shows four qubits that encode the same proportions of 0-ness and 1-ness but have four different phase factors.

We can easily generalize the superposition principle beyond a 2-state system. Mathematically, if a quantum system can exist in any one of n eigenstates $|\psi_0\rangle$ or $|\psi_1\rangle$ or $|\psi_2\rangle$ or ... $|\psi_{n-1}\rangle$, it can also exist in the "superposed" state

$$|\psi\rangle = \sum_{i=0}^{n-1} |\psi_i\rangle \omega_i$$

Unfortunately, it becomes difficult to draw diagrams to illustrate such superpositions.

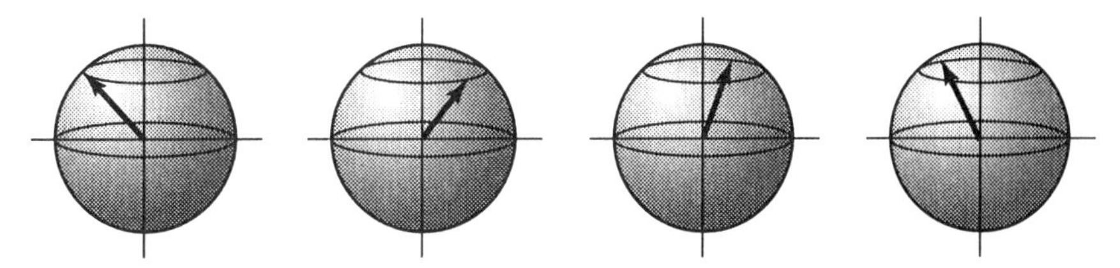

Fig. 3.4 Each sphere represents a qubit with the same proportions of the $|0\rangle$ and $|1\rangle$ states but with different phase factors, depicted as different values of the rotation about the vertical axis.

3.6 Probability Interpretation

What meaning can we attach to the weighting coefficients ω_i in the expansion of a state vector of an n-state quantum system $|\psi\rangle = \sum_{i=0}^{n-1} |\psi_i\rangle \omega_i$? Intuitively, the "larger" ω_i, relative to the other weighting coefficients, the more the eigenstate $|\psi_i\rangle$ contributes to the state vector $|\psi\rangle$. So we might expect that ω_i is related somehow to the probability of the system being in state $|\psi_i\rangle$. Unfortunately, the weighting coefficients are all complex numbers and we know that probabilities must be real numbers between zero and one. Fortunately, for any complex number, $z = x + iy$, the product of z with its complex conjugate is always a positive real number; that is, $z^* z = (x - iy)(x + iy) = x^2 + y^2 = |z|^2$ is a positive real number. In particular, this means that $\omega_i^* \omega_i = |\omega_i|^2$ is guaranteed to be a positive real number. So perhaps $|\omega_i|^2$ is the probability of the system being in state $|\psi_i\rangle$? Well, it could be, except that there is no guarantee that $|\omega_i|^2$ is between zero and one, which it has to be if it is a probability. Moreover, as the system can only exist in n possible states, $|\psi_0\rangle, |\psi_1\rangle, \ldots, |\psi_n\rangle$, the sum of the probabilities of it being in each state must add up to one. Given these two requirements, we are led to think of rescaling the weighting coefficients by dividing them by $\sum_{i=0}^{n-1} |\omega_i|^2$. Hence, for an n-state quantum system the probability of the system being in state $|\psi_i\rangle$ is given by

$$\Pr(\text{system in state } |\psi_i\rangle) = \frac{|\omega_i|^2}{\sum_{i=0}^{n-1} |\omega_i|^2}$$

The weighting coefficients ω_i are known as *probability amplitudes* (rather than probabilities). However, to simplify calculations it is customary to work with normalized state vectors, that is, state vectors for which the weighting coefficients have been rescaled so that the sum of their square moduli is equal to one.

3.7 Alternative Bases

The choice of a particular set of orthogonal axes is somewhat arbitrary. Different choices define what are called different "bases." For example, you could obtain a new set of axes by rotating any given set of axes through a fixed angle. Although the state vector does not change in this process, its projections onto the various axes do change. Consequently, the same state vector can assume a superficially different appearance if it is expressed in a different coordinate basis. In Fig. 3.5 we show the same state vector we drew earlier, represented in a different basis.

The significance of such basis transformations is this: each choice of basis defines a different set of eigenstates corresponding to the possible states in which a system may be found upon being measured. Thus the choice of what measurement you make on a quantum system is intimately tied to the possible outcomes you will obtain.

3.8 Eigenstates

A 2-state system has 2 eigenstates which we call $|\psi_0\rangle$ and $|\psi_1\rangle$. Thus the general state of a 2-state system can be written as a superposition of these eigenstates:

$$|\psi\rangle = \omega_0 |\psi_0\rangle + \omega_1 |\psi_1\rangle$$

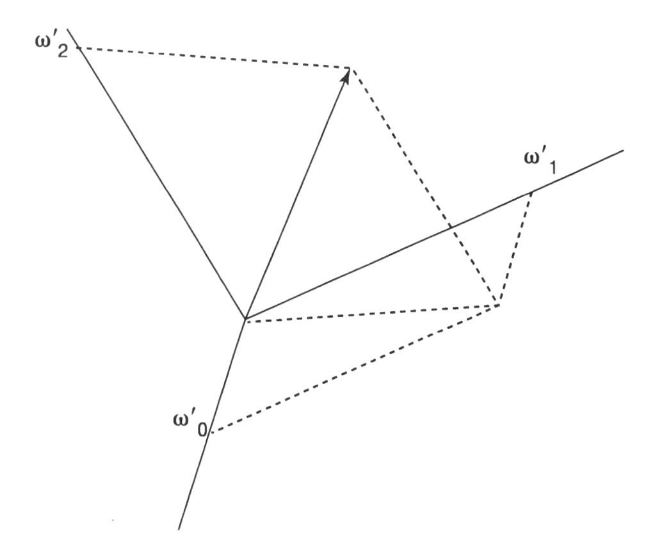

Fig. 3.5 The same state vector as shown in Fig. 3.2 in a different coordinate basis.

Once you specify a basis, in this case the basis $\{|\psi_0\rangle, |\psi_1\rangle\}$, the entire state vector is really just specified by the two amplitudes ω_0 and ω_1. If the system is wholly in the $|\psi_0\rangle$ state, representing the binary digit 0, then $\omega_0 = 1$ and $\omega_1 = 0$. Conversely, if the system is wholly in the state $|\psi_1\rangle$, representing the binary digit 1, $\omega_0 = 0$ and $\omega_1 = 1$. We can represent these situations conveniently by defining $|\psi_0\rangle$ and $|\psi_1\rangle$ to be equivalent to the column vectors $\begin{pmatrix}1\\0\end{pmatrix}$ and $\begin{pmatrix}0\\1\end{pmatrix}$, respectively. Thus the binary value 0 is represented by the state $|\psi_0\rangle = \begin{pmatrix}1\\0\end{pmatrix}$ and the binary value 1 is represented by the state $|\psi_1\rangle = \begin{pmatrix}0\\1\end{pmatrix}$. The superposition (which has no classical analogue) is represented by the state $|\psi\rangle$ given by

$$|\psi\rangle = \omega_0|\psi_0\rangle + \omega_1|\psi_1\rangle = \omega_0\begin{pmatrix}1\\0\end{pmatrix} + \omega_1\begin{pmatrix}0\\1\end{pmatrix} = \begin{pmatrix}\omega_0\\\omega_1\end{pmatrix}$$

To remind us that the eigenstates stand for bits we contract the notation and write $|0\rangle$ for $|\psi_0\rangle$ and $|1\rangle$ for $|\psi_1\rangle$. This substitution has done nothing other than to change the notation to a form closer to classical bits.

In terms of our vector-in-a-sphere picture of a qubit, the eigenstates $|0\rangle$ and $|1\rangle$ (corresponding to classical bits) can be pictured as shown in Fig. 3.6.

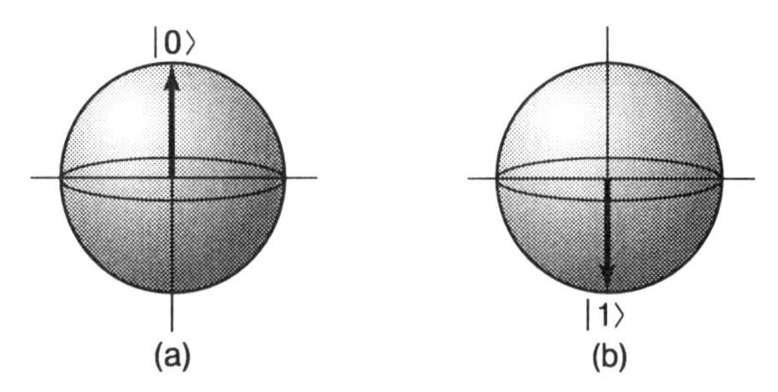

Fig. 3.6 A qubit can be pictured as a vector protruding from the center of a sphere. The south pole represents the eigenstate $|0\rangle$ and the north pole represents the eigenstate $|1\rangle$. Thus (a) shows a qubit for the binary value 0 and (b) shows a qubit for the binary value 1.

3.9 State of a Quantum Memory Register

So far we have described the state of a single 2-state system. But a memory register will consist of many such systems. How do we describe its state?

Quantum mechanics says you write it as the direct product:

$$\left|\psi^{(1)}\right\rangle \otimes \left|\psi^{(2)}\right\rangle = \left|\psi^{(1,2)}\right\rangle$$

where the superscripts merely indicate that the two qubits are associated with different Hilbert spaces (initially). Expanding the two state vectors into their column vector equivalents we have:

$$\left|\psi^{(1)}\right\rangle = \begin{pmatrix} \omega_0^{(1)} \\ \omega_1^{(1)} \end{pmatrix} \text{ and } \left|\psi^{(2)}\right\rangle = \begin{pmatrix} \omega_0^{(2)} \\ \omega_1^{(2)} \end{pmatrix}$$

In each eigenbasis we have:

$$\left|\psi^{(1)}\right\rangle = \left|\psi_0^{(1)}\right\rangle \omega_0^{(1)} + \left|\psi_1^{(1)}\right\rangle \omega_1^{(1)}$$

$$\left|\psi^{(2)}\right\rangle = \left|\psi_0^{(2)}\right\rangle \omega_0^{(2)} + \left|\psi_1^{(2)}\right\rangle \omega_1^{(2)}$$

where $\left|\psi_0^{(1)}\right\rangle = \begin{pmatrix} 1 \\ 0 \end{pmatrix}$, $\left|\psi_1^{(1)}\right\rangle = \begin{pmatrix} 0 \\ 1 \end{pmatrix}$, $\left|\psi_0^{(2)}\right\rangle = \begin{pmatrix} 1 \\ 0 \end{pmatrix}$ and $\left|\psi_1^{(2)}\right\rangle = \begin{pmatrix} 0 \\ 1 \end{pmatrix}$.

Multiplying out we obtain:

$$\left|\psi^{(1)}\right\rangle \otimes \left|\psi^{(2)}\right\rangle = \begin{pmatrix} \omega_0^{(1)} \\ \omega_1^{(1)} \end{pmatrix} \otimes \begin{pmatrix} \omega_0^{(2)} \\ \omega_1^{(2)} \end{pmatrix} = \begin{pmatrix} \omega_0^{(1)}\omega_0^{(2)} \\ \omega_0^{(1)}\omega_1^{(2)} \\ \omega_1^{(1)}\omega_0^{(2)} \\ \omega_1^{(1)}\omega_1^{(2)} \end{pmatrix} = \left|\psi^{(1,2)}\right\rangle = \begin{pmatrix} \omega_{00} \\ \omega_{01} \\ \omega_{10} \\ \omega_{11} \end{pmatrix}$$

The new basis states of the composite system are $|00\rangle$, $|01\rangle$, $|10\rangle$, and $|11\rangle$ where

$$|00\rangle = \begin{pmatrix} 1 \\ 0 \\ 0 \\ 0 \end{pmatrix}, \quad |01\rangle = \begin{pmatrix} 0 \\ 1 \\ 0 \\ 0 \end{pmatrix}$$

$$|10\rangle = \begin{pmatrix} 0 \\ 0 \\ 1 \\ 0 \end{pmatrix}, \quad |11\rangle = \begin{pmatrix} 0 \\ 0 \\ 0 \\ 1 \end{pmatrix}$$

and a general state of a 2-bit memory register is therefore

$$\left|\psi^{(1,2)}\right\rangle = \omega_{00}|00\rangle + \omega_{01}|01\rangle + \omega_{10}|10\rangle + \omega_{11}|11\rangle.$$

The generalization to an n-qubit quantum memory register is straightforward. The electronic supplement contains software that allows you to create, manipulate, and analyze general states of n-qubit quantum memory registers. For example, to create the general state of a 3-qubit memory register, $|\psi^{(1)}\rangle \otimes |\psi^{(2)}\rangle \otimes |\psi^{(3)}\rangle$, where each $|\psi^{(i)}\rangle$ has the form $|\psi^{(i)}\rangle = \omega_0^{(i)}|0\rangle + \omega_1^{(i)}|1\rangle$, you could enter the command:

In[]:=

```
Direct[w0[1] ket[0] + w1[1] ket[1],
       w0[2] ket[0] + w1[2] ket[1],
       w0[3] ket[0] + w1[3] ket[1] ]
```

which is a fairly literal translation of the standard equation, except that we use "ket[x]" to represent the ket vector, $|x\rangle$.

One of the most important properties of quantum memory registers is their ability to store an exponential amount of classical information in only a polynomial number of qubits by exploiting the Principle of Superposition. In Figure 3.7, for example, are two classical memory registers storing complementary sequences of bits. However, a single quantum memory register can store both sequences simultaneously in an equally weighted superposition of the two states representing each classical input. That is, the quantum register stores the state $\frac{1}{\sqrt{2}}(|1011001\rangle + |0100110\rangle)$.

In general, the amplitudes and phase factors may conspire to produce some very complicated superpositions, such as those in Figure 3.8.

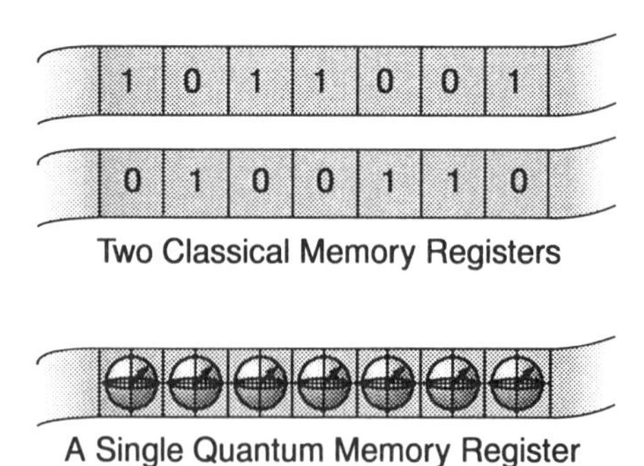

Fig. 3.7 A quantum memory register can store multiple sequences of classical bits in superposition.

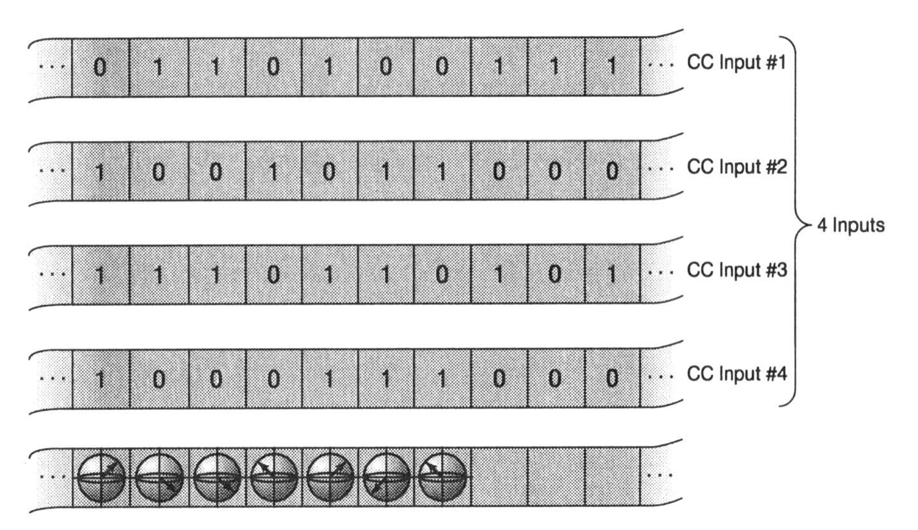

Fig. 3.8 A quantum memory register can store an exponential number of inputs using only a polynomial number of qubits.

In fact, it is also possible to evolve the set of qubits in a quantum memory register into a state that cannot be expressed, mathematically, as the direct product of states of the individual qubits that comprise the register. In such an entangled state the memory register as a whole has a well-defined quantum state even though its component qubits do not!

If you were to attempt to read a quantum memory register, that was in such an entangled state, by measuring the state of each qubit in turn, then the values that you will obtain for later qubits in your sequence will be determined by the outcomes of the measurements made on the earlier qubits in the sequence. Thus if a quantum memory register exists in an entangled state, you can change the state of one part of the register simply by measuring another part of the register. This phenomenon is important in many quantum algorithms and has no parallel in classical physics. Later in the book we show how entangled states can be put to work in teleportation, information transmission, and error-correcting codes. It is really entanglement that distinguishes quantum computing from classical computing.

3.10 Unitary Operators

The mathematical description of how something changes in the quantum world is given by the application of an operator. An operator is represented by a square matrix. If the quantum system consists of n 2-state quantum systems (as would the memory register of a quantum computer), then any operator that acts on the memory register would be represented by a $2^n \times 2^n$ dimensional matrix. Thus the general form for an operator that acts on a single qubit would be a 2×2 matrix:

$$\begin{pmatrix} a & b \\ c & d \end{pmatrix}\begin{pmatrix} \omega_0 \\ \omega_1 \end{pmatrix} = \begin{pmatrix} a\omega_0 + b\omega_1 \\ c\omega_0 + d\omega_1 \end{pmatrix}$$

Here's an operator that maps eigenstates into superpositions:

$$\hat{U}(\theta) = \begin{pmatrix} \cos\theta & -\sin\theta \\ \sin\theta & \cos\theta \end{pmatrix}$$

To get an equally weighted superposition, we rotate a state vector through $\pi/4$

$$\hat{U}\left(\frac{\pi}{4}\right)|0\rangle = \hat{U}\left(\frac{\pi}{4}\right)\begin{pmatrix} 1 \\ 0 \end{pmatrix} = \begin{pmatrix} \frac{1}{\sqrt{2}} \\ \frac{1}{\sqrt{2}} \end{pmatrix} = \frac{1}{\sqrt{2}}(|0\rangle + |1\rangle)$$

Similarly, you can think of logic gates as operators. The NOT operator defined as $\text{NOT} \equiv \begin{pmatrix} 0 & 1 \\ 1 & 0 \end{pmatrix}$, flips the state of its input, like so:

$$\text{NOT}|0\rangle = \begin{pmatrix} 0 & 1 \\ 1 & 0 \end{pmatrix}\begin{pmatrix} 1 \\ 0 \end{pmatrix} = \begin{pmatrix} 0 \\ 1 \end{pmatrix} = |1\rangle$$

$$\text{NOT}|1\rangle = \begin{pmatrix} 0 & 1 \\ 1 & 0 \end{pmatrix}\begin{pmatrix} 0 \\ 1 \end{pmatrix} = \begin{pmatrix} 1 \\ 0 \end{pmatrix} = |0\rangle$$

You will notice that the $U(\theta)$ and the NOT operators are invertible in the sense that you can infer the output from the input and vice versa. Indeed, as we shall see shortly, the evolution of any isolated unobserved quantum system is always described by such operators. In mathematical terms, these operators are described by unitary matrices. The defining property of a unitary matrix is that its conjugate transpose is equal to its inverse.

Let us turn back to computation. A set of logic gates is said to be "universal" if any feasible computation can be accomplished in a circuit comprising only gates of the type found in that set. In classical computation, it is well known that there are many universal sets of gates. For example, any classical computation can be accomplished using a circuit built out of AND and NOT gates. A similar result holds for quantum computation. Adriano Barenco, and collaborators, at Oxford University, have shown that a 2-qubit gate that has the form:

$$\hat{A}(\phi,\alpha,\theta)=\begin{pmatrix}1 & 0 & 0 & 0\\ 0 & 1 & 0 & 0\\ 0 & 0 & e^{i\alpha}\cos(\theta) & -ie^{i(\alpha-\phi)}\sin(\theta)\\ 0 & 0 & -ie^{i(\alpha+\phi)}\sin(\theta) & e^{i\alpha}\cos(\theta)\end{pmatrix}$$

where ϕ, α, and θ are constant irrational multiples of π and one another, is universal for quantum computation[Barenco95]. David DiVincenzo of IBM arrived at a similar conclusion, independently[DiVincenzo95a]. Thus, any feasible quantum computation can be accomplished by connecting together gates of the form $\hat{A}(\phi,\alpha,\theta)$ into a quantum circuit.

3.11 Schrödinger's Equation

Although the operator level description is adequate for characterizing *what* computation is effected under the action of a particular logic gate, it tells us nothing about the dynamical evolution that transforms the input state to a logic gate into the output state of that gate. To understand this process, we need to understand the more general question of how a state vector evolves in time. As the state of a quantum memory register is described by some state vector, this amounts to asking by what rule does the memory register of a quantum computer evolve?

Fortunately, physicist Erwin Schrödinger already gave us an answer to this question back in 1926, long before quantum computers were ever imagined. The orthodox view says that the state vector of an isolated quantum system evolves in two radically different ways depending on whether it is being observed. If *unobserved*, the system will undergo a smooth continuous evolution governed by Schrödinger's equation:

$$i\hbar\frac{\partial|\psi(t)\rangle}{\partial t}=\hat{H}(t)|\psi(t)\rangle$$

where $i=\sqrt{-1}$, $\hbar=1.0545\times10^{-34}$Js and $\hat{H}(t)$ is the Hamiltonian operator (which is related to the total energy of the system). Equivalently, in terms of the temporal evolution of the amplitudes we have

$$i\hbar\frac{\partial\omega_i(t)}{\partial t}=\sum_j H_{ij}(t)\omega_j(t)$$

However, if the quantum system is observed it will appear to undergo a sudden, discontinuous, and unpredictable jump into an eigenstate that depends upon what measurement was being made.

Recently, physicists Nicolas Cerf and Chris Adami of Caltech have resolved the apparent dichotomy between the smooth continuous unitary evolution described by Schrödinger's equation and the sudden discontinuous quantum jumps that appear to occur during measurements, in a very elegant and compelling manner. Cerf and Adami point out that, in fact, no quantum system is truly "isolated" from its environment and if you track the changes in the joint state of the quantum system and its environment, carefully, you will see that the state vector does not actually "collapse."[Cerf96a] Similar insights were arrived at earlier by Zurek, Zeh and Joos[Joos95, Zeh95, Zurek81, Zurek82, Zurek86, Zurek91, Zurek93]. Cerf and Adami improved upon these ideas by introducing a quantum version of information theory to describe the exchange of information between a system and its environment.

3.12 What Does the Hamiltonian Mean Physically and Computationally?

The Schrödinger equation contains the "Hamiltonian." The Hamiltonian is an operator (in fact, an Hermitian matrix) related to the total energy of the system. Its form is determined by the specific arrangement of atoms, molecules, and charges that constitute the computer.

You can think of the Hamiltonian as being analogous to the "hardware" of a conventional computer and the initial state of the quantum memory register as being analogous to the "data" fed into a conventional computer. Of course, for specialized quantum computers that only perform a single type of computation, the "program" is essentially folded into the definition of the Hamiltonian too, like the way ultrafast chips implement specialized programs in hardware. However, for the hypothetical universal quantum computer envisaged by Deutsch, the program would be part of the input state in which the machine was started.

The first actual quantum computer is likely to be specialized to perform a specific task. At this point in time, the most likely applications seem to be testing quantum mechanics, simulating quantum systems, teleporting information, finding a record in an unsorted database or factoring large integers.

3.13 Unitary Evolution

If the Hamiltonian is time independent, and the computer is started off with its memory register in the state $|\psi(0)\rangle$, then we can write the general solution of the Schrödinger equation as:

$$|\psi(t)\rangle = e^{-i\hat{H}t/\hbar}|\psi(0)\rangle = \hat{U}(t)|\psi(0)\rangle$$

where $\hat{U}(t) \equiv e^{-i\hat{H}t/\hbar}$ is called the evolution operator. The operator $\hat{U}(t)$ is always a unitary matrix. This means that the conjugate transpose of $\hat{U}(t)$ is equal to the inverse of $\hat{U}(t)$. This superficially innocuous property harbors an important implication. It means that the evolution operator of an ideal quantum computer, isolated from its environment, is reversible. Hence any ideal quantum computer must also be a reversible computer. We can now see why the results on reversible computing that began with the work of Bennett in 1973 were so important in the development of quantum computing. Quantum physicists knew all along that Schrödinger's equation gives rise to a unitary (and hence reversible) evolution. Thus if a quantum system had any chance of serving as a computer, it had to be possible to make computers that operated reversibly. This is exactly what Charles Bennett showed in 1973, which opened the way for others to speculate on a fully quantum model of a reversible computer. This line of thinking culminated in Benioff's reversible Turing machine, which operated using quantum mechanical interactions.

3.14 Interference

Quantum interference can occur whenever there is more than one way to obtain a particular result. In the case of the famous "double-slit" experiment, a particle shot towards a double slit has two possible pathways to transit the slit. These pathways interfere with one another quantum mechanically leading to a wave-like pattern of probability amplitude for finding the particle in various locations on the far side of the slit. This effect is illustrated in Fig. 3.9 which shows four snapshots, taken from an animation created by Terry Robb, of a particle passing through a double slit[Robb93]. The full animation can be found on the CD-ROM that accompanies this book.

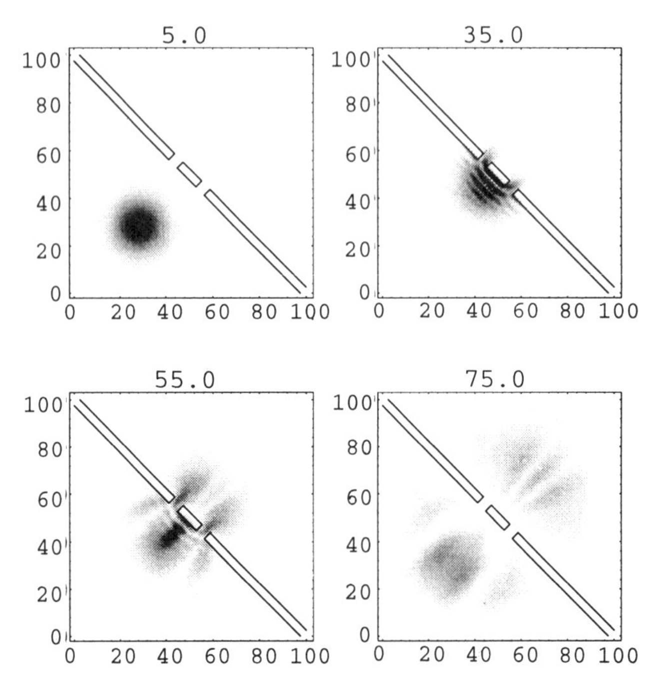

Fig. 3.9 A particle impinging on a double slit. Notice the interference pattern beyond the double slit (upper right quadrant of the lower right frame). Several other quantum animations can be found in [Robb93].

To obtain the effect of multiple pathways available to a quantum system, one must *first* add together the state vectors corresponding to each alternative and *then* calculate the net probability of obtaining the result from the square moduli of the weights in the composite state vector. Because the amplitudes are complex numbers with a wide variety of phases, these sums can reinforce or cancel out, leading to the enhancement or suppression of certain eigenstates in the output state vector.

Here is a simple example that illustrates the effects of interference on the outcome of a computation.

Let us suppose that we are dealing with the computational problem of generating a random bit, that is, 0 or 1 (or True/False, Up/Down, Left/Right, etc.). Suppose that there are two methods for computing the bit, which we call method A and method B. Suppose that the methods are a little biased so that, using method A, you obtain the answer 0 with probability 7/20. However, using method B, you obtain the answer 0 with probability 4/5. Now suppose that a classical computer for solving your problem works by first picking the method (A or B) and then using its selection to solve the problem. Let us assume that the computer picks method A with probability p (and picks method B with probability $1 - p$).

Mathematically, we can compute the overall probability of obtaining the answer 0 as follows:

$$P_{\text{classical}} = p\,p_A + (1-p)p_B$$
$$= \cos^2(\phi)p_A + \sin^2(\phi)p_B$$
$$P_{\text{classical}} = \frac{7}{20}p + \frac{4}{5}(1-p) = \frac{4}{5} - \frac{9}{20}p$$

The first term corresponds to picking method A and running it and the second term corresponds to picking method B and running it.

Now let us imagine a quantum approach to solving the same problem. As probability is a real number bounded between 0 and 1, we can, without loss of generality, write $p = \cos^2(\phi)$. Next we create the state, $|\psi\rangle$, that allows us to pick method A or method B in a manner that reflects the classical probabilities. That is, we define,

$$|\psi\rangle = \cos(\phi)|0\rangle + \sin(\phi)|1\rangle$$

where $p = \cos^2(\phi)$. We say we pick method A if we measure the state $|\psi\rangle$ and find it in the $|0\rangle$ state. This event this happens with probability p because $\cos^2(\phi) = p$.

Now, let us create the states $|A\rangle$ and $|B\rangle$ that are the quantum analogues of the methods A and B as follows.

$$|A\rangle = \sqrt{\frac{7}{20}}|0\rangle + \sqrt{1-\frac{7}{20}}|1\rangle$$
$$|B\rangle = -\frac{2}{\sqrt{5}}|0\rangle + \sqrt{1-\frac{4}{5}}|1\rangle$$

Thus if we measure state $|A\rangle$ (in the $\{|0\rangle,|1\rangle\}$ basis) let us say we get the answer "0" if we find the system in state $|0\rangle$ and let us say we get a "1" if we find the system in state $|1\rangle$. Using the definitions given, if we measure state $|A\rangle$ we obtain a "0" with probability 7/20. Likewise, if we measure state $|B\rangle$ we would find it in the $|0\rangle$ state with probability 4/5.

Therefore, the quantum probability of obtaining a "0" is given by

$$P_{\text{quantum}} = \frac{\left|\cos(\phi)\omega_A^{(0)} + \sin(\phi)\omega_B^{(0)}\right|^2}{\left|\cos(\phi)\omega_A^{(0)} + \sin(\phi)\omega_B^{(0)}\right|^2 + \left|\cos(\phi)\omega_A^{(1)} + \sin(\phi)\omega_B^{(1)}\right|^2}$$

$$p_{\text{quantum}} = \frac{23 - 9\cos(2\phi) - 8\sqrt{7}\sin(2\phi)}{40 - \left(8\sqrt{7} - 4\sqrt{13}\right)\sin(2\phi)}.$$

But the classical probability of obtaining a "0" is

$$P_{\text{classical}} = \frac{4}{5} - \frac{9}{20}\cos^2(\phi) = \frac{23 - 9\cos(2\phi)}{40}.$$

Hence the quantum probability can be equal to, greater than, or less than the classical probability depending on the choice of ϕ. This is the hallmark of interference wherein the probability amplitudes of accomplishing a given task are added first and then the resulting probabilities computed from the moduli squared of the final amplitudes. An alternative way of visualizing this data is to plot the pairs of corresponding classical and quantum probabilities for each possible angle ϕ. Data points above the 45° line correspond to regions where the quantum probability of obtaining the "0" outcome exceeds the classical probability of obtaining that outcome. For points below the 45° line the quantum probability of obtaining a "0" is less than the classical probability of obtaining a "0".

More extreme cases are revealing. For example, if Method A gives a 0 with probability 0, and Method B gives a zero with probability 2/5, then the corresponding probability diagrams show that the quantum method can have near certainty of yielding a zero at a particular value of ϕ.

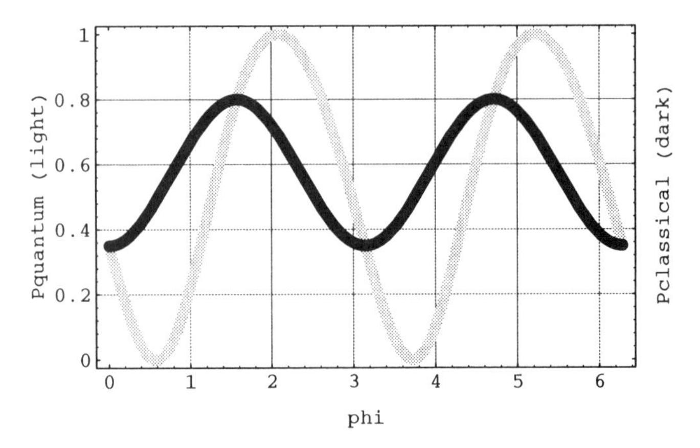

Fig. 3.10 Classical (black) and quantum (gray) probabilities of obtaining a "0" outcome for the random bit as a function of the angle ϕ. Method A gives a 0 with probability 7/20. Method B gives a 0 with probability 4/5.

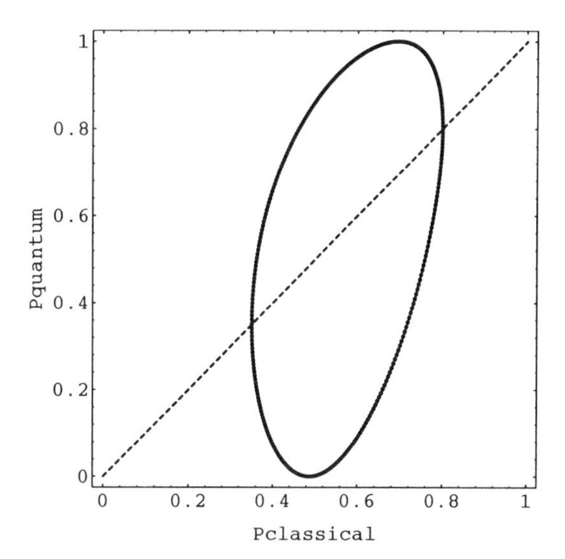

Fig. 3.11 The corresponding values of the classical and quantum probabilities of obtaining a "0" outcome for the random bit. Method A gives a 0 with probability 7/20. Method B gives a 0 with probability 4/5.

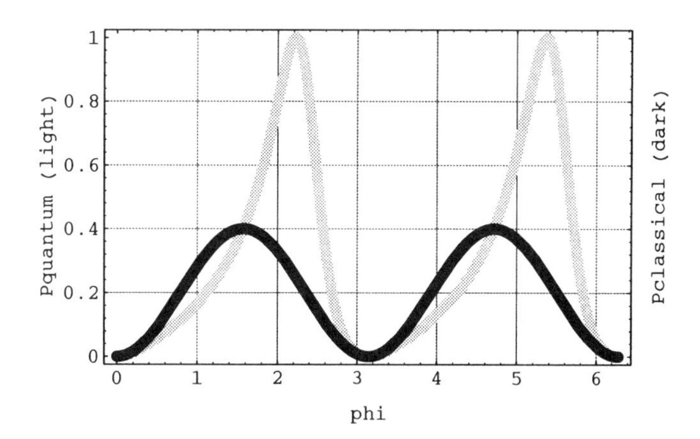

Fig. 3.12 Classical (black) and quantum (gray) probabilities of obtaining a "0" outcome for the random bit as a function of the angle ϕ. Method A gives a 0 with probability 0 (i.e., never). Method B gives a 0 with probability 2/5 (a 40% chance). Note that the quantum probability is hardly ever lower than the classical probability.

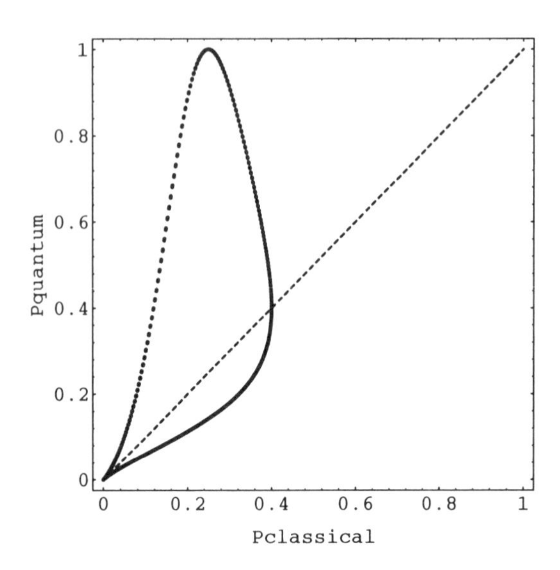

Fig. 3.13 The corresponding values of the classical and quantum probabilities of obtaining a "0" outcome for the random bit. Method A gives a 0 with probability 0. Method B gives a 0 with probability 2/5. The quantum probability can be much greater than the classical probability.

Similar interference effects can arise when the state of a quantum memory register is allowed to evolve unobserved. Bernstein and Vazirani provide the following example. Suppose memory states $|\psi^{(1)}(t)\rangle$ and $|\psi^{(2)}(t)\rangle$ both lead to memory state $|\psi(t+\Delta t)\rangle$ with amplitude $\sqrt{p}$. If we start the machine in either $|\psi^{(1)}\rangle$ or $|\psi^{(2)}\rangle$ and observe one time-step later, we see $|\psi\rangle$ with probability p. But if we start the machine in a superposition of $|\psi^{(1)}\rangle$ and $|\psi^{(2)}\rangle$, $\alpha_1|\psi^{(1)}\rangle+\alpha_2|\psi^{(2)}\rangle$, then how we observe it can greatly influence the outcome.

If we observe the superposed input at time 0 and then again one time-step later, we will see configuration $|\psi\rangle$ with probability $p\left(|\alpha_1|^2+|\alpha_2|^2\right)$. But if instead we do not observe the superposed input at time 0 and allow the computation to evolve in superposition, then we see memory state $|\psi\rangle$ one time-step later with probability $\left|\sqrt{p}\,\alpha_1+\sqrt{p}\,\alpha_2\right|^2=p|\alpha_1+\alpha_2|^2$. This can be greater than or less than $p\left(|\alpha_1|^2+|\alpha_2|^2\right)$. If $|\alpha_1+\alpha_2|^2>|\alpha_1|^2+|\alpha_2|^2$, we get constructive interference. Whereas if $|\alpha_1+\alpha_2|^2<|\alpha_1|^2+|\alpha_2|^2$, we get destructive interference. There will be extreme cancellation if $\alpha_1=-\alpha_2$.

3.15 Observables as Hermitian Operators

When certain operators act on eigenstates they reproduce the state multiplied by a real number.

$$\hat{A}|\psi_i\rangle = a_i|\psi_i\rangle$$

This captures, in some sense, what happens when we make an idealized measurement. We obtain a number (the eigenvalue) corresponding to the state (the eigenstate). Consider the most general 2 × 2 operator:

$$\hat{A} = \begin{pmatrix} a & b \\ c & d \end{pmatrix}$$

Its eigenstates (also called "eigenvectors") are:

$$|\psi_0\rangle = \begin{pmatrix} -\frac{-a+d+\sqrt{a^2+4bc-2ad+d^2}}{2c} \\ 1 \end{pmatrix}$$

$$|\psi_1\rangle = \begin{pmatrix} -\frac{-a+d-\sqrt{a^2+4bc-2ad+d^2}}{2c} \\ 1 \end{pmatrix}$$

When the operator acts on its eigenstates, it simply multiplies the eigenvector by a number.

$$\hat{A}|\psi_0\rangle = \begin{pmatrix} a & b \\ c & d \end{pmatrix}\begin{pmatrix} -\frac{-a+d+\sqrt{a^2+4bc-2ad+d^2}}{2c} \\ 1 \end{pmatrix}$$

$$= \tfrac{1}{2}\left(a + d - \sqrt{a^2 + 4bc - 2ad + d^2}\right)\begin{pmatrix} -\frac{-a+d+\sqrt{a^2+4bc-2ad+d^2}}{2c} \\ 1 \end{pmatrix}$$

$$= a_0 |\psi_0/$$

similarly for $|\psi_1\rangle$.

The number that multiplies the eigenvector will be real whenever A is an Hermitian operator (i.e., a square matrix equal to its own conjugate transpose).

Thus the model of observables in quantum mechanics uses Hermitian operators. If the system is not in an eigenstate, then the outcome of a measurement will be one of the observables of the operator, but which observable cannot be predicted with certainty in advance. However, we can ascribe various probabilities to obtaining each of the possible outcomes.

3.16 Measurement: Extracting Answers From Quantum Computers

The result of any measurement of a quantum system described by the state vector $|\psi\rangle$ is always one of the eigenvalues of the operator $\hat{A}$, corresponding to the observable being measured. If the system is in an eigenstate of the operator (i.e., if $|\psi\rangle$ happens to be an eigenstate of $\hat{A}$), then the outcome of the measurement will be one of the eigenvalues corresponding to this eigenstate. If the system is in a superposition of states, the state vector can be expressed as a weighted sum of the eigenstates of the operator, such that the weights are complex numbers, whose square moduli add up to 1. If a system, whose state is described by a superposition, is measured, the probability of each possible outcome (i.e., each possible eigenvalue) is given by $|\omega_i|^2$ where ω_i is the contribution of the ith eigenstate to the superposition.

Thus if the memory register of a quantum computer is in a superposition of possible states, then the result of a measurement of the state of the register will not be completely predictable. However, you can predict the probability of obtaining each of the possible results using:

$$\Pr(\text{system in state } |\psi_i\rangle) = \frac{|\omega_i|^2}{\sum_{i=0}^{n-1} |\omega_i|^2}$$

Realize that this is the probability of finding the *whole* memory in a particular configuration (a particular sequence of bits) when it is finally observed. Moreover, measurements of a subset of the qubits in the register project out the state of the whole register into a subset of eigenstates consistent with the answers obtained for the measured qubits.

Sometimes you might be more interested in the *average value* of some observable. If the operator corresponding to the observable is called $\hat{A}$, then the average value of this observable is denoted as $\langle\hat{A}\rangle$. The average value of the observable is the average of the results you would obtain from repeated measurements of a collection of identically prepared quantum systems. If each of these systems is in the normalized state $|\psi\rangle$ initially, then $\langle\hat{A}\rangle$ is computed from:

$$\langle\hat{A}\rangle = \langle\psi|\hat{A}|\psi\rangle$$

where $\langle\psi|$ is the "bra" corresponding to the "ket" $|\psi\rangle$ (see Section 3.4). In terms of vectors, the bra $\langle\psi|$ is the row vector given by taking the conjugate transpose of the ket $|\psi\rangle$. Thus the average value of the observable is given by the dot product of the bra vector, the observable operator, and the state (ket) vector, in that order. We show examples of average values of observables in the next chapter and when we discuss NMR-based quantum computers in Chapter 11.

The foregoing sections have covered a lot of quantum mechanics notation. We are now going to apply this notation to the quantum description of computers. As you will see, at a sufficiently abstract level the quantum level description is quite simple. Of course, fleshing out these abstract ideas in terms of real quantum devices is considerably more difficult.

3.17 Benioff's Quantum Computer

Benioff introduced the idea that the "tape" of the Turing machine could be replaced by a sequence of simple 2-state quantum systems[Benioff80]. This provided a primitive way of encoding a sequence of binary digits. Similarly, the "head" of the Turing machine was replaced by a quantum mechanical interaction that could read and/or reset the value of the spin state. The "rules" of the Turing machine were replaced by a carefully designed Schrödinger equation so that an initial configuration of spins would evolve into a final set of spins which when decoded as bits would be interpretable as having performed some useful calculation. Thus the "program" the computer was executing was implicitly contained in the details of the Schrödinger equation. The machine evolved in steps of fixed duration such that, at the end of each step, the tape was always back in one of its fundamental states in which each spin was either totally "up" (representing a 1) or totally "down" representing a 0. However, *during* a step the machine could temporarily go into superpositions of spin states.

Thus Benioff's computer failed to make full use of the potential for superposed computations because at the conclusion of each computational step the head measured the state of the tape which collapsed any superpositions on the tape. Consequently quantum coherence, and hence interference between superposed computations, the hallmark of true quantum computation, was destroyed at the end of every step. Nevertheless, it was the first time anyone had demonstrated that quantum mechanical interactions could be made to mimic the operations of a Turing machine.

Benioff's computer is impractical as a design for any real computer because it relies upon interactions between the "head" and spins on the "tape" that could be very far apart. In a real computer it would be better to only have to deal with nearby spins. Worse still, to actually design the Hamiltonian needed to mimic a particular Turing machine, one would have to know, not only the program, but also the complete set of computational orbits. In other words, you would need to know the answer to the problem you were trying to compute in order to design the quantum computer to solve it! Fortunately, through refinements to his original model, Benioff showed that this problem can be fixed by including a time-dependent Hamiltonian, but the issue of remote spin interactions cannot[Benioff81], [Benioff82a], [Benioff82], [Benioff86].

In 1985 Richard Feynman of the California Institute of Technology discovered a general way to turn any Benioff-like model into one that only needed local spin interactions[Feynman85].

3.18 Feynman's Quantum Computer

The Turing machine model is one way to describe an abstract computer. Another is as a circuit built out of a set of primitive logic gates. The two approaches are equivalent: any computation that can be done efficiently with a quantum Turing machine can also be done by a quantum circuit and vice versa[Yao93].

Feynman's model of a quantum computer is rather like a quantum version of a combinational logic circuit. One starts with a circuit level description of the computation to be performed. The circuit is built out of reversible quantum gates, such as Barenco's 2-qubit gate specified earlier. In general we can regard the circuit as consisting of k logic gates acting on m qubits. The overall unitary transformation achieved by the circuit can thus be written as $A_k \cdot A_{k-1} \cdot \mathrm{K} \cdot A_1$ where A_i is the operator describing the action of the ith gate.

Thus the question is what Hamiltonian $\hat{H}$ in the expression

$$\hat{U}(t) = \exp\left(-\frac{i}{\mathrm{h}}\hat{H}\,t\right) = \left(\exp\left(-\frac{i}{\mathrm{h}}\hat{H}\right)\right)^t,$$

will give rise to a dynamical evolution that implements $A_k \cdot A_{k-1} \cdot \mathrm{K} \cdot A_1$. This is the question to which Feynman found a general answer. He did this by augmenting the set of m qubits (used to represent the inputs and outputs of the circuit) with a set of k+1 extra qubits. These qubits were used merely to track the progress of the computation; that is, how many gates had thus far been applied. They therefore constituted a kind of program step counter. Mathematically, this can be accomplished by writing the Hamiltonian as

$$\hat{H} = \sum_{i=0}^{k-1} c_{i+1} \cdot a_i \cdot A_{i+1} + \left(c_{i+1} \cdot a_i \cdot A_{i+1}\right)^{\dagger}$$

where c and a are creation and annihilation operators, respectively (see next chapter).

Only one of the program counter sites, called the "cursor," is ever occupied. Thus by periodically measuring the cursor we can determine when all k gate operations have been applied. As soon as we find the cursor at the kth site we know that the entire circuit has been applied to the input. At this time the state of the m qubits is guaranteed to contain a valid answer to the computation embodied in the given circuit.

In this form the Hamiltonian will evolve the computer into a superposition of states representing different numbers of gate applications. Thus the cursor, too, occupies a superposed state and the completion time of the computer is indefinite.

The computer can be made ballistic by propelling the qubits into the circuit using a spin wave with well-defined velocity.

There is, however, an important connection between the time independence of the Hamiltonian and the locality of the unitary evolution operator. As $\hat{H}$ and $\hat{U}(t)$ are related via an exponential, you can imagine expanding the exponential as a power series in $\hat{H}$. Thus $\hat{U}(t)$ really involves a sum of matrix powers of $\hat{H}$. Hence if $\hat{H}$ involves interactions between only adjacent qubits in the memory register, then $\hat{U}(t)$ will involve interactions between qubits that are arbitrarily far apart as the effect of the matrix powers is to increase couplings to more remote qubits. This is considered undesirable as it might be difficult to arrange physically.

3.19 Deutsch's Quantum Computer

The first true Quantum Turing Machine was devised by David Deutsch of Oxford University in 1985[Deutsch85]. Deutsch's model differed from Benioff's in that it maintained the quantum memory register (the "tape") in a superposition of computational states throughout the entire operations of the computer.

In Deutsch's quantum computer, the time evolution operator $\hat{U}$ is desired to be local (i.e., only involve interactions between adjacent qubits) and time-independent. This ensures that, at each computational step, only adjacent qubits interact and that the computation has a well-defined finishing time. But a local and fixed unitary evolution operator cannot be achieved with a local and fixed Hamiltonian. It is necessary to have a time-dependent Hamiltonian instead. The solution of the Schrödinger equation having a time-dependent Hamiltonian can be written as:

$$|\psi(t)\rangle = T\left\{\exp\left(-\frac{i}{h}\int_0^t \hat{H}(\tau)\,d\tau\right)\right\}|\psi(0)\rangle$$

where T is called the "time-ordered integral" defined[Shankar94] by:

$$T\left\{\exp\left(-\frac{i}{h}\int_0^t \hat{H}(\tau)\,d\tau\right)\right\} = \lim_{N\to\infty}\prod_{n=0}^{N-1}\exp\left(-\frac{i}{h}\hat{H}(n\Delta)\Delta\right)$$

where $\Delta = t/N$. The unitary evolution operator $\hat{U}$ can be made local and time-independent by choosing the Hamiltonian to be of the form:

$$\hat{H}(t) = \frac{-\pi f(t)}{2T}\begin{pmatrix} 0 &&&&&&& \\ & 0 &&&&&& \\ && 0 &&&&& \\ &&& 0 &&&& \\ &&&& 0 &&& \\ &&&&& 0 && \\ &&&&&& 1 & \alpha \\ &&&&&& -\alpha & 1 \end{pmatrix}$$

where f is any function such that:

$$\frac{1}{T}\int_0^T f(t)\,dt = 1$$

$$\lim_{t\to 0} f(t) = 0$$

$$\lim_{t\to T} f(t) = 0$$

where the limits must be approached smoothly.

In the next chapter we look at a simulation of Feynman's model of a quantum computer in complete detail.

CHAPTER 4

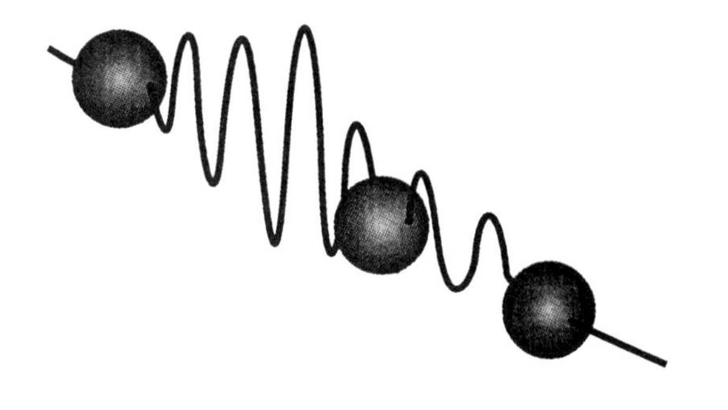

Simulating a Simple Quantum Computer

"Feynman noted that simulating quantum systems on classical computers is hard [...] it is an exponentially difficult problem to merely record the state of a quantum system, let alone integrate its equations of motion."
— Seth Lloyd

In the last chapter, we described mathematical models of various kinds of quantum computers, ranging from quantum implementations of classical reversible Turing machines to quantum circuits and true universal quantum computers. These models have proven to be very useful in anticipating the computational capabilities of hypothetical quantum devices. Nevertheless, it can be quite hard to see beyond the equations to get an intuition of how quantum computers would actually work. To remedy this problem we describe in this chapter a *simulation* of a simple quantum computer that is available in the software supplement that comes with this book. The tools in the supplement allow you to play with the simulator and perhaps try out novel quantum computations of your own.

Of course, as Richard Feynman pointed out, we currently do not know of an efficient way to simulate a quantum computer on a classical computer[Feynman82]. In particular, as quantum computers manipulate, in general, a superposition of an exponential number of possible inputs, and any classical computer must represent each of these inputs explicitly, it is easy to see that a classical computer will have a hard time keeping track of the complete details of a general quantum mechanical evolution. Consequently, we limit ourselves to simulating a simple quantum computer doing simple computations. Nevertheless, the simulations do provide a more intuitive appreciation of how a real quantum computer might operate.

Of the model quantum computers outlined in the previous chapter, Feynman's is the simplest to simulate as it can be specified using finite-dimensional matrices and involves only time-independent Hamiltonians. So let us take a look at the steps required to build a simulation of Feynman's quantum computer performing a particular computation[Feynman85]. In all cases, the basic sequence of steps is as follows:

1. Choose the computation that you want the computer to perform.
2. Represent that computation as a circuit built out of quantum logic gates.
3. Compute the Hamiltonian H that achieves this circuit.
4. Compute the Unitary evolution operator for H.
5. Determine the size of the memory register (to accommodate cursor + answer qubits).
6. Initialize the memory register (with input to the circuit).
7. Evolve the computer for some time.
8. Test whether the computation is done (read cursor qubits).
9. If so, extract the answer by reading the answer qubits. If not, allow further evolution (from the state projected when the cursor was last read).
10. Put it all together — bundle the previous steps into the full simulator function EvolveQC.

Let us flesh out each of these steps in the context of a fairly simple, but inherently quantum, computation.

4.1 What Computation Are We Going to Simulate?

The first step is to choose the computation that you want to simulate. Feynman used binary addition with carry in his original paper. We are going to use something simpler and more intrinsically "quantum." We choose to compute the logical NOT of a single bit, using two special quantum logic gates. You can think of the individual gates as each being "square-root-of-NOT" gates[Deutsch92a], which we denote as $\sqrt{\text{NOT}}$. The name comes from the fact that consecutive applications of two quantum logic gates can be understood mathematically as the dot product of the operators corresponding to those gates. Thus the computation that we want to simulate can be characterized as $\sqrt{\text{NOT}} \cdot \sqrt{\text{NOT}} \equiv \text{NOT}$.

What makes this computation "quantum" is the fact that it is impossible to have a single-input/single-output classical binary logic gate that works this way. Any classical binary $\sqrt{\text{NOT}}$ gate is going to have to output either a 0 or a 1 for each possible input 0 or 1. Suppose you defined the action of a classical $\sqrt{\text{NOT}}$ gate as the pair of transformations,

$$\sqrt{\text{NOT}}_{\text{classical}}(0) = 1$$
$$\sqrt{\text{NOT}}_{\text{classical}}(1) = 1$$

Then two consecutive applications of such a gate could invert a 0 successfully but not a 1. Similarly, if you defined a classical $\sqrt{\text{NOT}}$ gate as the pair of transformations,

$$\sqrt{\text{NOT}}_{\text{classical}}(0) = 1$$
$$\sqrt{\text{NOT}}_{\text{classical}}(1) = 0$$

then two consecutive applications of such a gate would not invert any input! In fact, there no way to define $\sqrt{\text{NOT}}$ classically, using binary logic, so that two consecutive applications of $\sqrt{\text{NOT}}$ reproduce the behavior of a NOT gate.

Thus, although computing NOT is a rather simple computation, the manner in which we do it illustrates several aspects of quantum computations including the use of qubits (superpositions of classical bits), the nature of quantum gates, and Feynman's method for converting a static, circuit-level, description of a computation to a dynamical Schrödinger evolution of a quantum computer that imitates the action of that circuit.

4.2 Representing a Computation as a Circuit

Once you have a particular computation in mind, the next question is how to design a quantum circuit that achieves it. To do so, you need to specify what quantum gates are to be used and how they are to be "wired" together.

We use the term "wired" rather loosely as there are no "wires," as such, at the quantum level. The "wiring" in a quantum circuit is merely a specification of which outputs (i.e., *logical* qubits) from one set of gates feed into which inputs (i.e., *logical* qubits) of another set of gates. Physically, the ports of two gates may "communicate" by either sharing the same *physical* qubit or via direct field interactions. Nevertheless, it is helpful to visualize a quantum circuit as if it were a classical combinatorial logic circuit. Thus a set of quantum gates and their associated wiring diagram define a quantum circuit.

The transformation of inputs to outputs of such circuits must be unitary because, to be realizable quantum mechanically, the circuit must cause the quantum state, that is, the input to the circuit, to evolve in accordance with Schrödinger's equation. However, Schrödinger's equation always predicts that an isolated quantum system will undergo a unitary evolution. Hence, for a hypothetical circuit to have any chance of being realizable quantum mechanically, it has to implement a unitary transformation on its inputs.

To ensure overall unitarity, this means that the quantum gates that comprise the circuit must, themselves, be unitary. However, unitarity implies reversibility. This means that the outputs of any quantum gate can be inferred, uniquely, from its inputs and vice versa. Mathematically, such gates are always describable by unitary matrices.

Just as you can pick different sets of classical logic gates in a classical circuit implementing a classical computation, so too can you pick different sets of quantum gates in a quantum circuit for a quantum computation. We now know that there are many possible choices for 2-input/2-output and 3-input/3-output universal, reversible, quantum logic gates[Barenco95a], [Barenco95]. You can build any quantum circuit out of any universal set of quantum gates. However, there are also other quantum gates, which although not forming a universal set, still provide useful (unitary) transformations for specific computations. This is the kind of gate we use in our $\sqrt{\text{NOT}} \cdot \sqrt{\text{NOT}} \equiv \text{NOT}$ circuit, namely, the square root of the NOT gate, $\sqrt{\text{NOT}}$.

Let us define the action of the $\sqrt{\text{NOT}}$ gate as follows:

$$\sqrt{\text{NOT}} \equiv \begin{pmatrix} \frac{1+i}{2} & \frac{1-i}{2} \\ \frac{1-i}{2} & \frac{1+i}{2} \end{pmatrix}$$

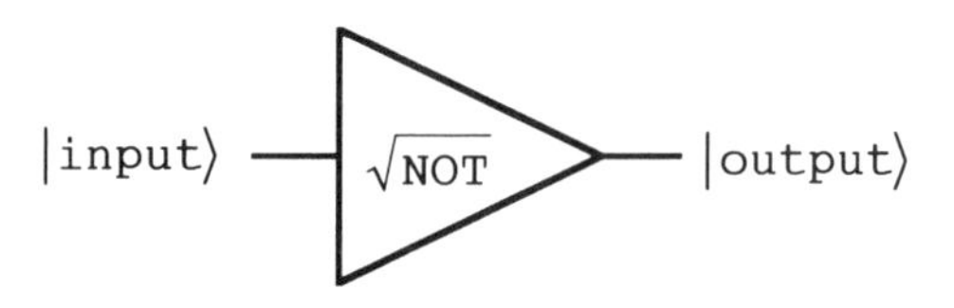

Fig. 4.1 Icon for a $\sqrt{\text{NOT}}$ gate.

This is a gate that acts on a single qubit. In a circuit diagram, we can represent the $\sqrt{\text{NOT}}$ gate as the icon shown in Fig. 4.1. Notice that there is one input and one output.

We can check that the matrix for $\sqrt{\text{NOT}}$ is unitary (and hence reversible) by looking at its product with its own conjugate transpose (denoted by †). For a unitary matrix, such a product ought to be the identity matrix.

$$\sqrt{\text{NOT}}\cdot\left(\sqrt{\text{NOT}}\right)^\dagger = \begin{pmatrix} \frac{1+i}{2} & \frac{1-i}{2} \\ \frac{1-i}{2} & \frac{1+i}{2} \end{pmatrix} \cdot \begin{pmatrix} \frac{1-i}{2} & \frac{1+i}{2} \\ \frac{1+i}{2} & \frac{1-i}{2} \end{pmatrix} = \begin{pmatrix} 1 & 0 \\ 0 & 1 \end{pmatrix}$$

So the $\sqrt{\text{NOT}}$ operator is, indeed, unitary.

You can see why we chose the name $\sqrt{\text{NOT}}$ by looking at the result of consecutive applications of this operator. We know that the NOT gate is represented by the (unitary) matrix:

$$\text{NOT} \equiv \begin{pmatrix} 0 & 1 \\ 1 & 0 \end{pmatrix}$$

The net effect of two $\sqrt{\text{NOT}}$ gates is given by the dot product of the matrices representing each $\sqrt{\text{NOT}}$ operation.

$$\sqrt{\text{NOT}}\cdot\sqrt{\text{NOT}} \equiv \begin{pmatrix} \frac{1+i}{2} & \frac{1-i}{2} \\ \frac{1-i}{2} & \frac{1+i}{2} \end{pmatrix} \cdot \begin{pmatrix} \frac{1+i}{2} & \frac{1-i}{2} \\ \frac{1-i}{2} & \frac{1+i}{2} \end{pmatrix} = \begin{pmatrix} 0 & 1 \\ 1 & 0 \end{pmatrix} \equiv \text{NOT}$$

Thus two consecutive applications of $\sqrt{\text{NOT}}$ achieve the same affect as the NOT operation.

However, the affect of a single $\sqrt{\text{NOT}}$ operator is quite unlike any classical logic gate. Here is the truth table for a single $\sqrt{\text{NOT}}$ gate:

In[]:=
```
TruthTable[ √NOT ]
```

Out[]=

$$\text{ket}[0] \to \left(\frac{1}{2}+\frac{i}{2}\right)\text{ket}[0]+\left(\frac{1}{2}-\frac{i}{2}\right)\text{ket}[1]$$

$$\text{ket}[1] \to \left(\frac{1}{2}-\frac{i}{2}\right)\text{ket}[0]+\left(\frac{1}{2}+\frac{i}{2}\right)\text{ket}[1]$$

Notice that the $\sqrt{\text{NOT}}$ gate has the effect of taking states representing classical bits ($|0\rangle$ and $|1\rangle$) into superpositions of states representing qubits ($\omega_0|0\rangle + \omega_1|1\rangle$). Nothing like this is possible using classical digital logic gates. Thus the $\sqrt{\text{NOT}}$ gate is an inherently quantum computation. Indeed, the $\sqrt{\text{NOT}}$ computation can be put to good use in true random number generation, as we show later in the book.

The $\sqrt{\text{NOT}}$ gate need not only be given inputs that are $|0\rangle$ or $|1\rangle$ however. The truth table for $\sqrt{\text{NOT}}$ also allows us, by the linearity of quantum mechanics, to predict how the gate will transform an input that is a superposition of $|0\rangle$ and $|1\rangle$ i.e. an input of the form $\omega_0|0\rangle + \omega_1|1\rangle$. This becomes important when we connect two $\sqrt{\text{NOT}}$ gates back to back to form a circuit for NOT. The first $\sqrt{\text{NOT}}$ gate converts a bit to a qubit, but the second gate can then convert a qubit back to a bit (albeit a different bit than the original). Thus, although a single $\sqrt{\text{NOT}}$ gate behaves non-classically, two consecutive applications of $\sqrt{\text{NOT}}$ behave as classical logic. For example, here is the truth table for $\sqrt{\text{NOT}} \cdot \sqrt{\text{NOT}}$:

In[]:=

```
TruthTable[ √NOT·√NOT ]
```

Out[]=

```
ket[0] -> ket[1]
ket[1] -> ket[0]
```

which is identical to the truth table for NOT.

In[]:=

```
TruthTable[ NOT ]
```

Out[]=

```
ket[0] → ket[1]
ket[1] → ket[0]
```

Thus our quantum circuit for computing NOT consists of two consecutive applications of $\sqrt{\text{NOT}}$. In other words we build the circuit shown in Fig. 4.2.

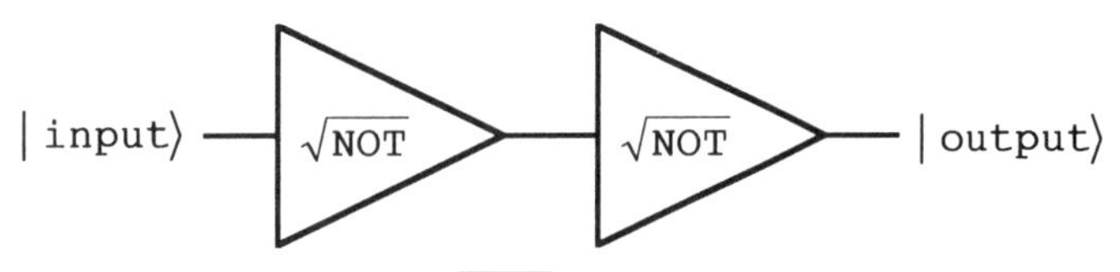

Fig. 4.2 A circuit for NOT as two $\sqrt{\text{NOT}}$ gates connected back to back.

Notice, however, that at the intermediate stage of the computation, after the action of just the first $\sqrt{\text{NOT}}$ gate, the state of the computation is a superposition of bits and hence is highly nonclassical.

4.3 Determining the Size of the Memory Register

At this point you know the computation that you want to perform and the circuit that will effect it. Next you must calculate how many qubits are needed to simulate the operation of this circuit.

The particles comprising the memory register of a (Feynman-like) quantum computer are divided into distinct sets. One set of particles is used to record the position of the cursor (which indicates how many steps of the computation have been performed), and the other set of particles is used to record the answer (or a superposition of answers) to the computation upon which the machine is working. If the computation requires the application of k logical operations (i.e., k gates) to some input pattern of m qubits, then there will be $k + 1$ qubits devoted to monitoring the cursor position and m qubits devoted to representing the answer (or superposition of answers). Thus if your circuit consists of k gate operations applied to a total of m input qubits, you will need $m + k + 1$ qubits to simulate the entire computer. Thus the complete memory register will look something like Fig. 4.3.

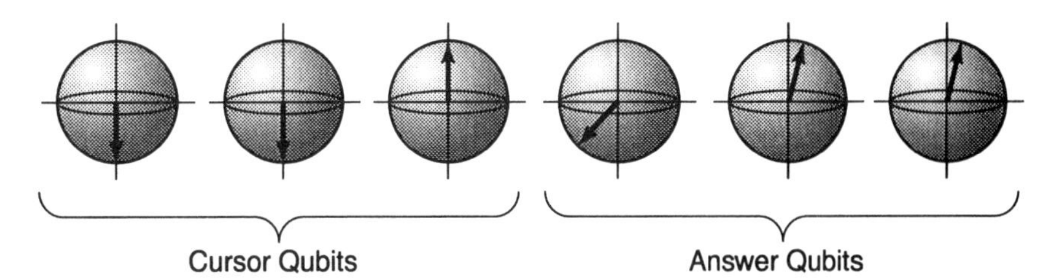

Fig. 4.3 The memory register of Feynman's quantum computer consists of a set of answer qubits that keep track of the evolving state of the computation which the circuit is performing, and a set of cursor qubits that keep track of the progress of the computation.

The states of the cursor qubits and answer qubits are coupled in the sense that, if you observe the position of the cursor and find it to be in its terminal position (i.e., at the $(k + 1)$-*th* site), then you can be sure that, at that moment, if you measured the answer qubits, they would reveal a guaranteed correct solution to the computation.

So what would be the required number of cursor bits and answer bits for the "square of the square-root-of-NOT" circuit? For the $\sqrt{\text{NOT}} \cdot \sqrt{\text{NOT}}$ circuit, there is only a single input qubit (so $m = 1$), and there are two quantum gates (so $k = 2$). Hence $m + k + 1 = 4$ and we can represent the complete state of the computer (cursor qubits and answer qubits) as a $2^4 \times 1$ dimensional column vector. Thus you can think of the complete memory register for the quantum computer to be composed of 4 qubits as shown in Fig. 4.4.

Without loss of generality we can suppose that the cursor uses the first three qubits in the register and the answer uses the fourth qubit in the register. Thus our unitary evolution operator is going to have to evolve the state of all four qubits simultaneously. Unfortunately, the $\sqrt{\text{NOT}}$ operator that we constructed earlier works on only a single qubit. So we are going to have to embed our $\sqrt{\text{NOT}}$ operator in an operator that works on four qubits at a time. How do we do this? Simple; just form the direct product of the desired operator with identity operators at the "dead" positions. For example, suppose that we want the $\sqrt{\text{NOT}}$ operator that acts on the 4th of 4 qubits. Let us call this $\sqrt{\text{NOT}}[4,4]$. We form this operator as follows.

$$\sqrt{\text{NOT}}[4,4] \equiv \hat{1} \otimes \hat{1} \otimes \hat{1} \otimes \sqrt{\text{NOT}} .$$

You can calculate such products using the function `Direct` found in the electronic supplement. The resulting operator is the following $2^4 \times 2^4$ dimensional tridiagonal matrix:

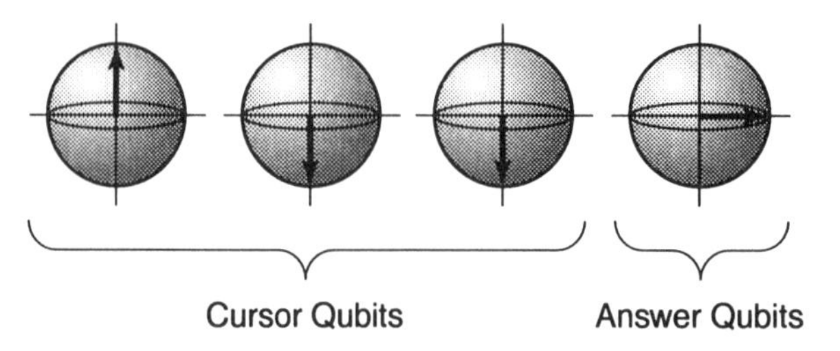

Fig. 4.4 A 4-qubit register sufficient for simulating the square of the square-root-of-NOT computation.

In[]:=

```
Direct[IdentityMatrix[2],
       IdentityMatrix[2],
       IdentityMatrix[2],
       √NOT ]
```

Out[]=

$$\begin{pmatrix}
\frac{1+i}{2} & \frac{1-i}{2} & 0 & 0 & 0 & 0 & 0 & 0 & 0 & 0 & 0 & 0 & 0 & 0 & 0 & 0 \\
\frac{1-i}{2} & \frac{1+i}{2} & 0 & 0 & 0 & 0 & 0 & 0 & 0 & 0 & 0 & 0 & 0 & 0 & 0 & 0 \\
0 & 0 & \frac{1+i}{2} & \frac{1-i}{2} & 0 & 0 & 0 & 0 & 0 & 0 & 0 & 0 & 0 & 0 & 0 & 0 \\
0 & 0 & \frac{1-i}{2} & \frac{1+i}{2} & 0 & 0 & 0 & 0 & 0 & 0 & 0 & 0 & 0 & 0 & 0 & 0 \\
0 & 0 & 0 & 0 & \frac{1+i}{2} & \frac{1-i}{2} & 0 & 0 & 0 & 0 & 0 & 0 & 0 & 0 & 0 & 0 \\
0 & 0 & 0 & 0 & \frac{1-i}{2} & \frac{1+i}{2} & 0 & 0 & 0 & 0 & 0 & 0 & 0 & 0 & 0 & 0 \\
0 & 0 & 0 & 0 & 0 & 0 & \frac{1+i}{2} & \frac{1-i}{2} & 0 & 0 & 0 & 0 & 0 & 0 & 0 & 0 \\
0 & 0 & 0 & 0 & 0 & 0 & \frac{1-i}{2} & \frac{1+i}{2} & 0 & 0 & 0 & 0 & 0 & 0 & 0 & 0 \\
0 & 0 & 0 & 0 & 0 & 0 & 0 & 0 & \frac{1+i}{2} & \frac{1-i}{2} & 0 & 0 & 0 & 0 & 0 & 0 \\
0 & 0 & 0 & 0 & 0 & 0 & 0 & 0 & \frac{1-i}{2} & \frac{1+i}{2} & 0 & 0 & 0 & 0 & 0 & 0 \\
0 & 0 & 0 & 0 & 0 & 0 & 0 & 0 & 0 & 0 & \frac{1+i}{2} & \frac{1-i}{2} & 0 & 0 & 0 & 0 \\
0 & 0 & 0 & 0 & 0 & 0 & 0 & 0 & 0 & 0 & \frac{1-i}{2} & \frac{1+i}{2} & 0 & 0 & 0 & 0 \\
0 & 0 & 0 & 0 & 0 & 0 & 0 & 0 & 0 & 0 & 0 & 0 & \frac{1+i}{2} & \frac{1-i}{2} & 0 & 0 \\
0 & 0 & 0 & 0 & 0 & 0 & 0 & 0 & 0 & 0 & 0 & 0 & \frac{1-i}{2} & \frac{1+i}{2} & 0 & 0 \\
0 & 0 & 0 & 0 & 0 & 0 & 0 & 0 & 0 & 0 & 0 & 0 & 0 & 0 & \frac{1+i}{2} & \frac{1-i}{2} \\
0 & 0 & 0 & 0 & 0 & 0 & 0 & 0 & 0 & 0 & 0 & 0 & 0 & 0 & \frac{1-i}{2} & \frac{1+i}{2}
\end{pmatrix}$$

As a check that this larger operator performs the $\sqrt{\text{NOT}}$ operation on the 4th of 4 qubits, we can compute the truth table for two consecutive applications of $\sqrt{\text{NOT}}[4,4]$.

In[]:=

```
TruthTable[ √NOT [4,4] . √NOT [4,4]]
```

Out[]=

```
ket[0, 0, 0, 0] -> ket[0, 0, 0, 1]
ket[0, 0, 0, 1] -> ket[0, 0, 0, 0]
ket[0, 0, 1, 0] -> ket[0, 0, 1, 1]
ket[0, 0, 1, 1] -> ket[0, 0, 1, 0]
ket[0, 1, 0, 0] -> ket[0, 1, 0, 1]
ket[0, 1, 0, 1] -> ket[0, 1, 0, 0]
ket[0, 1, 1, 0] -> ket[0, 1, 1, 1]
ket[0, 1, 1, 1] -> ket[0, 1, 1, 0]
ket[1, 0, 0, 0] -> ket[1, 0, 0, 1]
ket[1, 0, 0, 1] -> ket[1, 0, 0, 0]
ket[1, 0, 1, 0] -> ket[1, 0, 1, 1]
ket[1, 0, 1, 1] -> ket[1, 0, 1, 0]
ket[1, 1, 0, 0] -> ket[1, 1, 0, 1]
ket[1, 1, 0, 1] -> ket[1, 1, 0, 0]
ket[1, 1, 1, 0] -> ket[1, 1, 1, 1]
ket[1, 1, 1, 1] -> ket[1, 1, 1, 0]
```

Notice that the first three qubits are mapped into themselves and the NOT operation is applied to the last qubit. Hence $\sqrt{\text{NOT}}[4,4]$ is applying the $\sqrt{\text{NOT}}$ operation to the 4th of 4 qubits.

4.4 Computing the Hamiltonian Operator

You might think that once you have the description of the quantum circuit you have everything you need to simulate the quantum computer. Certainly the circuit level description tells you *what* transformation will be effected on any given input state. However, it does not embody any *dynamics*. A real quantum computer is nothing more than a physical system whose evolution over time can be interpreted as performing some particular computation. But how you arrange for this dynamical evolution is an important issue. The circuit level description tells you what the evolution needs to look like, but you still need to embody the computation in some dynamical process. How do you do this?

Recall that the time evolution of a quantum system is described by Schrödinger's equation. For the case of a time-independent Hamiltonian $\hat{H}$, the Schrödinger equation takes the form

$$i\hbar \frac{\partial |\psi(t)\rangle}{\partial t} = \hat{H} \cdot |\psi(t)\rangle$$

where $|\psi(t)\rangle$ describes the state of the physical system, or, in our case, the state of the memory register of a quantum computer, at time t, $i = \sqrt{-1}$, and $\hbar$ is Planck's constant divided by 2π equal to 1.0545×10^{-34} Js. Thus for a quantum system to perform as a computer, or more specifically a particular quantum circuit, you need to design a Hamiltonian that will cause a given initial state of the memory register $|\psi(0)\rangle$ to evolve in a way that mimics the action of the circuit.

You can push the problem back one step by looking at the solution of Schrödinger's equation, viz.:

$$|\psi(t)\rangle = e^{-i\hat{H}t/\hbar}|\psi(0)\rangle = \hat{U}(t)|\psi(0)\rangle .$$

This implies that, in time t the initial state of the memory register $|\psi(0)\rangle$ evolves into the state $|\psi(t)\rangle$ under the action of the operator $\hat{U}(t) = \exp(-i\hat{H}t/\hbar)$. Thus if $|\psi(0)\rangle$ represents some input to the circuit we are trying to simulate, then $|\psi(t)\rangle$ represents the output from the circuit at time t. Clearly, we would like the evolution operator $\hat{U}(t)$ to cause the state to evolve in a way that mimics the action of our desired circuit. So our task becomes that of finding a time-independent Hamiltonian operator $\hat{H}$ such that, when inserted into the matrix exponential, $\exp\left(-i\hat{H}t/\mathrm{h}\right)$ results in a unitary matrix $\hat{U}(t)$ that mimics the action of our desired circuit.

This is very difficult in general. However, physicist Richard Feynman found a way of making the computation piggyback on another dynamical system, the cursor system.

If the circuit, consisting of m inputs and m outputs, is composed of k logical operations (k gates), then Feynman augmented the input/output qubits with an additional set of $k + 1$ "cursor" qubits. The cursor serves as a kind of program step counter. If the cursor is ever found in the $(k + 1)$th site, then the memory register (of m qubits) is guaranteed to contain a valid answer to the computation at that time.

The form for the Hamiltonian that achieves this coupling between the cursor system and the computation is given by a sum of the (dot) product of creation and annihilation operators with each gate operator and the Hermitian conjugate of this. The idea came from work Feynman had done on the dynamics of spin waves traveling up and down a chain of coupled particles.

In general, a circuit will consist of a set of k gates. Its overall effect can be described by the dot product of the operators for each gate applied in sequence once the gate operators, M, have been embedded within the required number of qubits. For example, M_1 might apply an operation to the 2nd and 3rd of 4 qubits. M_2 might apply an operation to the 4th of 4 qubits, and so on.

$$M = M_k \cdot M_{k-1} \cdot \mathrm{K} \cdot M_1.$$

For the $\sqrt{\mathrm{NOT}} \cdot \sqrt{\mathrm{NOT}}$ circuit, there are two gates, M_1 and M_2 that can both be described by operators that act on the 4th of 4 qubits:

$$\hat{M}_2 \cdot \hat{M}_1 \equiv \sqrt{\mathrm{NOT}}[4,4] \cdot \sqrt{\mathrm{NOT}}[4,4].$$

The creation and annihilation operators are used to move the cursor forwards and backwards. To move the cursor from the *ith* site to the $(i + 1)$th site, we first have to annihilate the cursor at the *ith* site and then recreate it at the $(i + 1)$th site. This guarantees that there is only ever one cursor bit set at 1 amongst the cursor sites. This cursor motion must then be coupled to the corresponding gate. So the term for moving the cursor from the *ith* site to the $(i + 1)$th site and coupling it to the application of the $(i + 1)$th gate operation is

$$c_{i+1} \cdot a_i \cdot \hat{M}_{i+1}.$$

Creation and Annihilation Operators

The creation operator c acting on a single spin state is

$$c = \begin{pmatrix} 0 & 0 \\ 1 & 0 \end{pmatrix}.$$

The effect of this operator is to convert a zero-state to a 1-state and to convert a 1-state to the null state (i.e., a column vector of all zeroes).

Similarly, the annihilation operator a acting on a single spin state is

$$a = \begin{pmatrix} 0 & 1 \\ 0 & 0 \end{pmatrix}.$$

The annihilation operator converts a 1-state to a 0-state and converts a 0-state to the null state.

Versions of creation and annihilation operators can be created that act on the ith of four qubits using the Direct product with identity matrices at the "dead" positions that we described earlier. For example, the creation operator that acts on the second of four qubits is given by

$$c_2 = \begin{pmatrix} 1 & 0 \\ 0 & 1 \end{pmatrix} \otimes \begin{pmatrix} 0 & 0 \\ 1 & 0 \end{pmatrix} \otimes \begin{pmatrix} 1 & 0 \\ 0 & 1 \end{pmatrix} \otimes \begin{pmatrix} 1 & 0 \\ 0 & 1 \end{pmatrix}.$$

The other operators are defined similarly. We use c_i and a_i to represent the creation and annihilation operators that act on the ith of k qubits.

Now define the Hamiltonian operator to be:

$$\hat{H} = \sum_{i=0}^{k-1} c_{i+1} \cdot a_i \cdot \hat{M}_{i+1} + \left(c_{i+1} \cdot a_i \cdot \hat{M}_{i+1}\right)^{\dagger},$$

where † denotes the conjugate transpose. Notice that this definition uses creation and annihilation operators to move the cursor forwards and backwards and to apply the corresponding gate operator when, and only when, the cursor is in the correct position. Thus the net effect of such a Hamiltonian is to put the quantum computer into a superposition of computations in which different numbers of gate operators have been applied. However, the position of the cursor and the number of operations so far applied are perfectly correlated.

For the SqrtNOT-Squared circuit, the Hamiltonian is given by

In[]:=

```
sqrtNOTcircuit = {√NOT[4, 4], √NOT[4, 4]};
Hamiltonian[1, 2, sqrtNOTcircuit];
```

Out[]=

$$\begin{pmatrix}
0 & 0 & 0 & 0 & 0 & 0 & 0 & 0 & 0 & 0 & 0 & 0 & 0 & 0 & 0 & 0 \\
0 & 0 & 0 & 0 & 0 & 0 & 0 & 0 & 0 & 0 & 0 & 0 & 0 & 0 & 0 & 0 \\
0 & 0 & 0 & 0 & \frac{1+i}{2} & \frac{1-i}{2} & 0 & 0 & 0 & 0 & 0 & 0 & 0 & 0 & 0 & 0 \\
0 & 0 & 0 & 0 & \frac{1-i}{2} & \frac{1+i}{2} & 0 & 0 & 0 & 0 & 0 & 0 & 0 & 0 & 0 & 0 \\
0 & 0 & \frac{1-i}{2} & \frac{1+i}{2} & 0 & 0 & 0 & 0 & \frac{1+i}{2} & \frac{1-i}{2} & 0 & 0 & 0 & 0 & 0 & 0 \\
0 & 0 & \frac{1+i}{2} & \frac{1-i}{2} & 0 & 0 & 0 & 0 & \frac{1-i}{2} & \frac{1+i}{2} & 0 & 0 & 0 & 0 & 0 & 0 \\
0 & 0 & 0 & 0 & 0 & 0 & 0 & 0 & 0 & 0 & \frac{1+i}{2} & \frac{1-i}{2} & 0 & 0 & 0 & 0 \\
0 & 0 & 0 & 0 & 0 & 0 & 0 & 0 & 0 & 0 & \frac{1-i}{2} & \frac{1+i}{2} & 0 & 0 & 0 & 0 \\
0 & 0 & 0 & 0 & \frac{1-i}{2} & \frac{1+i}{2} & 0 & 0 & 0 & 0 & 0 & 0 & 0 & 0 & 0 & 0 \\
0 & 0 & 0 & 0 & \frac{1+i}{2} & \frac{1-i}{2} & 0 & 0 & 0 & 0 & 0 & 0 & 0 & 0 & 0 & 0 \\
0 & 0 & 0 & 0 & 0 & 0 & \frac{1-i}{2} & \frac{1+i}{2} & 0 & 0 & 0 & 0 & \frac{1+i}{2} & \frac{1-i}{2} & 0 & 0 \\
0 & 0 & 0 & 0 & 0 & 0 & \frac{1+i}{2} & \frac{1-i}{2} & 0 & 0 & 0 & 0 & \frac{1-i}{2} & \frac{1+i}{2} & 0 & 0 \\
0 & 0 & 0 & 0 & 0 & 0 & 0 & 0 & 0 & 0 & \frac{1-i}{2} & \frac{1+i}{2} & 0 & 0 & 0 & 0 \\
0 & 0 & 0 & 0 & 0 & 0 & 0 & 0 & 0 & 0 & \frac{1+i}{2} & \frac{1-i}{2} & 0 & 0 & 0 & 0 \\
0 & 0 & 0 & 0 & 0 & 0 & 0 & 0 & 0 & 0 & 0 & 0 & 0 & 0 & 0 & 0 \\
0 & 0 & 0 & 0 & 0 & 0 & 0 & 0 & 0 & 0 & 0 & 0 & 0 & 0 & 0 & 0
\end{pmatrix}$$

4.5 Computing the Unitary Evolution Operator

Given the preceeding form for the Hamiltonian, we can compute the evolution operator $\hat{U}(t)$ from

$$\hat{U}(t) = \left(e^{-i\,\hat{H}/h}\right)^{t}.$$

In[]:=

```
U[1, 2, sqrtNOTcircuit, t]
```

Out[]=

$$\hat{U}(t) = \begin{pmatrix} 1 & 0 & 0 & 0 & 0 & 0 & 0 & 0 & 0 & 0 & 0 & 0 & 0 & 0 & 0 & 0 \\ 0 & 1 & 0 & 0 & 0 & 0 & 0 & 0 & 0 & 0 & 0 & 0 & 0 & 0 & 0 & 0 \\ 0 & 0 & \beta & 0 & \delta & \varepsilon & 0 & 0 & 0 & \gamma & 0 & 0 & 0 & 0 & 0 & 0 \\ 0 & 0 & 0 & \beta & \varepsilon & \delta & 0 & 0 & \gamma & 0 & 0 & 0 & 0 & 0 & 0 & 0 \\ 0 & 0 & \varepsilon & \delta & \alpha & 0 & 0 & 0 & \delta & \varepsilon & 0 & 0 & 0 & 0 & 0 & 0 \\ 0 & 0 & \delta & \varepsilon & 0 & \alpha & 0 & 0 & \varepsilon & \delta & 0 & 0 & 0 & 0 & 0 & 0 \\ 0 & 0 & 0 & 0 & 0 & 0 & \beta & 0 & 0 & 0 & \delta & \varepsilon & 0 & \gamma & 0 & 0 \\ 0 & 0 & 0 & 0 & 0 & 0 & 0 & \beta & 0 & 0 & \varepsilon & \delta & \gamma & 0 & 0 & 0 \\ 0 & 0 & 0 & \gamma & \varepsilon & \delta & 0 & 0 & \beta & 0 & 0 & 0 & 0 & 0 & 0 & 0 \\ 0 & 0 & \gamma & 0 & \delta & \varepsilon & 0 & 0 & 0 & \beta & 0 & 0 & 0 & 0 & 0 & 0 \\ 0 & 0 & 0 & 0 & 0 & 0 & \varepsilon & \delta & 0 & 0 & \alpha & 0 & \delta & \varepsilon & 0 & 0 \\ 0 & 0 & 0 & 0 & 0 & 0 & \delta & \varepsilon & 0 & 0 & 0 & \alpha & \varepsilon & \delta & 0 & 0 \\ 0 & 0 & 0 & 0 & 0 & 0 & 0 & \gamma & 0 & 0 & \varepsilon & \delta & \beta & 0 & 0 & 0 \\ 0 & 0 & 0 & 0 & 0 & 0 & \gamma & 0 & 0 & 0 & \delta & \varepsilon & 0 & \beta & 0 & 0 \\ 0 & 0 & 0 & 0 & 0 & 0 & 0 & 0 & 0 & 0 & 0 & 0 & 0 & 0 & 1 & 0 \\ 0 & 0 & 0 & 0 & 0 & 0 & 0 & 0 & 0 & 0 & 0 & 0 & 0 & 0 & 0 & 1 \end{pmatrix}^{t}$$

where

$$\alpha = \cos\left(\sqrt{2}\right),$$

$$\beta = \frac{1}{2} + \frac{\alpha}{2},$$

$$\gamma = -\frac{1}{2} + \frac{\alpha}{2},$$

$$\delta = \frac{1}{2\sqrt{2}}(1-i)\ \sin\left(\sqrt{2}\right),$$

$$\varepsilon = \frac{1}{2\sqrt{2}}(-1-i)\ \sin\left(\sqrt{2}\right).$$

4.6 Running the Quantum Computer for a Fixed Length of Time

Once the unitary time evolution operator $\hat{U}(t)$ has been determined and a particular initial state of the memory register of the quantum computer $|\psi(0)\rangle$ has been selected, you can calculate the state of the memory register at any future time t from the equation

$$|\psi(t)\rangle = \hat{U}(t) \cdot |\psi(0)\rangle.$$

The initial state specifies a state for each qubit in the memory register. Recall that the memory register consists of two sets of qubits; the cursor qubits and the answer qubits. The initial state of the $k+1$ cursor qubits is obligatory: the first cursor qubit must always be set to state $|1\rangle$ and the rest to state $|0\rangle$. However, the state of the m answer qubits can be initialized to be any valid superposition of m qubits. For the square-root-of-NOT-squared circuit, there is only one answer qubit ($m = 1$) and, as the whole circuit is supposed to perform the NOT operation, it is sensible to set this answer qubit (which also serves as the input to the circuit) to either $|0\rangle$ or $|1\rangle$ initially. In which case we expect, after the computation has taken place, to observe the answer qubit and find it in state $|1\rangle$ or state $|0\rangle$, respectively.

Thus if the input to the circuit is $|0\rangle$, the initial state of the memory register will be $|1\rangle \otimes |0\rangle \otimes |0\rangle \otimes |0\rangle \equiv |1000\rangle$. Likewise, if the input to the circuit is $|1\rangle$, the initial state of the memory register will be $|1\rangle \otimes |0\rangle \otimes |0\rangle \otimes |1\rangle \equiv |1001\rangle$.

The preceding evolution equation is valid for any quantum computation, provided no observations of the state of the memory register are made in the interval $(0, t)$, the computer works perfectly, and the computer does not interact with its environment. The effects of imperfections ("internal errors") are considered in the next chapter and the effects of stray couplings to the environment ("external errors") are considered in Chapter 10. Combating errors is a crucial issue for quantum computing as it is necessary to keep the memory register in a coherent superposition of states long enough for computation to take place.

In the electronic supplement, the preceding operations, which took us from the (timeless) quantum circuit description to the (time-independent) Hamiltonian and hence the (time-dependent) unitary evolution operator, are bundled together in the function SchrodingerEvolution.

In[]:=

```
?SchrodingerEvolution
```

Out[]=

```
SchrodingerEvolution[psi0, m, k, circuit, t] evolves
the given circuit for time t from the initial
configuration psi0 (a ket vector e.g.,
ket[1,0,0,0]). You must specify the numbers of gates
and inputs in your circuit. In general, if your
circuit contains k gates, there will be k+1 cursor
qubits. If your circuit has m inputs and m outputs,
there will be m answer qubits.
```

The first argument to SchrodingerEvolution is the initial state of the memory register. If we specify this state as ket[1,0,0,0] the 4th bit (corresponding to the memory register bit) is set to 0, we are saying that we want to compute $\sqrt{\text{NOT}} \cdot \sqrt{\text{NOT}} \cdot |0\rangle$ (which ought to be

NOT $\cdot|0\rangle=|1\rangle$). Also, we start the cursor off at the first cursor site. As time evolves, the cursor runs back and forth along its possible positions and the probability of finding the cursor at the $(k+1)$th cursor position rises and falls repeatedly. Let us simulate this circuit for increasing evolution times (specified in the last argument of SchrodingerEvolution). For simplicity we set `h` equal to 1 and we evolve the computer for time 0.5 units.

In[]:=

```
sqrtNOTcircuit = {√NOT[4, 4], √NOT[4, 4]};
evoln1 =
    SchrodingerEvolution[ket[1,0,0,0],
                          1, 2, sqrtNOTcircuit, 0.5]
```

Out[]=

```
-0.119878 ket[0, 0, 1, 1] +
(0.229681 - 0.229681 I) ket[0, 1, 0, 0] +
(-0.229681 - 0.229681 I) ket[0, 1, 0, 1] +
0.880122 ket[1, 0, 0, 0]
```

SchrodingerEvolution returned the state of the computer 0.5 time units after the initial state. At this time, the probabilities of finding the memory register (cursor bits and program bit) in each possible configuration are given by the function Probabilities.

In[]:=

```
?Probabilities
```

Out[]=

```
Probabilities[superposition] returns the
probabilities of finding a system in a state
described by superposition in each of its possible
eigenstates upon being measured (observed). If
Probabilities is given the option ShowEigenstates ->
True the function returns a list of {eigenstate,
probability} pairs.
```

In[]:=

```
Probabilities[evoln1, ShowEigenstates->True] //
ColumnForm
```

Out[]=
```
{ket[0, 0, 0, 0], 0}
{ket[0, 0, 0, 1], 0}
{ket[0, 0, 1, 0], 0}
{ket[0, 0, 1, 1], 0.0143707}
{ket[0, 1, 0, 0], 0.105507}
{ket[0, 1, 0, 1], 0.105507}
{ket[0, 1, 1, 0], 0}
{ket[0, 1, 1, 1], 0}
{ket[1, 0, 0, 0], 0.774615}
{ket[1, 0, 0, 1], 0}
{ket[1, 0, 1, 0], 0}
{ket[1, 0, 1, 1], 0}
{ket[1, 1, 0, 0], 0}
{ket[1, 1, 0, 1], 0}
{ket[1, 1, 1, 0], 0}
{ket[1, 1, 1, 1], 0}
```

On the left we see the possible states in which the memory register can be found. On the right, we see the corresponding probability of finding the register in that state at that time if (and we emphasize if) the complete register (cursor qubits and answer qubits) were observed at that time. If we were to run the simulation again for a longer period of time, we would see that the probabilities change, indicating that the odds of finding the register in one state or another change over time.

In[]:=
```
sqrtNOTcircuit = {√NOT[4, 4], √NOT[4, 4]};
evoln2 =
    SchrodingerEvolution[ket[1,0,0,0],
                              1, 2, sqrtNOTcircuit, 2]
```
Out[]=
```
-0.975682 ket[0, 0, 1, 1] +
(0.10892 - 0.10892 I) ket[0, 1, 0, 0] +
(-0.10892 - 0.10892 I) ket[0, 1, 0, 1] +
0.0243184 ket[1, 0, 0, 0]
```

In[]:=
```
Probabilities[evoln2, ShowEigenstates->True] //
ColumnForm
```

Out[]=
```
{ket[0, 0, 0, 0], 0}
{ket[0, 0, 0, 1], 0}
{ket[0, 0, 1, 0], 0}
{ket[0, 0, 1, 1], 0.951955}
{ket[0, 1, 0, 0], 0.023727}
{ket[0, 1, 0, 1], 0.023727}
{ket[0, 1, 1, 0], 0}
{ket[0, 1, 1, 1], 0}
{ket[1, 0, 0, 0], 0.000591386}
{ket[1, 0, 0, 1], 0}
{ket[1, 0, 1, 0], 0}
{ket[1, 0, 1, 1], 0}
{ket[1, 1, 0, 0], 0}
{ket[1, 1, 0, 1], 0}
{ket[1, 1, 1, 0], 0}
{ket[1, 1, 1, 1], 0}
```

The output indicates that, after the initial state has been allowed to evolve (without any measurements) for 2 time units, there is a 95% chance of finding the cursor at the $(k + 1)$th site (the third qubit set to 1 in ket[0,0,1,1]). Notice also that there is zero probability of finding the register in certain other configurations (such as ket[1,0,1,1]), indicating that certain configurations are forbidden. For example, any configuration that has more than one 1 bit amongst the cursor qubits is forbidden because there can only ever be a single 1 amongst these qubits.

As you can see, the probabilities of obtaining the correct answer rise and fall over time. In Feynman's model of a quantum computer there is uncertainty as to when the computation finishes. However, notice also that whenever the cursor is observed at the $(k + 1)$th position, the answer then in the memory register (the 4th bit) is *always* correct. Hence by periodically observing the state of the $(k + 1)$th bit, as soon as it is found to be in the 1-state, the answer is then known to be in the memory register of the computer.

It is important to understand that when the cursor position is observed, the state of the answer qubits is projected into a state consistent with the observed cursor position. However, the answer bit is not observed in this process, so it is possible for the answer qubits to remain in a superposed (but projected) state after the cursor has been observed.

4.7 Running the Quantum Computer Until the Computation Is Done

In practice, the fact that we cannot predict, with certainty, the times at which the computation will be completed, means that we must periodically observe the cursor position to see if the computation has finished. Although we do not directly observe the state of the answer bits during such observations of the cursor bits, these observations do affect the state of the whole memory register (cursor bits and answer bits). Once we find the cursor at a particular site, say, the ith cursor site, the state of the complete register is projected into a superposition of states consistent with this observation, that is, a superposition of states in which the ith cursor bit is a 1.

If the cursor is found at the $(k + 1)$th site, then further evolution is prevented and the state of the complete memory register (cursor bits plus answer bits) is measured. If the cursor is not yet at the $(k + 1)$th site, the Schrödinger evolution is allowed to resume taking the new projected state as the initial state. Thus the whole computer evolves in a punctuated fashion.

To test for completion, we must measure just the qubits used to encode the cursor in the memory register. We can simulate the act of reading the cursor bits of the memory register using the function ReadCursorBits. This takes two arguments, the number of bits being used to keep track of the cursor position and the superposition of eigenstates that represents the complete state of the memory register at a given time. It returns the position of the cursor and the new state into which the memory register is projected.

In[]:=

```
?ReadCursorBits
```

Out[]=

```
ReadCursorBits[numCursorBits, superposition] reads
the state of the cursor bits of a Feynman-like quan-
tum computer. If the computation can be accomplished
in k+1 (logic gate) operations, the cursor will con-
sist of a chain of k+1 atoms, only one of which can
ever be in the |1> state. The cursor keeps track of
how many logical operations have been applied to the
program bits thus far. The state of the program bits
of the computer are unaffected by measuring just the
cursor bits. If the cursor is ever found at the
(k+1)th site, then, if you measured the program bits
at that moment, they would be guaranteed to contain
a valid answer to the computation on which the quan-
tum computer was working.
```

For example, if the state of the complete memory register is ket[1,0,0,1] and you at that moment read the cursor position, the answer you would obtain would be {1,0,0} (because the cursor is certainly at site 1).

In[]:=
```
ReadCursorBits[3, ket[1,0,0,1]]
```
Out[]=
```
{{1, 0, 0}, ket[1, 0, 0, 1]}
```

The function ReadCursorBits also returns the state of the register that is projected out after the measurement of the cursor position has been made. You can see this better by supposing that the memory register contains a superposition of possible computational states. For example, assume that the register is in the state 1/2 ket[1,0,0,0] - (1/4 + I/3) ket[0,1,0,0] + Sqrt[83/144] ket[0,0,1,0]. What might we get if we measure the cursor position?

In[]:=
```
ReadCursorBits[3, 1/2             ket[1,0,0,0] -
                  (1/4 + 1/3 I) ket[0,1,0,0] +
                  Sqrt[83/144]  ket[0,0,1,0]]
```
Out[]=
```
                    3     4 I
{{0, 1, 0}, (-(-) - ---) ket[0, 1, 0, 0]}
                    5      5
```

This time when we measured the cursor position we found it at the second of the three possible sites, {0,1,0}. After the measurement was made the state of the memory register was projected into the state (-3/5 - 4/5 I) ket[0,1,0,0].

Now in order to measure the cursor position, you need to measure the state of each cursor bit in turn. If you call ReadCursorBits with the option TraceProgress->True you can see the sequence of steps that takes place during this process.

In[]:=
```
ReadCursorBits[3, 1/2             ket[1,0,0,0] -
                  (1/4 + 1/3 I) ket[0,1,0,0] +
                  Sqrt[83/144]  ket[0,0,1,0],
              TraceProgress->True]
```

Out[]=

$$\{\text{BeforeAnyMeasurements}, \frac{\text{Sqrt}[83]\,\text{ket}[0,0,1,0]}{12} + \left(-\frac{1}{4} - \frac{I}{3}\right)\text{ket}[0,1,0,0] + \frac{\text{ket}[1,0,0,0]}{2}\}$$

$$\{0, \frac{\text{Sqrt}\left[\frac{83}{3}\right]\text{ket}[0,0,1,0]}{6} + \left(-\frac{1}{6} - \frac{2I}{9}\right)\text{Sqrt}[3]\,\text{ket}[0,1,0,0]\}$$

$$\{1, \left(-\frac{3}{5} - \frac{4I}{5}\right)\text{ket}[0,1,0,0]\}$$

$$\{0, \left(-\frac{3}{5} - \frac{4I}{5}\right)\text{ket}[0,1,0,0]\}$$

"BeforeAnyMeasurements" corresponds to knowing nothing about the cursor position. At that time the state of the computer is a superposition of three states corresponding to the three possible cursor positions. Once the first cursor bit is measured we start to gain information about the cursor location. In the present example we obtain a "0" indicating that the cursor is not at the first location. A side effect of this measurement is that the state of the computer is projected into a state consistent with not finding the cursor at the first position. Next we measure the cursor to see if it is at its second position. This time we read a "1", indicating that the cursor is indeed at the second position, and project the computer into a state consistent with finding the cursor at this location. The third and final measurement is done purely to check that there is no error as there should only ever by a single "1" amongst the cursor qubits. The projected state does not change because the state was already consistent with not finding the cursor at the third location (because we already know it is at the second location). Hence the last measurement yields no new information and so the projected state after the third cursor measurement is identical to the state after the second cursor measurement.

4.8 Extracting the Answer

If the cursor is ever found at the $(k + 1)$th site, then the computation is halted and the answer is read off from amongst the program bits (at sites $k + 2$ through $k + 1 + m$). If the cursor is not at the $(k + 1)$th site, then the computer is allowed to evolve again according to Schrödinger's equation for some other length of time (which may or may not be fixed) and then the cursor is again observed and the whole cycle repeats itself. Note that although the cursor bits are observed to determine whether the machine has finished, the program bits are not observed until the very end of the computation. Thus the state of the program bits is never scrambled during the evolution of the quantum computer even though the cursor position is observed. So, in the Feynman model of a quantum computer, there is no doubt as to the correctness of the answer, merely the time at which the answer is available.

4.9 Putting It All Together

Now that we know how to measure the cursor position without disturbing the program bits, we can envisage a method of evolving the quantum computer whilst periodically checking whether the computation has finished. The CD-ROM contains a simulator that allows you to trace through the steps such a quantum computer would follow. The simulator is invoked using the function EvolveQC.

In[]:=

```
?EvolveQC
```

Out[]=

```
The function EvolveQC[initState, circuit] evolves a
Feynman-like quantum computer, specified as a cir-
cuit of interconnected quantum logic gates, from
some initial state until the computation is com-
plete. The output is a list of snapshots of the
state of the QC at successive cursor-measurement
times. Each snapshot consists of a 4-element list
whose elements are the time at which the cursor is
measured, the state of the register immediately be-
fore the cursor is measured, the result of the meas-
urement, and the state of the register immediately
after the cursor position is measured. The last is
the projected state of the register. EvolveQC can
take the optional argument TimeBetweenObservations
which can be set to a number or a probability dis-
tribution. The default time between observations is
1 time unit.
```

To be concrete let us pick up the computation that we began in the last chapter, and actually see how a quantum computer would perform this calculation.

In[]:=

```
SeedRandom[13377];   (* seed to ensure reproducible
results *)
EvolveQC[ket[1,0,0,0], sqrtNOTcircuit]
```

Out[]=

```
{{0, ket[1, 0, 0, 0],
     {1, 0, 0},
     ket[1, 0, 0, 0]},

  {1, -0.422028 ket[0, 0, 1, 1] +
      (0.349228 - 0.349228 I) ket[0, 1, 0, 0] +
      (-0.349228 - 0.349228 I) ket[0, 1, 0, 1] +
      0.577972 ket[1, 0, 0, 0],

      {0, 1, 0},

      (0.5 - 0.5 I) ket[0, 1, 0, 0] +
      (-0.5 - 0.5 I) ket[0, 1, 0, 1]},

  {2, -0.698456 ket[0, 0, 1, 1] +
      (0.0779718 - 0.0779718 I) ket[0, 1, 0, 0] +
      (-0.0779718 - 0.0779718 I) ket[0, 1, 0, 1] -
      0.698456 ket[1, 0, 0, 0],

      {0, 0, 1},

      -1. ket[0, 0, 1, 1]}}
```

We can give EvolveQC the option Explain->True to produce a more descriptive explanation of the progress of the computation.

In[]:=

```
SeedRandom[13377];   (* same seed gives same result
as above *)
EvolveQC[ket[1,0,0,0], sqrtNOTcircuit,
         Explain->True]
```

Out[]=

```
Time t=0
State of QC = ket[1, 0, 0, 0]
Cursor observed at position = {1, 0, 0}
Projected state of QC = ket[1, 0, 0, 0]
```

```
Time t=1
State of QC =                -0.422028 ket[0, 0, 1, 1] +
           (0.349228 - 0.349228 I) ket[0, 1, 0, 0] +
          (-0.349228 - 0.349228 I) ket[0, 1, 0, 1] +
                            0.577972 ket[1, 0, 0, 0]
Cursor observed at position = {0, 1, 0}
Projected state of QC =
                     (0.5 - 0.5 I) ket[0, 1, 0, 0] +
                    (-0.5 - 0.5 I) ket[0, 1, 0, 1]


Time t=2
State of QC =                -0.698456 ket[0, 0, 1, 1] +
         (0.0779718 - 0.0779718 I) ket[0, 1, 0, 0] +
        (-0.0779718 - 0.0779718 I) ket[0, 1, 0, 1] -
                            0.698456 ket[1, 0, 0, 0]
Cursor observed at position = {0, 0, 1}
Projected state of QC = -1. ket[0, 0, 1, 1]
```

If you run the same calculation multiple times, without seeding the random number generator, you will find that it is possible to get different computational results each time due to the unavoidable indeterminism of the quantum measurement process used to find the cursor position.

In[]:=

```
EvolveQC[ket[1,0,0,0], sqrtNOTcircuit,
     Explain->True]
```

Out[]=

```
Time t=0
State of QC = ket[1, 0, 0, 0]
Cursor observed at position = {1, 0, 0}
Projected state of QC = ket[1, 0, 0, 0]


Time t=1
State of QC =                -0.422028 ket[0, 0, 1, 1] +
           (0.349228 - 0.349228 I) ket[0, 1, 0, 0] +
          (-0.349228 - 0.349228 I) ket[0, 1, 0, 1] +
                            0.577972 ket[1, 0, 0, 0]
Cursor observed at position = {1, 0, 0}
Projected state of QC = 1. ket[1, 0, 0, 0]


Time t=2
State of QC =                -0.422028 ket[0, 0, 1, 1] +
           (0.349228 - 0.349228 I) ket[0, 1, 0, 0] +
          (-0.349228 - 0.349228 I) ket[0, 1, 0, 1] +
                            0.577972 ket[1, 0, 0, 0]
```

```
Cursor observed at position = {1, 0, 0}
Projected state of QC = 1. ket[1, 0, 0, 0]


Time t=3
State of QC =             -0.422028 ket[0, 0, 1, 1] +
           (0.349228 - 0.349228 I) ket[0, 1, 0, 0] +
          (-0.349228 - 0.349228 I) ket[0, 1, 0, 1] +
                          0.577972 ket[1, 0, 0, 0]
Cursor observed at position = {0, 1, 0}
Projected state of QC =
                     (0.5 - 0.5 I) ket[0, 1, 0, 0] +
                    (-0.5 - 0.5 I) ket[0, 1, 0, 1]


Time t=4
State of QC =             -0.698456 ket[0, 0, 1, 1] +
         (0.0779718 - 0.0779718 I) ket[0, 1, 0, 0] +
        (-0.0779718 - 0.0779718 I) ket[0, 1, 0, 1] -
                          0.698456 ket[1, 0, 0, 0]
Cursor observed at position = {1, 0, 0}
Projected state of QC = -1. ket[1, 0, 0, 0]


Time t=5
State of QC =              0.422028 ket[0, 0, 1, 1] +
          (-0.349228 + 0.349228 I) ket[0, 1, 0, 0] +
           (0.349228 + 0.349228 I) ket[0, 1, 0, 1] -
                          0.577972 ket[1, 0, 0, 0]
Cursor observed at position = {0, 1, 0}
Projected state of QC =
                    (-0.5 + 0.5 I) ket[0, 1, 0, 0] +
                     (0.5 + 0.5 I) ket[0, 1, 0, 1]


Time t=6
State of QC =              0.698456 ket[0, 0, 1, 1] +
        (-0.0779718 + 0.0779718 I) ket[0, 1, 0, 0] +
         (0.0779718 + 0.0779718 I) ket[0, 1, 0, 1] +
                          0.698456 ket[1, 0, 0, 0]
Cursor observed at position = {0, 0, 1}
Projected state of QC = 1. ket[0, 0, 1, 1]
```

Here is the output from a different run.

In[]:=
```
EvolveQC[ket[1,0,0,0], sqrtNOTcircuit,
Explain->True]
```
Out[]=
```
Time t=0
State of QC = ket[1, 0, 0, 0]
Cursor observed at position = {1, 0, 0}
Projected state of QC = ket[1, 0, 0, 0]


Time t=1
State of QC =               -0.422028 ket[0, 0, 1, 1] +
           (0.349228 - 0.349228 I) ket[0, 1, 0, 0] +
          (-0.349228 - 0.349228 I) ket[0, 1, 0, 1] +
                          0.577972 ket[1, 0, 0, 0]
Cursor observed at position = {1, 0, 0}
Projected state of QC = 1. ket[1, 0, 0, 0]


Time t=2
State of QC =               -0.422028 ket[0, 0, 1, 1] +
           (0.349228 - 0.349228 I) ket[0, 1, 0, 0] +
          (-0.349228 - 0.349228 I) ket[0, 1, 0, 1] +
                          0.577972 ket[1, 0, 0, 0]
Cursor observed at position = {1, 0, 0}
Projected state of QC = 1. ket[1, 0, 0, 0]


Time t=3
State of QC =               -0.422028 ket[0, 0, 1, 1] +
           (0.349228 - 0.349228 I) ket[0, 1, 0, 0] +
          (-0.349228 - 0.349228 I) ket[0, 1, 0, 1] +
                          0.577972 ket[1, 0, 0, 0]
Cursor observed at position = {0, 0, 1}
Projected state of QC = -1. ket[0, 0, 1, 1]
```

In all three cases, the correct answer was returned, but it was returned at different times. Thus in the Feynman quantum computer, the answer returned is always guaranteed to be correct but there is uncertainty as to when the answer will be available.

Graphical Illustration of the Evolution of the Memory Register

We can create a graphical illustration of how the state of the memory register evolves over time using the function PlotEvolution. This shows the probability with which each eigenstate contributes to the superposition of states in the memory register at the successive times at which the cursor is observed. PlotEvolution takes a single input, which is the output of the function EvolveQC, and returns a graphic that shows the probability (i.e., modulus amplitude squared) of each eigenstate in the superposition of eigenstates at the times when the cursor is observed. For compactness of notation we label the eigenstates of the (in this case 4-bit) memory register, |i>, in base 10 notation; for example, |5> corresponds to the eigenstate of the memory register that is really |0101> and |15> corresponds to the eigenstate of the memory register that is really |1111>. By default, PlotEvolution shows the composition of the superposition of states in the memory register immediately before the cursor is observed (Fig. 4.5).

In[]:=

```
SeedRandom[123457];
evoln = EvolveQC[ket[1,0,0,0], sqrtNOTcircuit];
PlotEvolution[evoln];
```

Out[]=

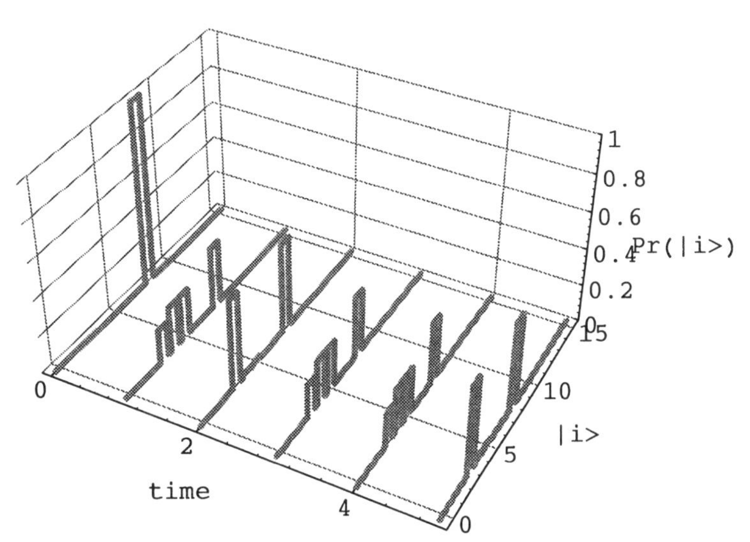

Fig. 4.5 Graphical illustration of the evolution of the state of the cursor and memory register in Feynman's quantum computer. The state, at the indicated times, is shown *before* the cursor has been read.

In[]:=
```
PlotEvolution[evoln, AfterObservingCursor->True];
```
Out[]=

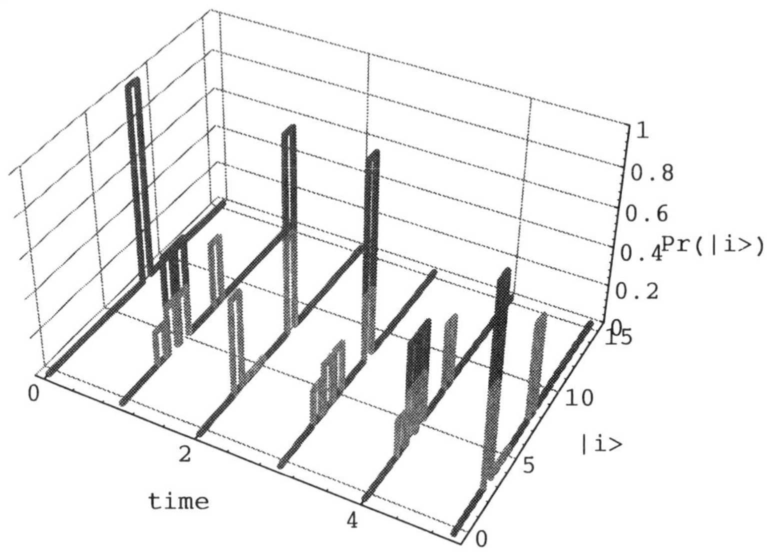

Fig. 4.6 The same evolution as shown in Fig. 4.5 with the introduction of the states the cursor and memory qubits are projected into *after* the cursor has been read.

However, by setting the option AfterObservingCursor->True you can see the effect that observing the cursor has on the composition of the superposition. The light lines in Fig. 4.6 refer to the state of the memory register immediately before the cursor position is observed and the darker lines refer to the state of the register immediately after the cursor has been observed. The projective effect of the cursor observations is quite noticeable. If you measure the cursor position too frequently, then you can sometimes prevent the system from evolving at all! Remember, the computer only evolves during the time it is not being observed (measured). If you do not let it evolve for long enough between measurements, then no appreciable probability amplitude will accumulate in those states representing the cursor anywhere other than at its initial state. In such a situation, each time you measure the cursor, you keep collapsing it back to its initial state and so the computation never advances. In the following example, we keep collapsing the cursor state back to $|100\rangle$. Eventually, we got lucky and projected the state into $|001\rangle$ (the terminal state showing that the computation was done).

In[]:=
```
longCalc = EvolveQC[ket[1,0,0,0], sqrtNOTcircuit,
                    TimeBetweenObservations->0.2];
```
The evolution recorded in longCalc reveals the following sequence of cursor positions (right column) at the indicated times (left column). The cursor result {1,0,0} means the cursor is at the first site (corresponding to state $|100\rangle$), {0,1,0} means it is at the second site (corresponding to state $|010\rangle$), and so on.

```
{0.0,  {1, 0, 0}}
{0.2,  {1, 0, 0}}
{0.4,  {1, 0, 0}}
{0.6,  {1, 0, 0}}
{0.8,  {1, 0, 0}}
{1.0,  {0, 1, 0}}
{1.2,  {0, 1, 0}}
{1.4,  {0, 1, 0}}
{1.6,  {0, 1, 0}}
{1.8,  {0, 1, 0}}
{2.0,  {0, 1, 0}}
{2.2,  {0, 1, 0}}
{2.4,  {0, 1, 0}}
{2.6,  {0, 1, 0}}
{2.8,  {1, 0, 0}}
{3.0,  {1, 0, 0}}
{3.2,  {1, 0, 0}}
{3.4,  {1, 0, 0}}
{3.6,  {1, 0, 0}}
{3.8,  {1, 0, 0}}
{4.0,  {1, 0, 0}}
{4.2,  {1, 0, 0}}
{4.4,  {1, 0, 0}}
{4.6,  {1, 0, 0}}
{4.8,  {1, 0, 0}}
{5.0,  {1, 0, 0}}
{5.2,  {1, 0, 0}}
{5.4,  {1, 0, 0}}
{5.6,  {1, 0, 0}}
{5.8,  {1, 0, 0}}
{6.0,  {1, 0, 0}}
{6.2,  {1, 0, 0}}
{6.4,  {1, 0, 0}}
{6.6,  {1, 0, 0}}
{6.8,  {1, 0, 0}}
{7.0,  {1, 0, 0}}
{7.2,  {1, 0, 0}}
{7.4,  {1, 0, 0}}
{7.6,  {1, 0, 0}}
{7.8,  {1, 0, 0}}
{8.0,  {1, 0, 0}}
{8.2,  {1, 0, 0}}
{8.4,  {1, 0, 0}}
{8.6,  {1, 0, 0}}
{8.8,  {1, 0, 0}}
{9.0,  {1, 0, 0}}
{9.2,  {1, 0, 0}}
{9.4,  {1, 0, 0}}
{9.6,  {1, 0, 0}}
{9.8,  {1, 0, 0}}
{10.0, {1, 0, 0}}
{10.2, {1, 0, 0}}
{10.4, {1, 0, 0}}
{10.6, {1, 0, 0}}
{10.8, {1, 0, 0}}
{11.0, {1, 0, 0}}
{11.2, {1, 0, 0}}
{11.4, {1, 0, 0}}
{11.6, {1, 0, 0}}
{11.8, {1, 0, 0}}
{12.0, {1, 0, 0}}
{12.2, {1, 0, 0}}
{12.4, {1, 0, 0}}
{12.6, {1, 0, 0}}
{12.8, {0, 1, 0}}
{13.0, {0, 0, 1}}
```

CHAPTER 5

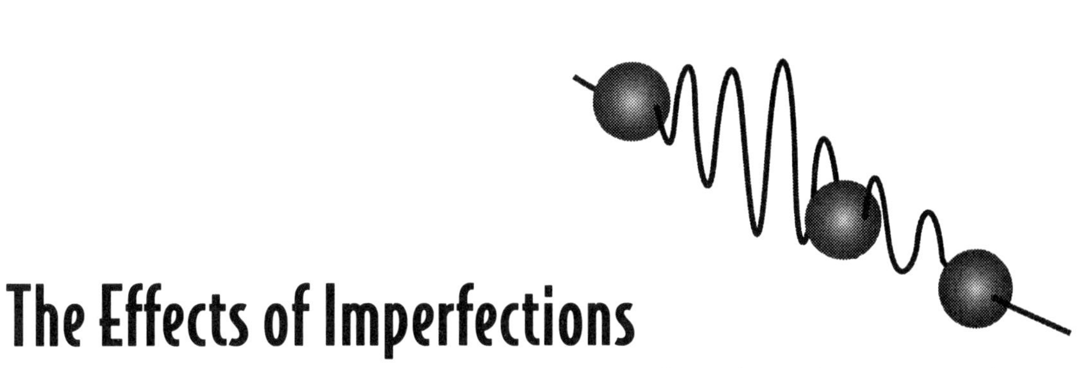

The Effects of Imperfections

"Nature knows no trifling; she is always sincere, always serious, always stern; she is always in the right, and the errors and mistakes are invariably ours."
— Goethe

Up to this point, we have considered the operation of an *ideal* quantum computer. In reality there will be unavoidable physical effects that conspire to make such ideal operation impossible. In this chapter we consider errors that occur during the prepare, evolve, and measure cycles of a quantum computation. We limit our attention to just those sources of error that are internal to the computer, that is, imperfections in the design of the machine, mistakes in the data fed into it, and imprecision in timing measurements. Of course there are many other sources of error, too, that come from the inevitable interactions between the computer and its environment. But for the moment we focus on just the internal errors and postpone our discussion of external error models and error correction until Chapter 10.

The errors in preparation and measurement are both analogous to "software" errors. The evolution errors are analogous to hardware errors as would occur in an incorrect circuit. As we show, the software errors in a quantum computer are conserved, that is, constant, throughout the computation. By contrast, the hardware errors grow quadratically in time, although this is much better than the classical case.

That errors occur is not surprising, although classically it is possible theoretically to have a computer that operates perfectly. With a quantum computer we cannot have perfect operation even theoretically. The Uncertainty Principle prevents us from making critical measurements with certainty so that we can never know for sure if, for example, we wanted to verify that the input state we desire is the one that is actually being used. Similarly, we cannot be sure that the answer that we measure is exactly the same one that the computer found.

To ground our discussion of errors it is helpful to know the physical sources of these errors. The errors come from a variety of sources depending in part on the embodiment of the qubits, for example, electron spin, photon polarization, or molecular energy level. In general, errors in the initial and final calibrations of the preparation and measuring devices are sources of error. Also, background effects that distort the Hamiltonian cause errors. For example, in the case of electron spin states representing qubits we must be careful to avoid the effects of stray magnetic fields that could tilt the spin by an angle and lead to a small chance of a subsequent measurement error. In the case of photon polarization we must be concerned with the properties of our polarizers and other types of photon analyzers such as birefringent crystals and Pockels cells, as discussed in Chapter 6. Are the crystals perfect and are they oriented properly? If not, the chances of an incorrect measurement will be increased. In the case of molecular energy levels we must be concerned with the pulse length of the light that goes into exciting the electrons because the wrong length pulse will change the probability of a transition to another state, and hence the "value" of the qubit, and thus affect the entire course of the computation.

With these effects in mind we can now look into the consequences of the inevitable errors on the prepare, evolve, and measure stages of quantum computation.

5.1 Imperfections in Preparation

In the preparation stage we prepare a register of qubits into a particular state that will serve as the initial state for our computation. The initial state may be a classical state or it may be a quantum mechanical superposition. But what if there is an error in the preparation of this initial data? That is, what if one or more of the initial state qubits was not in the state we expected it to be? How trustworthy would our answer be at the end of the computation? Classically, this would result in errors that grow exponentially with the number of steps performed. Quantum mechanically there is no growth of the errors as the computation progresses[Zurek84]. This can be seen straightforwardly by considering an initial state with a small error in one of the bits. For example, suppose the perfect (i.e., error-free) preparation and subsequent evolution is given by

$$|\psi_{\text{out}}\rangle = \exp\left[-(i/\hbar)\int_0^T \hat{H}(t)\,dt\right]\cdot|\psi_{\text{in}}\rangle = \hat{U}(T)\cdot|\psi_{\text{in}}\rangle.$$

Note that we are using a time-dependent Hamiltonian which means that our results are general and work for the time-independent case as well. The evolution with error is then given by

$$|\psi_{\varepsilon,\text{out}}\rangle = \exp\left[-\frac{i}{\hbar}\int_0^T \hat{H}(t)dt\right]\cdot|\psi_{\varepsilon,\text{in}}\rangle = \hat{U}(T)\cdot|\psi_{\varepsilon,\text{in}}\rangle,$$

where $|\psi_{\varepsilon,\text{in}}\rangle$ is the input state corrupted by some error ε and $|\psi_{\varepsilon,\text{out}}\rangle$ is the resulting output state. Now suppose we choose a particular type of error for ε. Consider an input state consisting of N qubits,

$$\begin{aligned}|\psi_{\varepsilon,\text{in}}\rangle &= |b_1\rangle\otimes|b_2\rangle\otimes\cdots\otimes\left[\sqrt{1-|\varepsilon|^2}\,|0_k\rangle+\varepsilon|1_k\rangle\right]\otimes\cdots\otimes|b_{N-1}\rangle\\ &= \sqrt{1-|\varepsilon|^2}\,|\psi_{\text{in}}\rangle+\varepsilon|\text{junk}\rangle\end{aligned}$$

where ε is a small preparation error of the kth bit and the "junk" ket is some state unrelated to the desired state that effectively leaks probability out of the correct computation. Making the appropriate substitutions we have:

$$|\psi_{\varepsilon,\text{out}}\rangle = \hat{U}|\psi_{\varepsilon,\text{in}}\rangle = \sqrt{1-|\varepsilon|^2}\,|\psi_{\text{out}}\rangle+\varepsilon|\text{junkier}\rangle,$$

which shows that the errors are constant, or conserved, in time and a huge improvement over the classical exponential growth of errors.

The preceding result can easily be generalized to the case where there is an error ε_i in each of the qubits and by the same linearity arguments given previously we have that the errors are conserved and do not grow with time. This important fact results from the linearity of quantum mechanics.

5.2 Imperfections in Evolution

We now consider errors that take place during the evolution of a quantum computation assuming that the preparation and measurement operations are perfect for the sake of simplicity. Again the reasoning is straightforward, except that this time the error is in the Hamiltonian and not in the initial state so that

$$\hat{H}_\varepsilon = \hat{H} + \hat{\eta},$$

where η is a small error in the Hamiltonian. The effect of this error manifests itself in the probabilities of the possible outcomes. That is, there will be a different chance for a particular outcome because of the presence of the error.

Up to now we have only dealt with the results of particular measurements of quantum states. Since the existence of an error is probabilistic we cannot tell ahead of time whether a particular computation will have an error. Thus, when discussing errors, we need to know about the "expected" value of the computation over a large number of trials. In quantum mechanical terms we are interested in the so-called "expectation value" of the computation. The expectation value is calculated from the probability weighted-sum of all the possible measurements. Suppose the eigenvectors of $|\psi\rangle$ are $|\phi_i\rangle$ with amplitudes ω_i and

$$|\psi\rangle = \sum_{i=1}^{n} \omega_i |\phi_i\rangle.$$

Suppose further that we intend to make a measurement with an operator $\hat{E}$ that has eigenvalues E_i. The expectation value of $\hat{E}$ is given by

$$\langle \hat{E} \rangle = \langle \psi | \hat{E} | \psi \rangle = \sum_{i=1}^{n} \omega_i^* \, \omega_i \, E_i.$$

The expectation value for an error, relative to the perfect output state, is then given by the amplitude of the expectation value of the "perfect" output state and the "error" output state, namely,

$$\begin{aligned} \langle \psi_{\text{out}} | \psi_{\varepsilon,\text{out}} \rangle &= \left(\langle \psi_{\text{in}} | e^{(i/\hbar)\hat{H}t} \right) \left(e^{-(i/\hbar)(\hat{H}+\hat{\eta})t} | \psi_{\text{in}} \rangle \right) \\ &= \langle \psi_{\text{in}} | e^{-(i/\hbar)\hat{\eta}t} | \psi_{\text{in}} \rangle \end{aligned}$$

where we have used the relations $|\psi_{\text{out}}\rangle = e^{-i\hat{H}t/\hbar} |\psi_{\text{in}}\rangle$ and

$$
\begin{aligned}
\left|\psi_{\varepsilon,\mathrm{out}}\right\rangle &= e^{-i\hat{H}_{\varepsilon}t/\mathrm{h}}\left|\psi_{\mathrm{in}}\right\rangle \\
&= e^{-i\eta t/\mathrm{h}}e^{-iHt/\mathrm{h}}\left|\psi_{\mathrm{in}}\right\rangle \\
&= e^{-i\eta t/\mathrm{h}}\left|\psi_{\mathrm{out}}\right\rangle.
\end{aligned}
$$

If η is small we can expand the exponential and only retain the linear term from which we obtain:

$$
\begin{aligned}
\left\langle\psi_{\mathrm{out}}\middle|\psi_{\varepsilon,\mathrm{out}}\right\rangle &= \left\langle\psi_{\mathrm{in}}\right|1-(i/\mathrm{h})\hat{\eta}t+\mathrm{L}\left|\psi_{\mathrm{in}}\right\rangle \\
&= \left\langle\psi_{\mathrm{in}}\middle|\psi_{\mathrm{in}}\right\rangle-(it/\mathrm{h})\left\langle\psi_{\mathrm{in}}\right|\hat{\eta}\left|\psi_{\mathrm{in}}\right\rangle \\
&= 1-(it/\mathrm{h})\left\langle\psi_{\mathrm{in}}\right|\hat{\eta}\left|\psi_{\mathrm{in}}\right\rangle
\end{aligned}
$$

If, in addition to being small, $\hat{\eta}$ is randomly distributed around 0, then the second term is zero and the figure of merit for determining the effect of the Hamiltonian error is given by the squared amplitude of the previous equation:

$$
\begin{aligned}
\left|\left\langle\psi_{\mathrm{out}}\middle|\psi_{\varepsilon,\mathrm{out}}\right\rangle\right|^{2} &= \left|\left\langle\psi_{\mathrm{in}}\right|1-(i/\mathrm{h})\hat{\eta}t+\mathrm{L}\left|\psi_{\mathrm{in}}\right\rangle\right|^{2} \\
&\approx \left\langle\psi_{\mathrm{in}}\right|1-(i/\mathrm{h})\hat{\eta}t\left|\psi_{\mathrm{in}}\right\rangle\left\langle\psi_{\mathrm{in}}\right|1+(i/\mathrm{h})\hat{\eta}^{*}t\left|\psi_{\mathrm{in}}\right\rangle \\
&= \left\langle\psi_{\mathrm{in}}\right|1-(i/\mathrm{h})\hat{\eta}t\left|\psi_{\mathrm{in}}\right\rangle\left\langle\psi_{\mathrm{in}}\right|1+(i/\mathrm{h})\hat{\eta}^{*}t\left|\psi_{\mathrm{in}}\right\rangle \\
&= \left\langle\psi_{\mathrm{in}}\right|1\left|\psi_{\mathrm{in}}\right\rangle\left\langle\psi_{\mathrm{in}}\right|1\left|\psi_{\mathrm{in}}\right\rangle-\left\langle\psi_{\mathrm{in}}\right|(i/\mathrm{h})\hat{\eta}t\left|\psi_{\mathrm{in}}\right\rangle+ \\
&\quad \left\langle\psi_{\mathrm{in}}\right|(i/\mathrm{h})\hat{\eta}t\left|\psi_{\mathrm{in}}\right\rangle+(t/\mathrm{h})^{2}\left\langle\psi_{\mathrm{in}}\right|\hat{\eta}\left|\psi_{\mathrm{in}}\right\rangle\left\langle\psi_{\mathrm{in}}\right|\hat{\eta}^{*}\left|\psi_{\mathrm{in}}\right\rangle \\
&= \left|\left\langle\psi_{\mathrm{in}}\middle|\psi_{\mathrm{in}}\right\rangle\right|^{2}-(t/\mathrm{h})^{2}\left|\left\langle\psi_{\mathrm{in}}\right|\hat{\eta}\left|\psi_{\mathrm{in}}\right\rangle\right|^{2}
\end{aligned}
$$

which finally reduces to:

$$
\left|\left\langle\psi_{\mathrm{out}}\middle|\psi_{\varepsilon,\mathrm{out}}\right\rangle\right|^{2}=1-\alpha^{2}t^{2}
$$

with

$$
\alpha^{2}=(1/\mathrm{h})^{2}\left|\left\langle\psi_{\mathrm{in}}\right|\hat{\eta}\left|\psi_{\mathrm{in}}\right\rangle\right|^{2}=(1/\mathrm{h})^{2}\left(\left\langle\psi_{\mathrm{in}}\right|\hat{\eta}^{2}\left|\psi_{\mathrm{in}}\right\rangle-\left|\left\langle\psi_{\mathrm{in}}\right|\hat{\eta}\left|\psi_{\mathrm{in}}\right\rangle\right|^{2}\right)
$$

to highlight the quadratic growth in time of the error. The inner product is the probability that the output due to an error in the Hamiltonian is different from the output using a perfect Hamiltonian. Note that the asymptotic behavior as the error η approaches zero is correct and consequently the inner product approaches 1 as we expect, since there is no difference between the two states (i.e., $\hat{\eta}=0$).

Fig. 5.1 shows the result of an analysis over many runs using different values for errors in the Hamiltonian. The figure plots the probability versus the magnitude of the Hamiltonian error. As can be seen in the figure, for small values of errors in the Hamiltonian (proportional to α) the behavior is indeed quadratic. As the magnitude of the error increases, the effect falls off much more slowly as higher order terms contribute as shown by the divergence of the solid and dashed curves.

Each data point in Fig. 5.1 represents the results of 500 runs on a Feynman quantum computer simulator of the $\sqrt{\text{NOT}} \cdot \sqrt{\text{NOT}} \cdot |0\rangle$ function, run until completion as defined by the cursor being measured in the stop location. Errors come in the following varieties. The first kind of error is a corruption of the cursor where, because of an error in the Hamiltonian, an extra cursor bit appears when the stop bit is checked. No matter what the data register reads, the answer is considered to be wrong if there is more than one cursor bit. The second kind of error occurs when the data register is corrupted and the answer is incorrect when the cursor is measured to be in the stop location.

Similarly, Fig. 5.2 shows that the performance, measured in terms of correctness of solution, drops off quadratically in time for small times and begins to deviate from the theoretical prediction as the time to finish increases. We used a small fixed value for the Hamiltonian error of $\alpha = 0.03$ so that the errors were influenced mostly by time effects rather than by the magnitude of α.

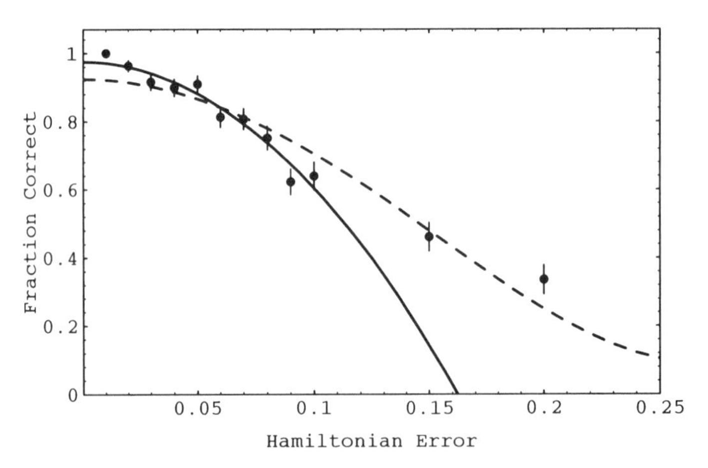

Fig. 5.1 Change in performance of $\sqrt{\text{NOT}} \cdot \sqrt{\text{NOT}} \cdot |0\rangle$ as a function of the error in the Hamiltonian (α). The Hamiltonian error function was defined to be $|N(0, \text{error})| + iN(0, \text{error})$ where $N(m, s)$ is the Normal distribution with mean m and standard deviation s. The data correspond to a time fixed at 2 time-steps from the start. The solid curve is the best-fit quadratic to the first 10 points. Each point corresponds to the runs finishing at time 2 and the error bars are the standard error of the mean of the runs assuming a binomial distribution (i.e., error $= \sqrt{(1-f)f/N}$ where N is the number of runs in the data point and f is the fraction that are correct). Note the divergence from the best-fit quadratic as the error becomes larger corresponding to the contribution of higher-order terms in the Hamiltonian error. The dashed curve is the best fit to a quadratic plus a fourth-order term using all the data points.

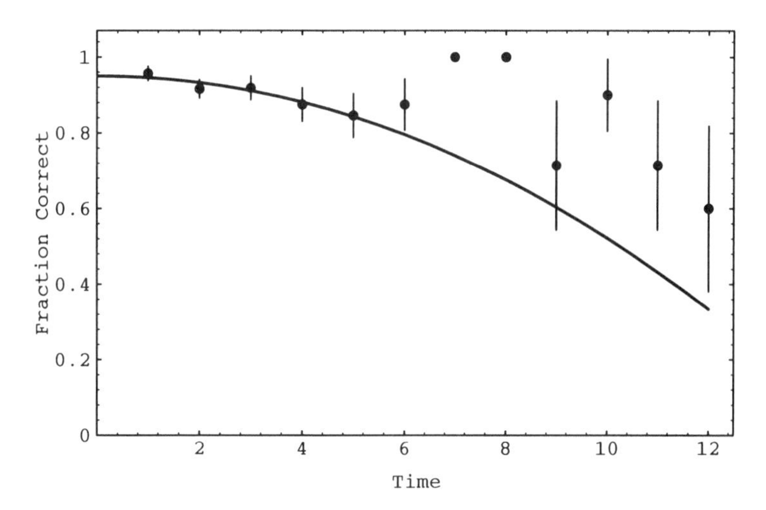

Fig. 5.2 Change in performance of $\sqrt{\text{NOT}} \cdot \sqrt{\text{NOT}} \cdot |0\rangle$ as a function of time where the error in the Hamiltonian was fixed at $|N(0, 0.03)| + iN(0, 0.03)$. The solid curve is the best-fit quadratic to the first 6 points. Each point corresponds to the runs that finished at that time and the error bars are the standard error of the mean of the runs as in Fig. 5.1. Again the error behaves quadratically for small times as per the theoretical prediction and begins to deviate for larger times as the higher-order terms begin to contribute. Note that for the larger times the error bars become larger because there are fewer runs that take a long time to run. For times 7 and 8 the performance is perfect so there is no statistical spread.

The results of the simulation runs in Figs. 5.1 and 5.2 show that quantum computers can be quite robust with respect to rather large errors in the circuit, which means that fairly high levels of errors can be tolerated in the manufacturing while still yielding a viable quantum computer.

5.3 Imperfections in Measurement

What happens if we make more than one measurement? That is, what if the computation is not done when we measure it and we need to make multiple measurements? Suppose that we make n measurements at equally spaced intervals in time. Then on average the error Hamiltonian will be $\hat{\eta}/n$ between measurements and we're left with a probability of an error of

$$\left|\langle \psi_{\text{out}} | \psi_{\varepsilon,\text{out}} \rangle\right|^2 = \left(1 - (\alpha t/n)^2\right)^n > 1 - (\alpha t)^2,$$

which is better than we had for the case with only one measurement. This means that more measurements will not make our computations less accurate.

Recall that Feynman's computer does not stop when it obtains an answer; it keeps going and it is up to the programmer to measure the system at a time when the computation happens to be complete. Suppose that the measurement takes place within a time τ centered around the correct time T which is a time when the Feynman computer is finished. Then the probability of getting the correct result is:

$$\left|\langle\psi|e^{-iH\tau/\mathrm{h}}|\psi\rangle\right|^2 \approx 1-\langle H\rangle(i\tau/\mathrm{h})-\left(\langle H^2\rangle-\langle H\rangle^2\right)\tau^2/\mathrm{h}^2+\mathrm{L}\ .$$

The average value of the Hamiltonian is zero ($\langle H\rangle = 0$) if the clock cursor waves moving forward are defined to have positive energy and the clock waves moving backwards are defined to have negative energy. That means that the probability of error is simply the term

$$\langle H^2\rangle\tau^2/\mathrm{h}^2,$$

which grows quadratically in the error in timing the measurement of the answer.

We are now at the point where we have established the basics for how quantum computing operates and how it can go wrong. In Chapter 10 we discuss how errors such as these and others can be corrected in some cases.

CHAPTER 6

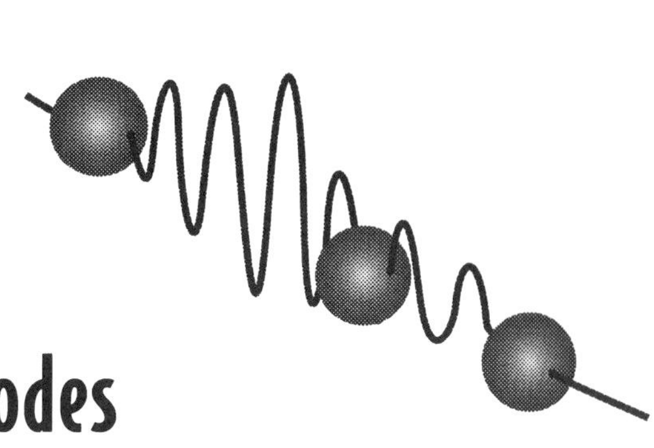

Breaking Unbreakable Codes

"There are three names in quantum computing: Shor, Alice and Bob."
— Michail Zak

In Chapters 3 and 4 we described a quantum computer that was specialized to do a simple calculation, namely, to compute the square of the square-root-of-NOT. This illustrated how quantum theory could be harnessed in the service of computation. However, you might be wondering whether a quantum computer can do anything more impressive? After all, the square of the square-root-of-NOT function is not exactly a vital computation. Is there a *significant* problem that a quantum computer can solve exponentially faster than any classical computer?

This question troubled many of the scientists engaged in quantum computing for the better part of a decade. For quantum computing to take off, someone needed to find at least one important application for which a quantum computer outperformed any classical computer. This application was discovered in 1994 when Peter Shor, of AT&T Bell Labs, discovered how to make a quantum computer factor a large integer superefficiently. Although this may sound like an esoteric problem, the ability to factor large integers efficiently is the key to breaking one of the most secure cryptographic schemes currently in use, the RSA-cryptosystem, invented by Rivest, Shamir, and Adleman in 1978[Rivest78].

In this chapter we describe how the RSA cryptosystem works and how you could try to break the RSA code using first a classical computer and then a quantum computer. We include output from a run of the RSA algorithm that shows you how to encrypt and decrypt a secret message and a simulator that mimics a quantum computer factoring an integer and, hence, breaking the RSA.

6.1 Codes and Code-Breakers

When people exchange sensitive information such as military plans, banking transactions, and confidential messages they routinely encrypt the information using a secret code. Such messages are secure as long as unscrambling the message is computationally intractable.

The need for coded messages is hardly new. The first known coded text is a small piece of Babylonian cuneiform dating from around 1500 BC It describes a secret method for making a glaze for pottery. By recording the instructions in a code, the potter could protect his secrets from the prying eyes of jealous competitors.

In more modern times more sophisticated codes have been developed that are very difficult to crack by hand.

In the Second World War, for example, the Germans used very elaborate codes to pass orders between the German High Command and officers in the field. The Enigma code was invented primarily for communicating with U-boats.

The breaking of such coded messages quickly became a military priority. The British and the Americans both established code-breaking centers staffed by some of the best mathematicians of their generation. Despite the skill of these individuals, the need to use a machine to break the codes quickly became apparent.

A machine that reads in a coded message and unscrambles it by applying some systematic algorithm is nothing other than a Turing machine. So it is not surprising that Alan Turing was co-opted into the British code breaking effort during the Second World War.

Turing joined the Code and Cipher School at Bletchley Park in England. This was an elite collection of mathematicians who were charged with the task of breaking the secret codes used by the Germans. Their first designs emulated the Turing machine model rather literally. They consisted of electromechanical contraptions that used tapes and elaborate pulley systems. The tapes broke frequently and the machines were often out of service.

Speed was crucial. A message had to be decoded before the event threatened in the message took place. The old electro-mechanical machines were simply too slow to get the job done in time. Consequently, British Intelligence commissioned the construction of an electronic code-breaking machine instead.

At the time, there was no computer industry as such, so the British had to make do with the best they had — the Post Office. The Post Office controlled the telephone system. They had more experience in the use of vacuum tube technology than anyone else.

At first the Bletchley Park scientists were skeptical that an electronic device would be sufficiently reliable. However, when the machine was first tried out it broke a coded message in 30 minutes that had taken several hours the day before.

In the summer of 1944, at the height of the Second World War, British military intelligence intercepted a coded message from the German high command to the commander of the Luftwaffe. The code that the Germans used was called Enigma. Unbeknownst to the Germans, the Allies had found a way of breaking the Enigma code. Their secret messages were not secret at all. This particular message outlined a plan for a bombing raid on the city of Coventry, an industrial center in the middle of England. The British commanders faced a dilemma: if they warned the people of Coventry about the impending raid they would signal the Germans that the Enigma code had been broken. On the other hand, to do nothing would inevitably mean the deaths of many civilians. With a heavy heart Winston Churchill, then Prime Minister, decided not to inform the citizenry in order to protect the future intelligence stream of the allies. The subsequent raid on Coventry killed many people but probably many more lives were saved from the knowledge gleaned from intercepted German messages in the months that followed.

Clearly the ability to make and break secret codes is a crucial skill in the modern world.

6.2 Code Making

There are numerous ways to make a coded message. Whatever method is employed, the idea is to scramble a message in such a way as to make it unintelligible to potential adversaries yet transparent to intended recipients. Typically, this involves converting the text of the message, which is usually given as a string of characters, into a sequence of numbers, and then mapping these numbers into other numbers by applying some "encryption algorithm" to the numerical encoding of the message text. Upon receipt, the recipient must decrypt the encrypted message back into the sequence of integers and reconvert these integers to the text string. Thus, in order to send and receive secure messages, you typically need to define two functions MessageToIntegers and Encrypt together with their inverses IntegersToMessage and Decrypt. In most cryptosystems, a sender encrypts a message *M* by computing:

Encrypt[MessageToIntegers[M]] -> E,

where the inner function MessageToIntegers is computed first, creating an encoding of a message in a set of "message integers" and then the encryption function is applied to these message integers. This process returns *E*, the encryption of *M*.

To decrypt *E*, a recipient must compute the inverse functions, that is,

IntegersToMessage[Decrypt[E]] -> M.

Before any secure messages can be exchanged, the parties who wish to communicate must agree upon the "alphabet" of symbols from which future messages will be composed. Typically, such an alphabet contains more than merely lower-case letters. It may also include punctuation marks, upper-case letters, numbers, and parentheses. The exact composition of the alphabet is unimportant. All that matters is that both parties agree upon the set, and alphabetic ordering, of symbols to be used.

An alphabet, sufficient for simple communications, is shown in the following. This particular alphabet, which we have called $Alphabet, contains 76 distinct symbols.

In[]:=

```
$Alphabet =
{"a", "b", "c", "d", "e", "f", "g", "h", "i", "j",
"k", "l", "m", "n", "o", "p", "q", "r", "s", "t",
"u", "v", "w", "x", "y", "z", "A", "B", "C", "D",
"E", "F", "G", "H", "I", "J", "K", "L", "M", "N",
"O", "P", "Q", "R", "S", "T", "U", "V", "W", "X",
"Y", "Z", "1", "2", "3", "4", "5", "6", "7", "8",
"9", "0", "!", "?", ".", ",", ";", ":", "(", ")",
"{", "}", "[", "]", " ", "'"};
```

Given such an alphabet, it is straightforward to define the MessageToIntegers function. If the alphabet contains `l` distinct symbols, you can convert the message to a sequence of integers by numbering the symbols in the alphabet from 0 to `l-1` and then substituting the appropriate integer for successive symbols appearing in the message. Thus the secret message, "My PIN number is 1234!" would, using $Alphabet, be converted into the sequence of integers {38, 24, 74, 41, 34, 39, 74, 13, 20, 12, 1, 4, 17, 74, 8, 18, 74, 52, 53, 54, 55,

62}. The code supplement to this chapter contains output from functions that will do this for you:

In[]:=

```
MessageToIntegers["My PIN number is 1234!",$Alphabet]
```

Out[]=

```
{38, 24, 74, 41, 34, 39, 74, 13, 20, 12, 1, 4, 17,
 74, 8, 18, 74, 52, 53, 54, 55, 62}.
```

Likewise, to decode a message given a sequence of such integers, you simply have to invert the substitution operation.

In[]:=

```
IntegersToMessage[{38, 24, 74, 41, 34, 39, 74, 13,
                   20, 12, 1, 4, 17, 74, 8, 18, 74,
                   52, 53, 54, 55, 62}, $Alphabet]
```

Out[]=

```
My PIN number is 1234!
```

Simple substitution alone, however, does not lead to a secure message. Substitution codes can be broken easily, by exploiting known statistical properties of the language in which the message is written. For example, in written English, different letters occur with a predictable frequency distribution. Knowledge of this distribution allows an adversary to guess large parts of a substitution code (see Fig. 6.1). To ensure that a message will be difficult for an adversary to read, you need to do more than merely substitute symbols; you need to encrypt the message in some way. Encrypt and Decrypt do the scrambling and unscrambling, respectively. To be an *effective* method of communication, both Encrypt and Decrypt must be reasonably *fast* computations. Unfortunately, this is unlikely to make for a secure communication because if decryption is fast, an

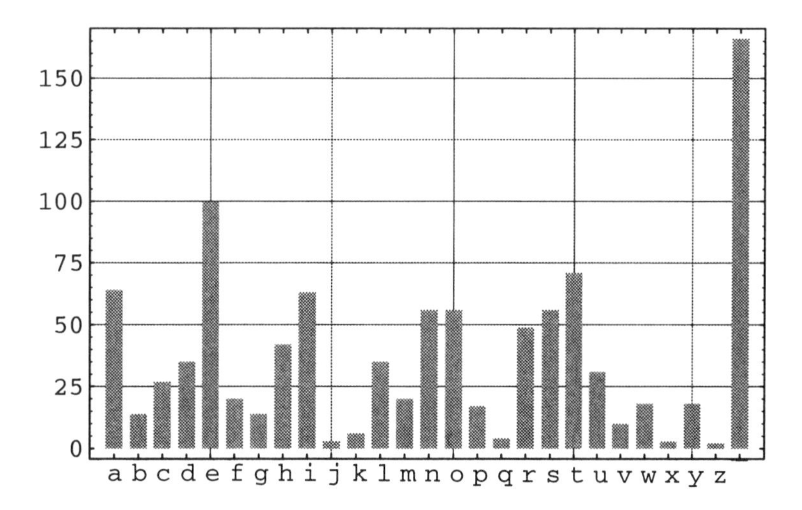

Fig. 6.1 Frequencies of letters in English per 1000 characters The blank space symbol is denoted as _.

adversary might be able to guess the definition of Decrypt through trial and error.

6.3 Trapdoor Functions

For the purpose of establishing *secure* communications, the functions Encrypt and Decrypt need to possess certain properties. In particular, the person sending a message would like to be able to compute Encrypt efficiently, but arrange for it to be almost impossible for an adversary to compute its inverse Decrypt in any reasonable length of time. There are many functions in mathematics that have this property of being easy to compute but hard to invert. For example, given two large prime numbers, p and q, say, it is very easy to compute their product $n = p \times q$. Here is a timing estimate.

```
In[]:=
    p = 15485863;
    q = 15485867;
    Timing[p q]
Out[]=
    {0. Second, 239812014798221}
```

The multiplication is done so quickly that the time taken is returned as approximately equal to zero! On the other hand, given a large integer n = 239812014798221, the task of finding two integers p and q such that $p \times q = n$ is a much more difficult problem. Here is a timing estimate for this problem.

```
In[]:=
    n = 239812014798221;
    Timing[FactorInteger[n]]
Out[]=
    {5.91667 Second, {{15485863, 1}, {15485867, 1}}}
```

Such "one-way" functions are used extensively in cryptography. Unfortunately, there is a slight drawback. If we rely purely on one-way functions, it is as hard for an intended recipient to decrypt the message as it is for an adversary. This calls for a minor refinement of the idea: This is the notion of a "trapdoor" function.

A trapdoor function is some mathematical procedure that is easy to compute but very hard to invert *unless you have access to a special "key."* The most powerful modern cryptographic techniques for making unbreakable codes use the concept of a trapdoor function.

To encrypt a message using a trapdoor function with a key *K*, the sender computes:

Encrypt[MessageToIntegers[M], K]] -> E

Upon receipt, a recipient unscrambles the encrypted message by computing:

IntegersToMessage[Decrypt[E], K] -> M.

Provided the parties know the key, both encrypting the message for the sender and decrypting the message for the recipient are easy computations. However, if an adversary who intercepts the encrypted message does not know the secret key, he will find it effectively impossible to unscramble the message.

6.4 One-Time Pads

The "one-time pad" is a cryptosystem based on a trapdoor function. This particular cryptosystem is so secure that it is rumored to be used for communicating diplomatic information between Washington and Moscow[Welsh88]. Despite its high level of security, the one-time pad cryptosystem is quite simple.

Suppose two parties, Alice and Bob, wish to send each other secure messages. Before any secure communications commence, Alice and Bob must meet covertly to create a huge number of secret keys in the form of random integers picked uniformly in the range 0 to `l`−1, where `l` is the number of symbols in the alphabet. Typically, these keys are printed in a booklet or "pad," which is where the word "pad" in the name "one-time pad" originates. You can simulate creating such a pad using the function MakeKeyPad in the electronic supplement.

In[]:=

```
?MakeKeyPad
```

Out[]=

```
MakeKeyPad[alphabet, n, k] creates a pad of keys
containing n pages and k keys per page. Each key is
a random integer picked uniformly and independently
in the range 0 to Length[alphabet]-1.
```

For example, to make a key pad containing 5 pages with 25 keys per page[1] enter:

In[]:=

[1] This is an unrealistically small pad of keys, of course. It is merely used to illustrate how the one-time pad works. You would typically create a pad with thousands of pages and hundreds of keys per page.

```
$Pad = MakeKeyPad[$Alphabet, 5, 25];
```

Out[]=

```
{
{37, 69, 40, 19, 17, 65, 34, 26, 62, 32, 29, 57,
 31, 27, 56, 53, 36, 15, 52, 72, 7, 48, 48, 19, 41},

{5, 75, 70, 18, 56, 15, 15, 9, 44, 41, 0, 72, 26,
 31, 20, 37, 36, 23, 41, 19, 38, 63, 1, 68, 18},

{30, 57, 26, 33, 36, 75, 52, 16, 1, 70, 48, 14, 42,
 23, 15, 20, 28, 45, 34, 51, 55, 37, 6, 8, 66},

{32, 73, 68, 22, 0, 70, 57, 0, 9, 24, 42, 26, 32,
 45, 46, 47, 14, 35, 10, 59, 35, 24, 62, 66, 13},

{54, 36, 71, 1, 28, 23, 26, 39, 4, 67, 23, 33, 7, 9,
 38, 37, 10, 32, 5, 64, 73, 63, 32, 20, 68}
}
```

Alice and Bob guard each of their copies of the key pad and return home. Now the parties are ready to communicate secret messages using the one-time pad cryptosystem. The steps that must be taken for Alice to send a secure message to Bob are shown in Table 6.1.

Here is an example of using the program in the electronic supplement to set up and use a one-time pad cryptosystem. We use the alphabet $Alphabet and the key pad $Pad that we created earlier.

Suppose you want to send someone the secret message "My PIN number is 1234!" You have already agreed with your intended recipient to use messages composed of symbols from $Alphabet and you each have a copy of $Pad.

Given your choice of which page to use from the key pad, the key pad itself, and the common alphabet, you can create a one-time pad encryption of a message using the function EncryptOTP.[†]

In[]:=

```
?EncryptOTP
```

Out[]=

```
EncryptOTP[message, page, pad, alphabet] computes an
encryption of the message string "message" using a
one-time pad cryptosystem based on the keys speci-
fied on page number "page" of the pad of keys,
"pad." All messages are presumed to be based on the
given alphabet. Note that "alphabet" contains sym-
bols such as punctuation marks, numbers, brackets,
```

[†] Unfortunately, this function is currently disabled to ensure compliance with U.S. export controls regarding the export of cryptographic software.

Table 6.1 One-Time Pad Cryptosystem

One-Time Pad Cryptosystem	
1.	Alice converts her message into a sequence of integers by using, for example, the function MessageToIntegers. Thus the original message text, consisting of N symbols, becomes the N message integers $M=\{m_1, m_2, \mathrm{K}\ , m_N\}$.
2.	Next Alice chooses a page P of keys from her copy of the key pad that she shares with Bob. This provides a supply of random integers $\{k_1, k_2, \mathrm{K}\ \}$. Alice only needs to use the first N such keys to encrypt M.
3.	To perform the encryption, Alice computes a sequence of N encrypted integers $E=\{e_1, e_2, \mathrm{K}\ , e_N\}$ by applying the rule $e_i = m_i + k_i \pmod{1}$; that is, the ith encrypted integer is the sum, modulo 1, of the ith message integer and ith key integer.
4.	Alice sends Bob (E,P) the encrypted message E and page number P of the keys that she used.
5.	Upon receipt, Bob looks up the keys on page P from his copy of the key pad. He finds the keys that Alice used, namely, $\{k_1, k_2, \mathrm{K}\ \}$. Using the encrypted message $E=\{e_1, e_2, \mathrm{K}\ , e_N\}$ Bob reconstructs the message integers $M=\{m_1, m_2, \mathrm{K}\ , m_N\}$ by computing $m_i = e_i - k_i + 1 \pmod{1}$.
6.	Finally, Bob converts M back into the original message using IntegersToMessage (the inverse of MessageToIntegers).

```
and upper-case letters in addition to the usual 26
lower-case letters. After the message has been sent
the page of keys that was used should be discarded
from the pad.
```

For example, the message "My PIN number is 1234!" becomes, using page 3 of the key pad:

In[]:=

```
EncryptOTP["My PIN number is 1234!",3,$Pad,$Alphabet]
```

Out[]=

```
{68, 5, 24, 74, 70, 38, 50, 29, 21, 6, 49, 18, 59,
 21, 23, 38, 26, 21, 11, 29, 34, 23}
```

A recipient of the encrypted message can decrypt it using the function DecryptOTP.[†]

In[]:=

```
?DecryptOTP
```

Out[]=

```
DecryptOTP[cipher, page, pad, alphabet] decrypts the
secret message "cipher" that was encrypted using a
one-time pad cryptosystem based on the keys on page
number "page" of the pad of keys, "pad." All mes-
sages are presumed to be based on the given alpha-
bet. After the message has been decrypted the page
of keys that was used should be erased from the pad.
```

In[]:=

```
DecryptOTP[{{68, 5, 24, 74, 70, 38, 50, 29, 21, 6,
             49, 18, 59, 21, 23, 38, 26, 21, 11, 29,
             34, 23}}, 3, $Pad, $Alphabet]
```

Out[]=

```
My PIN number is 1234!
```

In order to crack a one-time pad, an adversary would be faced with a daunting task. He would have to try out all possible keys. For an N symbol message, and an `l` symbol alphabet, there are l^N possible keys. Even for our simple alphabet, which contains 76 symbols, and our simple message, "My PIN number is 1234!" which contains 22 symbols, this amounts to 2×10^{41} possible keys! This makes one-time pads extremely secure.

Unfortunately, one-time pads consume vast numbers of keys and require the sender and receiver to conspire to exchange key sets covertly prior to sending each other secure messages. Moreover, the integrity of the entire cryptosystem rests on maintaining the secrecy of the key pads. Should one of the key pads ever fall into the wrong hands, the messages could be decrypted easily by an adversary. These factors make one-time pads of limited utility. A more practical scheme needs a way to *distribute* the keys, in a secure fashion, without the sender and recipient having to meet face to face.

6.5 The RSA Public Key Cryptography Scheme

The RSA system, invented by Ronald Rivest, Adi Shamir, and Leonard Adleman in 1978[Rivest78], is a cryptosystem that solves the key distribution problem. Unlike the one-time pad scheme, in RSA the sender and recipient do not need to meet beforehand to

[†] Unfortunately, this function is currently disabled to ensure compliance with U.S. export controls regarding the export of cryptographic software.

exchange secret keys. Instead, the sender and receiver use *different* keys to encrypt and decrypt a message. This makes it significantly more practicable than the one-time pad scheme. In fact, today most secure communications are based on the RSA cryptosystem.

The basic idea is as follows. A person wishing to *receive* secret messages, using RSA, creates his own pair of keys, consisting of a "public key" and a "private key." He publicizes the public key, but keeps the private key hidden. When someone wants to send him a secret message, the sender obtains the public key of the intended recipient and uses it to encrypt his message. Upon receiving the scrambled message, the recipient uses his private key to decrypt the message. The trick is to understand how the public key and private key need to be related to make the scheme work in an efficient, yet secure, fashion.

To be an *efficient* cryptographic scheme, it must be easy for a sender to compute the encryption of a message; that is,

Encrypt[MessageToIntegers[M], $PublicKey] = E

Similarly, it must be easy for the intended recipient to compute the decryption of an encrypted message; that is,

IntegersToMessage[Decrypt[E, $PrivateKey]] = M.

Moreover, it must be easy to generate the public key/private key pairs.

To be a *secure* cryptographic scheme, it must be extremely difficult to determine the message M given only knowledge of E and the public key $PublicKey. Also, it must be extremely difficult to guess the correct key pair.

Such a dual-key scheme is called a "public key cryptosystem." It is possible to have different cryptosystems by choosing different mathematical functions for creating the key pairs and encrypting and decrypting the messages.

The RSA system is just such a special case. It relies on the presumed difficulty of factoring large integers. Suppose Alice wants to receive secret messages from other people. To create a public key/private key pair, Alice picks two large prime numbers, p and q, and computes their product, $n = pq$. She then finds two special integers, d and e, that are related to p and q. The integer d is a random integer that is coprime to $(p - 1)(q - 1)$ (i.e., the greatest common divisor of d and $(p - 1)(q - 1)$ is 1). The integer e is the modular inverse of d; that is, $ed \equiv 1 \bmod (p - 1)(q - 1)$. Alice uses these special integers to create a public key consisting of the pair of numbers $\{e, n\}$ and a private key consisting of the pair of numbers $\{d, n\}$. Alice broadcasts her public key but keeps her private key hidden.

Now suppose Bob wishes to send Alice a secret message. Even though Bob and Alice have not conspired beforehand to exchange key pads, Bob can still send a message to Alice that only she can unscramble. To do so, Bob looks up the public key that Alice has advertised and represents his text message M_{text} as a sequence of integers in the range *1* to n. Let us call these message integers $M_{integers}$. Now Bob creates his encrypted message E by applying the rule:

$$E_i = M_i^e \bmod n$$

for each of the integers $M_i \in M_{integers}$.

Upon receipt of these integers, Alice decrypts the message using the rule:

$$M_i = E_i^d \bmod n.$$

The final step is then to reconvert the message integers to the corresponding text characters. Thus the RSA cryptosystem can be summarized as in Table 6.2.

Here is an example of how it all works. First check how to use the key creation function by calling ?CreatePublicKeyAndPrivateKey.[†]

In[]:=

```
?CreatePublicKeyAndPrivateKey
CreatePublicKeyAndPrivateKey[n] creates a matching
public key and private key pair based on an integer
with roughly n digits. These keys can then be used
as the basis for establishing secure communications
over a public channel using the RSA cryptosystem.
```

Table 6.2 Summary of the RSA public key cryptosystem.

RSA Public Key Cryptography

1. Find two large primes p and q and compute their product $n = pq$.
2. Find an integer d that is coprime to $(p - 1)(q - 1)$.
3. Compute e from $ed \equiv 1 \bmod (p - 1)(q - 1)$.
4. Broadcast the public key, that is, the pair of numbers (e, n).
5. Represent the message to be transmitted, M_{text}, say, as a sequence of integers $\{M_i\}$ each in the range *1* to n.
6. Encrypt each M_i using the public key by applying the rule: $$E_i = M_i^e \bmod n.$$
7. The receiver decrypts the message using the rule: $$M_i = E_i^d \bmod n.$$
8. Reconvert the $\{M_i\}$ back to the original message M_{text}.

[†] Unfortunately, this function is currently disabled to ensure compliance with U.S. export controls regarding the export of cryptographic software.

Now suppose Alice wants to set up a public key and private key pair based on a 20-digit integer. Alice would enter:

In[]:=

```
{$PublicKey, $PrivateKey} =
      CreatePublicKeyAndPrivateKey[20]
```

Out[]=

```
Picking p: p = 6257493337
Picking q: q = 6356046119

Hence n = p q = 39772916239307209103

Picking large integer d, coprime to (p-1)(q-1):
d = 53809585979820802 31

Computing modular inverse, e, from
e d = 1 mod (p-1)(q-1): e = 34928543677329462263

Public  Key is {e, n} =
{34928543677329462263, 39772916239307209103}

Private Key is {d, n} =
{5380958597982080231, 39772916239307209103}

{{34928543677329462263, 39772916239307209103},
 {5380958597982080231, 39772916239307209103}}
```

Next Alice broadcasts her public key $PublicKey, but keeps the private key $PrivateKey secret. Hence the public key is available to everyone but the private key is known only to Alice. If someone wants to send Alice an encrypted message that only Alice can decode, he or she simply looks up Alice's public key, converts the text message to a sequence of integers and encrypts each of the message integers using the rule $E_i = M_i^e \bmod n$ as previously explained. These steps are bundled together in the function EncryptRSA.[†]

In[]:=

```
?EncryptRSA

EncryptRSA[message, $PublicKey] creates an RSA en-
cryption of the given message string using the pub-
lic key, $PublicKey. $PublicKey, and its matching
private key, $PrivateKey, are created using the
function CreatePublicKeyAndPrivateKey.
```

[†] Unfortunately, this function is currently disabled to ensure compliance with U.S. export controls regarding the export of cryptographic software.

```
In[]:=
   cipherText =
   EncryptRSA["Buy 1000 shares in Netscape now!",
              $PublicKey]
Out[]=
   {1523508400427731756 3, 1926391541109234134 7,
    3087004823872457017 3, 1861921174133240880 4,
    2028580298523594573,  1978791198995175914 5,
    6735419423900161318,  3783643979652554421 5,
    1762513366632835095 4, 2935932729727566345,
    9984261459086080108,  1281853053122136453 1,
    2846177880364009532,  7785078759135308564,
    2330216654924735484,  1326881231667055862 2}
```

This is the message that is sent over the public channel. To decode the message the recipient, Alice, uses her unique knowledge of her private key $\{d, n\}$. Alice first unscrambles each cipher integer using the rule $M_i = E_i^d \bmod n$ and then maps the message integers back into characters. These steps are bundled into the function DecryptRSA.

```
In[]:=
   ?DecryptRSA

   DecryptRSA[cipher, $PrivateKey] decrypts the en-
   crypted message cipher using the private key,
   $PrivateKey. $PrivateKey, and its matching public
   key, $PublicKey, are created using the function Cre-
   atePublicKeyAndPrivateKey.

In[]:=
   DecryptRSA[cipherText, $PrivateKey]
Out[]=
   Buy 1000 shares in Netscape now!
```

Hence we have succeeded in encrypting and decrypting a secret message using the RSA protocol.

What makes RSA so useful is not merely the fact that there is an algorithm by which messages can be encrypted and decrypted but rather that the algorithm can be computed efficiently. Speed is vital if a cryptosystem is to provide a viable means of secure communication. Fortunately, each step in the RSA procedure can be done quickly. This may surprise you because it is not immediately obvious that finding large prime numbers, finding relatively prime numbers, and computing exponentials modulo an integer can all be done efficiently. However, Dominic Welsh, of Oxford University, suggests that large prime numbers can be found quickly using the Rabin-

Solovay-Strassen algorithm; that an integer d relatively prime to $(p - 1)(q - 1)$ can be found efficiently by finding a prime number that is larger than either p or q and that the integer e in the congruence $ed \equiv 1(\mathrm{mod}(p - 1)(q - 1))$ can be found quickly using Euclid's algorithm[Welsh88]. Moreover, exponentials modulo an integer can be computed in polynomial time. Thus every step that the RSA procedure requires can be computed efficiently making it a viable cryptosystem overall.

Does the ease of the computations underlying RSA mean that RSA is vulnerable to attack? To answer this, let us take a look at what an adversary would have to do to crack RSA encoded messages.

6.6 Code-Breaking on a Classical Computer

In order for an adversary to break the RSA code, it is necessary to find the private key $\{d, n\}$ given knowledge of the public key $\{e, n\}$. Once an adversary obtained the private key, intercepted messages could be decoded using exactly the same decoding procedure as that used by the intended recipient. So how is an adversary to find the private key given knowledge of the public key?

If an adversary could find the two factors of n, that is, p and q such that $n = p \times q$, he could easily compute their product $(p - 1)(q - 1)$. Then, as the adversary also knows e, he could compute d from the congruence $ed \equiv 1(\mathrm{mod}(p - 1)(q - 1))$. Once d is known, an intercepted coded message E_i could be decoded by computing $E_i^d \mathrm{mod} n$. So breaking RSA will be difficult only if factoring large composite integers is also difficult.

Fortunately, the computing time needed to factor an integer rises sharply with the "size" of the integer. You can think of the "size" of an integer as the number of bits needed to specify the integer. To date, the best known classical algorithm for factoring large integers is the "Number Field Sieve"[Lenstra93]. This has a running time that is proportional to $\exp(c\ \mathrm{size}^{1/3})$ where c is a constant. This is super-polynomial and is generally considered to be intractable.

Fig. 6.2 shows a plot of how the actual CPU-time needed to factor a number in Mathematica increases with the number of binary digits in the number. Notice that the plot uses a log scale, so the increase in CPU time with the size of the integer is quite dramatic. On a logarithmic scale, a straight line indicates an exponential trend. Notice, however, that our timing data shows a slight downwards bend in the sequence of data points for larger integers. This confirms

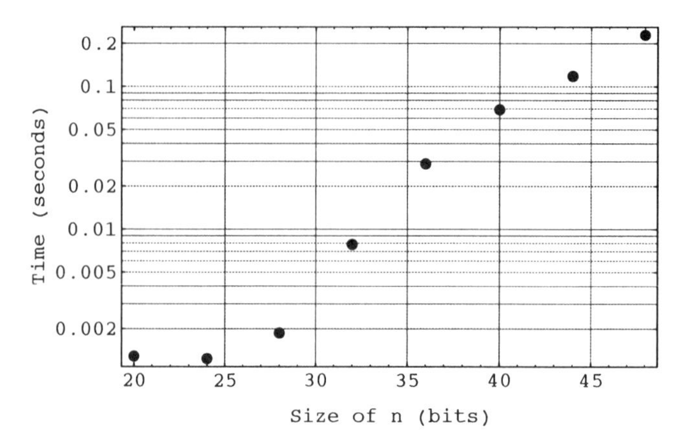

Fig. 6.2 Log plot of the average time taken to factor an integer as a function of the number of binary digits needed to specify the integer.

that the analytic form for the CPU-time as a function of the size of the integer is slightly sub-exponential. This is in agreement with the theoretical scaling for the best known factoring algorithms that we mentioned in Chapter 2 (see Fig. 2.5). In particular, if we fit the data to a function of the form $\exp\left(\alpha x^{1/3}(\ln(x))^{2/3}\right)$, we obtain very good agreement. Although the time needed to factor an integer seems to scale slightly less than exponential it is still super-polynomial and hence appears to be a difficult problem for a classical computer.

Thus the security of the RSA cryptosystem rests upon the presumption that factoring large integers is computationally intractable.

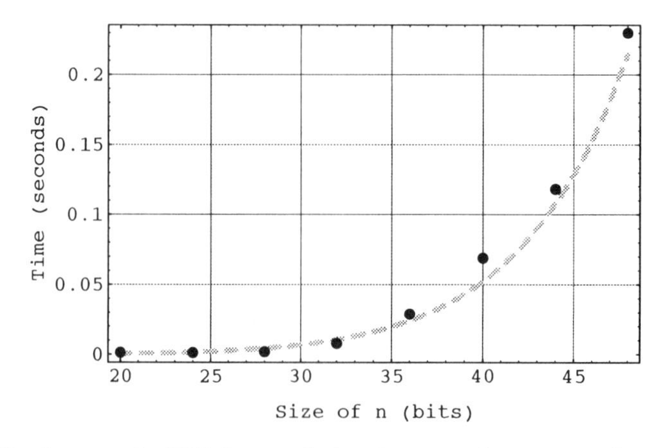

Fig. 6.3 The increase in CPU-time needed to factor an integer as a function of the size of the integer. The dashed line shows the best fit of a sub-exponential to the data. Each data point has 10,000 samples.

We say "presumption" because no one has yet *proved* that factoring is a truly intractable problem. There is a possibility that someone may discover a polynomial-time classical algorithm. However, as yet, no-one has. Umesh Vazirani, a computer scientist at the University of California Berkeley, has characterized the difficulty of factoring a 2,000-digit number rather graphically. Vazirani states[Vazirani94], "It's not just a case that all the computers in the world today would be unable to factor that number. It's really much more dramatic... Even if you imagine that every particle in the Universe was a [classical] computer and was computing at full speed for the entire life of the Universe, that would be insufficient to factor that number."

Thus messages encrypted using the RSA scheme with a key containing 2,000 digits should be quite impossible to break. Hence if someone wanted to make a secure message, all they would have to do is to pick a large enough value for n in the public key so that such a number could not be factored within a reasonable time.

How big does n have to be given the current state of (classical) computer technology? When Rivest, Shamir, and Adleman first devised their cryptosystem back in 1977 they challenged anyone to factor a particular 129-digit number that came to be known as RSA-129. The challenge stood for 17 years. Over that time period, computers became about 2,000 times faster. Even then, RSA-129 was only factored after a Herculean effort involving a network of some 1,600 computers.

Given the difficulty of factoring a 129-digit number, it is easy to see that by basing an RSA cryptosystem on a number that involved, say, 150 digits, any encrypted messages would be utterly secure for the foreseeable future. Secure, that is, with respect to the capabilities of hordes of classical computers. But what about quantum computers? Could a quantum computer factor large numbers fast enough to break the supposedly invulnerable RSA cryptosystem?

When we compare the efficiency of classical computers and quantum computers we are comparing how the number of computational steps needed to perform a computation scales up as larger instances of the problem are considered. We are not comparing execution times directly.

It is conventional, in computer science, to declare a computation tractable if, in the worst case, the required number of steps scales up as a polynomial in the "size" of the input problem. Otherwise the problem is deemed intractable. Factoring large integers is an intractable problem for any classical computer, but, as we show, is tractable for a quantum computer.

There is no difference in terms of which functions a classical computer and a quantum computer can compute. The class of computable functions is exactly the same on both kinds of machines. In particular, both classical and quantum computers can factor integers.

Nevertheless, quantum computers can compute certain functions more efficiently than classical computers because they can exploit certain quantum phenomena that have no classical analogues.

6.7 Code-Breaking on a Quantum Computer

All that you need to do to factor a large integer, n, say, is to calculate the product of all ordered pairs of integers that are less than $\sqrt{n}$ until either you find a pair of integers whose product is exactly n or else you run out of integers to try; that is, you reach $\sqrt{n}$. Although crude, if n is factorizable, this algorithm is guaranteed to find a pair of factors.

This algorithm would seem to lend itself to quantum parallelism, the technique introduced by David Deutsch in his original paper on universal quantum computers. Quantum parallelism allows you to conduct many similar calculations in parallel superposition in the time it takes to do just one of the calculations classically. You could imagine loading two quantum memory registers with a superposition of inputs representing all integers between 0 and $\lceil \sqrt{n} \rceil$ (i.e., the next integer larger than $\sqrt{n}$). Then you could imagine computing, in quantum parallel, the product of all pairs of such numbers and placing the result in a third quantum register. If you sampled from this final register you might find a case in which the two factors multiplied to n.

Unfortunately, this direct approach to finding the factors, that relies on superposition alone, is naive. Although such a scheme would certainly create the sought after factorization somewhere in the midst of the superposition, the probability of you obtaining that answer when you measured the output register would be as great as that of obtaining any other answer. Consequently, as there are many more nonsolutions to a given factorization problem than there are solutions, you would be much more likely to obtain a nonsolution than a solution when you measured the output register. In fact, you would have to repeat the whole calculation exponentially many times to have a high enough chance of finding the factors. Thus although superposition enables you to test all possible factorizations in the time needed to test just one of them classically, quantum measurement severely limits your ability to extract a useful answer.

What is missing from the preceding algorithm is quantum interference among the possible computations. For quantum parallelism to be useful, you need to arrange for the nonsolutions to interfere destructively and the solutions to interfere constructively in the state of the output memory register. If this can be done, the output register will contain a superposition of all possible answers (both solutions and nonsolutions) but with different amplitudes, with the squares of the absolute values of the amplitudes of the solution states being much higher than those of the nonsolution states. Consequently, when you make a measurement to "read" the state output register the solutions have a higher probability of being selected than the nonsolutions. It is such interference effects, rather than superposition alone, that imbues quantum parallelism with its power.

In the years that followed Deutsch's introduction of quantum parallelism, scientists devised a few computational problems for which the quantum computer would, theoretically, outperform first a deterministic and then a probabilistic Turing machine (see Chapter 2 for a complete history). Unfortunately, all of these problems were rather contrived; they were invented solely for the purpose of showing that a quantum computer could outperform a classical computer on at least *some* problems. But what was needed was for someone to find a *significant* problem, one that really mattered, for which the quantum computer outperformed the classical computer.

The "killer ap" came in 1994 when Peter Shor, of AT&T Bell Labs, showed how a quantum computer could be used to factor a large integer superefficiently[Shor94]. This was big news. Immediately the interest of security agencies and banks was piqued. If a quantum computer could factor an integer efficiently it could also break the RSA cryptosystem, and then all forms of secret information would become vulnerable. Let us take a look at how this quantum algorithm works.

6.8 A Trick From Number Theory

Shor's quantum algorithm for factoring relies upon a result from classical number theory that relates the period of a particular periodic function to the factors of an integer.

Given an integer n (the number to be factored), construct a function $f_n(a) = x^a \bmod n$ where x is an integer chosen at random that is coprime to n. Being "coprime" means that the greatest common divisor of x and n is 1 (i.e., $\gcd(x,n) = 1$). The "mod" function, on the right-hand side of the definition, calculates the remainder after dividing x^a by n. However, finding $f_n(a)$ for an exponential number of input, a, is expensive.

Why is this remotely interesting with respect to the problem of factoring n? It turns out that the new function that we just constructed, $f_n(a)$, is periodic. This means that as you increase the argument of the function, taking $a = 0, 1, 2, \mathrm{K}$, the values of the function $f_n(0), f_n(1), f_n(2), \mathrm{K}$ fall into a repeating pattern eventually. Different values of x give rise to different patterns. The number of values in between the repeating pattern, for a particular value of x, is called "the period of x modulo n" which is usually given the label r. Every rth element in the sequence $f_n(0), f_n(1), f_n(2), \mathrm{K}$ is therefore the same. In particular, this implies that:

$$x^r \equiv 1 \bmod n.$$

If the period r is an even number, then we can rewrite this equation as the difference between two squares using the following sequence of steps:

$$\left(x^{r/2}\right)^2 \equiv 1 \bmod n$$

$$\left(x^{r/2}\right)^2 - 1 \equiv 0 \bmod n$$

$$\left(x^{r/2}\right)^2 - 1^2 \equiv 0 \bmod n$$

$$\left(x^{r/2} - 1\right)\left(x^{r/2} + 1\right) \equiv 0 \bmod n$$

Table 6.3 Demonstration of the periodicity of x^a mod n.

Input (a)	Calling $f_{n,x}(a)$	Simplifies to...
0	$x^0 \bmod n$	1
1	$x^1 \bmod n$	x
M	M	M
r	$x^r \bmod n$	1
$r+1$	$x^{r+1} \bmod n$	x
M	M	M
$2r$	$x^{2r} \bmod n$	1
$2r+1$	$x^{2r+1} \bmod n$	x
M	M	M

The final form reveals that the product $(x^{r/2}-1)(x^{r/2}+1)$ is some integer multiple of n. That is, dividing $(x^{r/2}-1)(x^{r/2}+1)$ by n results in a remainder of zero. So, unless $x^{r/2} \equiv \pm 1 \bmod n$, at least one of the terms on the left-hand side must have a nontrivial factor in common with n. Thus we have a good chance of finding a factor of n by computing $\gcd(x^{r/2}-1,n)$ and $\gcd(x^{r/2}+1,n)$.

The upshot of this diversion into number theory has shown that it is possible to find the factors of a number n, by finding the period of a function $f_{x,n}(a)=x^a \bmod n$ where x is a random integer coprime to n.

Unfortunately, there is no known way of calculating the required period *efficiently* on any classical computer. Shor, however, discovered a way to calculate the period efficiently using a quantum computer.

6.9 Shor's Algorithm for Factoring on a Quantum Computer

The crux of Shor's algorithm for factoring a large integer is a *quantum* method for computing the period r of the function $f_{x,n}(a)=x^a \bmod n$ that is exponentially more efficient than any classical method. This is significant because the result from number theory that we just described tells us that once we know r, we can obtain the factors of n from $\gcd(x^{r/2}-1,n)$ and $\gcd(x^{r/2}+1,n)$. However, calculating the greatest common divisors is easy computationally. In fact a fast algorithm for calculating greatest common divisors has been known since the time of Euclid. So Shor's hybrid algorithm, which uses a fast quantum method for calculating the period and a fast classical method for calculating the factors from the period, is extremely efficient overall.

So how does Shor's algorithm work? Let us start off by giving the big picture and focus on the details later. Our goal is to find the period of the function $f_{x,n}(a)=x^a \bmod n$. To do so, we create a single quantum memory register that we partition into two parts called Register1 and Register2. Although the complete register consists of a chain of qubits, we are going to use a more compact notation for representing its contents. If Register1 is holding the (base 10) number a and Register2 is holding the (base 10) number b we represent the state of the complete register as $|a,b\rangle$.

Next we create, in Register1, a superposition of the integers $a = 0,1,2,3,\ldots$ that will become the arguments of the function $f_{x,n}(a)$. Then we evaluate, in quantum parallel, $f_{x,n}(a)$ evaluated on each input a and place the result in Register2. This creates a superposition, in Register2, of the function evaluations $f_{x,n}(a)$ in the time it takes to compute just one such function evaluation classically. Next we measure the state of Register2. This collapses the superposition stored in Register2 and we obtain some answer, k say. This means that there was some value of a such that $x^a \bmod n = k$.

However, this simple act of measuring the state of Register2 has a critical side-effect on the state (i.e., the contents) of Register1. Remember that Register1 and Register2 are simply two parts of a single quantum register. Just as we saw in the Feynman quantum computer simulator in Chapter 3, measurements made on one part of a quantum register have the effect of projecting out the states of other parts of the register. For the case of the simulator, we projected the state of the qubits used to represent the state of the computation by observing the qubits used to represent the cursor position. Similarly, by observing Register2 we actually change the contents of Register1 to be a superposition of just those values of a, such that $x^a \bmod n = k$. The values now stored in Register1 are of the form $\{a, a+r, a+2r, K\}$ encoded as a superposition of states of the form $\omega|a\rangle + \omega|a+r\rangle + \omega|a+2r\rangle + K$. Notice that, at the moment, the amplitudes with which each state appears are all equal.

Next we compute the discrete Fourier transform of the contents of Register1 and put the result back into Register1. As Register1 contains a periodic function, its Fourier transform will be peaked[2] at multiples of the inverse period $1/r$. What this means, physically, is that the Fourier transform of the superposition stored in Register1 will be some new superposition. However, now the amplitudes with which various states appear are no longer equal. States corresponding to integer multiples of the inverse period, and those close to them, appear with greater amplitudes than those that do not correspond to integer multiples of the inverse period. Thus, if you now measure the state (i.e., "read the contents") of Register1, you would be highly likely to obtain a result that is close to some multiple of the inverse period.

If you repeat this whole process just a few times, you will quickly obtain enough samples of integer multiples of the inverse period to be able to determine the period unambiguously.

[2] Fourier transforms map functions in the time domain to functions in the frequency domain. The frequency is the inverse of the period.

Thus Shor's quantum algorithm for finding the period of $f_{x,n}(a) = x^a \bmod n$ relies upon quantum parallelism to create a superposition of values of the periodic function $f_{x,n}(a)$ upon measurement to project out a periodic function in Register1 and upon a discrete Fourier transform to bring about the desired interference effect between solutions (integer multiples of the inverse period) and nonsolutions (numbers that are not integer multiples of the inverse period).

Having found the period, the factors of n are found from $\gcd(x^{r/2} - 1, n)$ and $\gcd(x^{r/2} + 1, n)$. As in the purely classical case, the method will find nontrivial factors of n provided r is even and $x^{r/2} \not\equiv \pm 1 \bmod n$. The restriction that q should have small prime factors is merely to ensure that it is physically possible to implement Shor's algorithm in a quantum circuit that is small in comparison to n.

The final step in Shor's algorithm makes use of a "continued fraction expansion." This is a purely classical computation and does not really involve the quantum computer. The continued fraction expansion is used to identify what multiple of the inverse period r is obtained.

A continued fraction is an expression of the form:

$$a_0 + \cfrac{1}{a_1 + \cfrac{1}{a_2 + \cfrac{1}{\mathrm{K} + \cfrac{1}{a_N}}}}$$

which is usually abbreviated as $x = [a_0, a_1, \mathrm{K}\ , a_N]$. A truncated continued fraction is called a "convergent." The nth convergent is written $[a_0, a_1, \mathrm{K}\ , a_n]$. There is a theorem on continued fractions that states: "If $\frac{p}{q}$ is any rational number satisfying $\left|\frac{p}{q} - x\right| < \frac{1}{2q^2}$ then $\frac{p}{q}$ is a convergent of the continued fraction. Moreover, the convergent is such that $\gcd(p,q) = 1$". We can use this theorem to find the closest rationals to the terms λ / r and hence find the period r.

Table 6.4 Shor's Quantum Algorithm for Factoring Composite Integers.

Shor's Algorithm for Factoring n

1. Pick a number q (with small prime factors) such that $2n^2 \leq q \leq 3n^2$.
2. Pick a random integer x that is coprime to n.
3. Repeat steps labeled (a) through (g) order $\log(q)$ times, *using the same random number* x each time:

(a) Create a quantum memory register and partition the qubits into two sets, called register 1 and register 2. If the qubits in register 1 are in the state reg1 and those in register 2 are in the state reg2, we represent the joint state of both parts of the register as $|\text{reg1}, \text{reg2}\rangle$.

(b) Load register 1 with all integers in the range 0 to q - 1 and load register 2 with all zeroes. The state of the complete register is given by:

$$|\psi\rangle = \frac{1}{\sqrt{q}} \sum_{a=0}^{q-1} |a, 0\rangle.$$

(c) Now apply (exploiting quantum parallelism) the transformation $x^a \bmod n$ to each number in register 1 and place the result in register 2. The state of the complete register becomes:

$$|\psi\rangle = \frac{1}{\sqrt{q}} \sum_{a=0}^{q-1} |a, x^a \bmod n\rangle.$$

(d) Measure the state of register 2, obtaining some result k. This has the effect of projecting out the state of register 1 to be a superposition of just those values of a such that $x^a \bmod n = k$. Hence the state of the complete register is:

$$|\psi\rangle = \frac{1}{\sqrt{\|A\|}} \sum_{a' \in A} |a', k_{/},$$

where $A = \{a' : x^{a'} \bmod n = k\}$ and $\|A\|$ is the number of elements in this set.

(e) Next compute the discrete Fourier transform of the projected state in register 1. The discrete Fourier transform maps each state $|a'\rangle$ into a superposition given by

$$|a'\rangle \text{ a } \frac{1}{\sqrt{q}} \sum_{c=0}^{q-1} e^{2\pi i a' c/q} |c\rangle.$$

Thus the net effect of the discrete Fourier transform is to map the projected state in register 1 into the superposition given by

$$|\psi\rangle = \frac{1}{\sqrt{\|A\|}} \sum_{a' \in A} \frac{1}{\sqrt{q}} \sum_{c=0}^{q-1} e^{2\pi i a' c/q} |c, k\rangle.$$

(f) Measure the state of register 1. This effectively samples from the discrete Fourier transform. This returns some number c' that is some multiple λ of q/r where r is the desired period; that is, $c'/q \approx \lambda/r$ for some positive integer λ.

Table 6.4 *continued*

(g) To determine the period r we need to estimate λ. This is accomplished by computing the continued fraction expansion c'/q as long as the denominator is less than n and retaining the closest such fraction to λ/r.

4. By repeating the steps (a) through (g) we create a set of samples of the discrete Fourier transform in register 1. This gives samples of multiples of $1/r$ as λ_1/r, λ_2/r, λ_3/r, ... for various integers λ_i. After a few repeats of the algorithm we have enough samples of register 1 to compute, using a continued fraction technique, what the λ_i could be and hence to guess r.
5. Once r is known, the factors of n can be obtained from $\gcd(x^{r/2}-1,n)$ and $\gcd(x^{r/2}+1,n)$.

6.10 Simulation of Shor's Algorithm

The electronic supplement contains a program that simulates the steps taken by a quantum computer factoring an integer using Shor's algorithm. To run the simulator, pick an integer to be factored and enter RunShorsAlgorithm[*integer*]. As the algorithm is probabilistic, it will usually not follow the same sequence of steps each time it is run. Moreover, for certain choices of the random number x, Shor's algorithm will fail, and, for others, lead to the trivial divisors 1 and n. However, a few retries are sufficient to find nontrivial factors of n. Here is the output from the program when asked to factor the integer 15.

In[]:=

```
RunShorsAlgorithm[15]
```

Out[]=

```
Picking x ... x=13
Picking q ... q = 243
Loading register 1 and register 2 with zeroes...
```

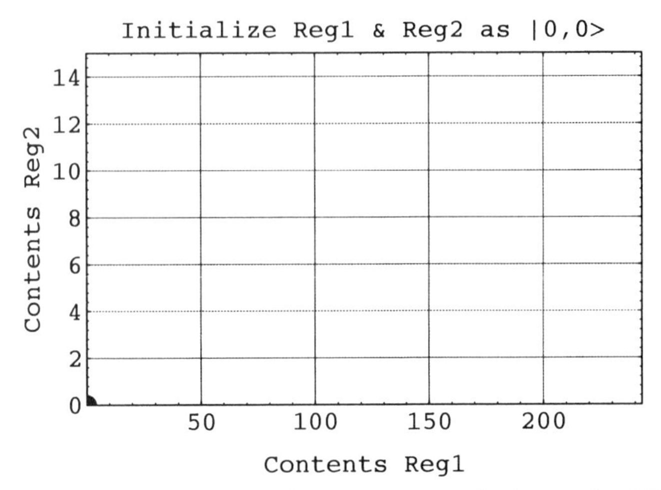

Loading register 1 with a superposition of all possible inputs

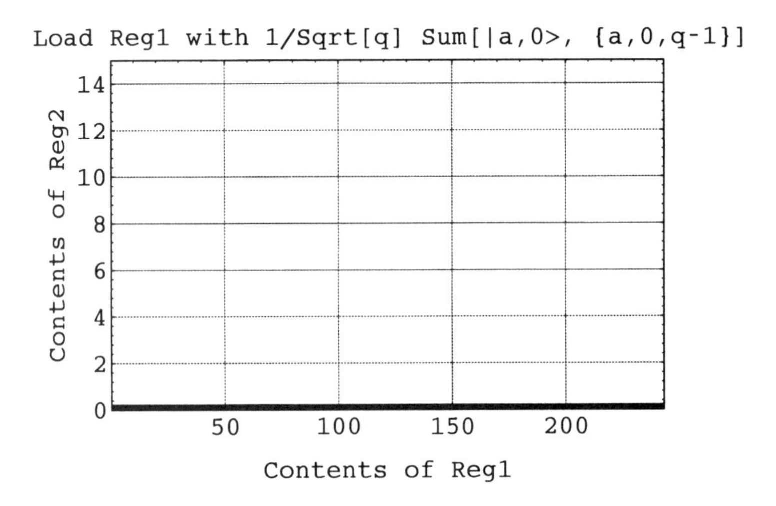

Now computing the period of x^a mod n

```
reg1 = 0.06415 ket[0] + 0.06415 ket[1] +
       0.06415 ket[2] + 0.06415 ket[3] +
       ...
       0.06415 ket[240] + 0.06415 ket[241] +
       0.06415 ket[242]

reg2 = {1, 13, 4, 7, 1, 13, 4, 7, 1, 13, 4, 7, 1,
        13, 4, 7, 1, 13, 4, 7, 1, 13, 4, 7, 1, 13,
        4, 7, 1, 13, 4, 7, 1, 13, 4, 7, 1, 13, 4, 7,
        ...,
        7, 1, 13, 4, 7, 1, 13, 4, 7, 1, 13, 4, 7, 1,
        13, 4, 7, 1, 13, 4}
```

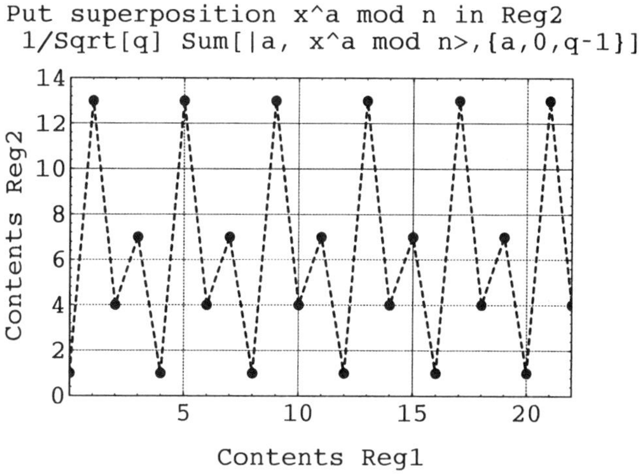
Put superposition x^a mod n in Reg2
1/Sqrt[q] Sum[|a, x^a mod n>,{a,0,q-1}]
14
12
10
8
6
4
2
0
Contents Reg2
5
10
15
20
Contents Reg1

Note that we added a dashed line to connect the values stored in register 2 to make it easier to see that they form a periodic function. However, this is not meant to imply that the function stored in register 2 is continuous. Also, the contents of register 1 still extend out to 242. We have abbreviated the horizontal axis in the figure to illustrate the periodicity better.

```
measureReg2 = 1
```

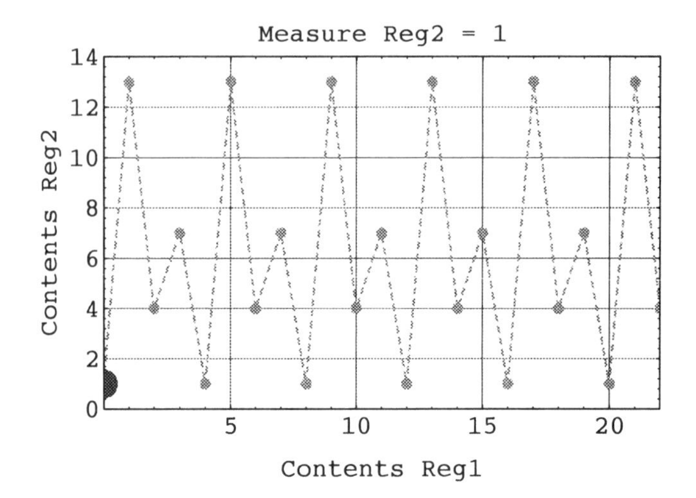


```
projectReg1 = 0.128037 ket[0] + 0.128037 ket[4] +
              0.128037 ket[8] + 0.128037 ket[12] +
              0.128037 ket[16] +
              ...
              0.128037 ket[232] + 0.128037 ket[236]
              + 0.128037 ket[240]
```

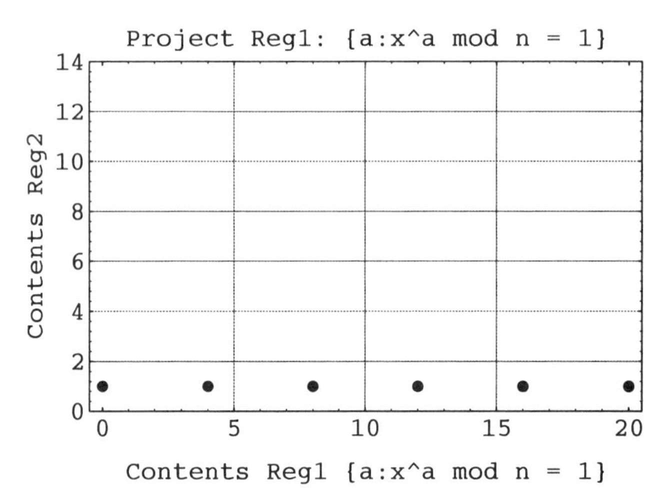


```
Now computing discrete Fourier transform ... this
takes a while

Fourier = 0.501028 ket[0] +
          (0.00205271 - 0.0000796543 I) ket[1] +
          (0.00205064 - 0.000159389 I) ket[2] +
          ...
          (0.00205064 + 0.000159389 I) ket[241] +
          (0.00205271 + 0.0000796543 I) ket[242]
```

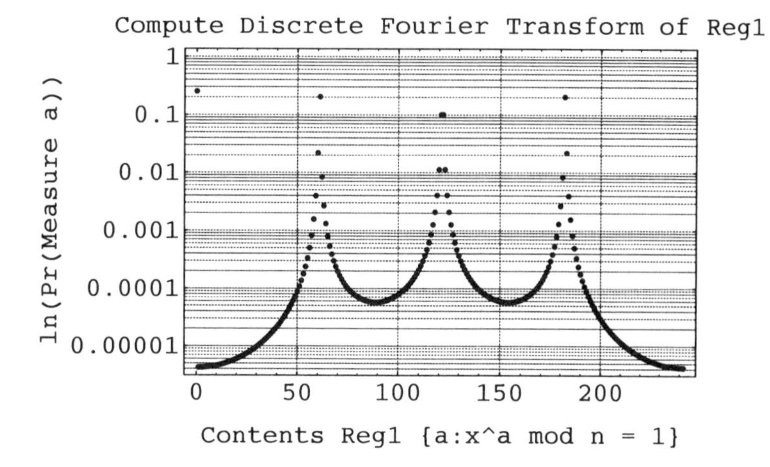

Upon observing the state of the reg1 (that now contains the Fourier transform) we obtain the result: ket[61]:

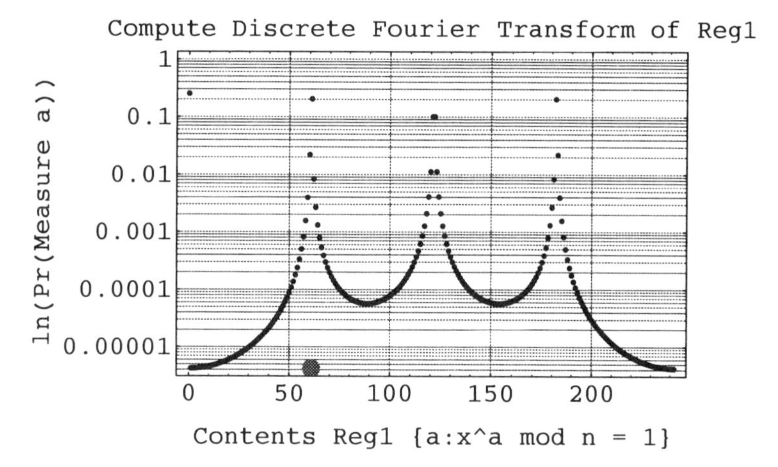

Now repeat the previous steps O(log(q)) times ... Obtain samples of the discrete Fourier transform of register 1:

Found samples{ket[121], ket[61], ket[184], ket[182], ket[61], ket[122], ket[0], ket[121], ket[0], ket[121], ket[181], ket[61]}

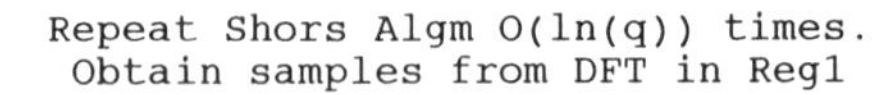

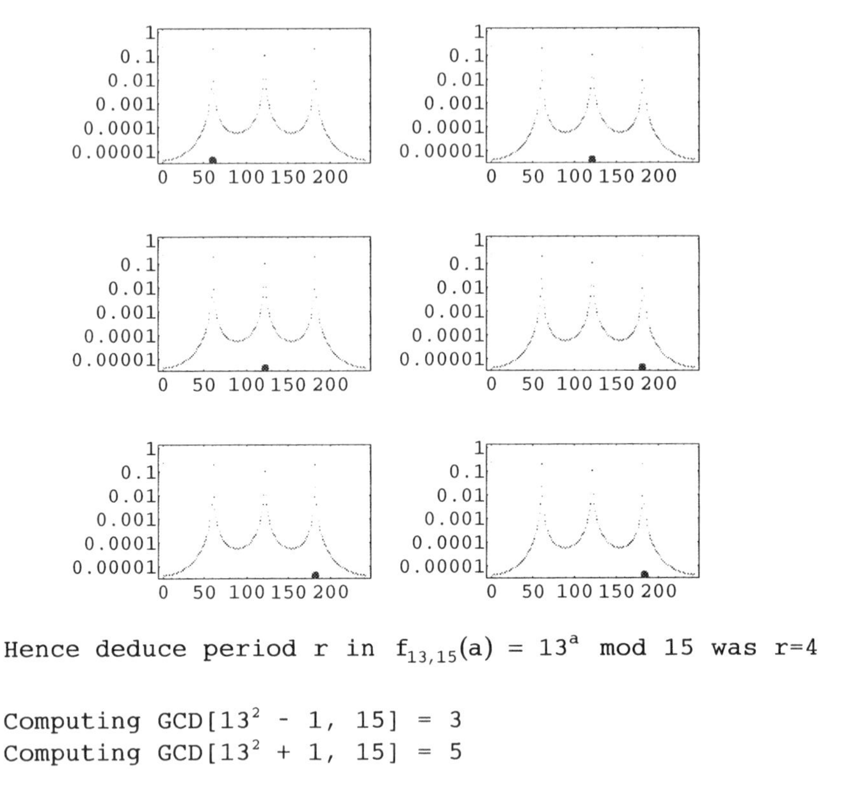


```
Hence deduce period r in f_{13,15}(a) = 13^a mod 15 was r=4

Computing GCD[13^2 - 1, 15] = 3
Computing GCD[13^2 + 1, 15] = 5

Conclusion: Factors of 15 are 3 and 5.

Period was r=4
```

6.11 Shor's Algorithm Can Sometimes Fail

Shor's algorithmis probabilistic: it does not always return a nontrivial factor. For example, here is a case in which the algorithm fails.

In[]:=

```
RunShorsAlgorithm[15]
```

Out[]=

```
Picking x ... x=14
Picking q ... q = 243
Now computing the period of x^a mod n

reg1 = 0.06415 ket[0] +
       0.06415 ket[1] +
```

```
       0.06415 ket[2] +
       0.06415 ket[3] +
       ....
       0.06415 ket[238] +
       0.06415 ket[239] +
       0.06415 ket[240] +
       0.06415 ket[241] +
       0.06415 ket[242]

reg2 = {1, 14, 1, 14, 1, 14, 1, 14, 1,..., 14, 1}

measureReg2 = 1

projectReg1 = 0.0905357 ket[0] +
              0.0905357 ket[2] +
              0.0905357 ket[4] +
              0.0905357 ket[6] +
              0.0905357 ket[8] +
              ...
              0.0905357 ket[234] +
              0.0905357 ket[236] +
              0.0905357 ket[238] +
              0.0905357 ket[240] +
              0.0905357 ket[242]
Now computing discrete Fourier transform ... this
takes a while

Fourier = 0.70856 ket[0] +
          (0.00290394 - 0.0000375452 I) ket[1] +
          (0.00290394 - 0.000075103 I) ket[2] +
          (0.00290394 - 0.000112686 I) ket[3] +
          (0.00290394 - 0.000150307 I) ket[4] +
          (0.00290394 - 0.000187978 I) ket[5] +
          ...
          (0.00290394 + 0.000187978 I) ket[238] +
          (0.00290394 + 0.000150307 I) ket[239] +
          (0.00290394 + 0.000112686 I) ket[240] +
          (0.00290394 + 0.000075103 I) ket[241] +
          (0.00290394 + 0.0000375452 I) ket[242]

Upon observing the state of the reg1 (that now con-
tains the Fourier transform) we obtain the value:
ket[0].

Repeating the whole process 12 times samples =
{ket[0], ket[125], ket[0], ket[0], ket[0], ket[121],
 ket[122], ket[122], ket[0], ket[121], ket[0],
ket[121]}
```

```
Hence deduce period r in f_{15,14}(a) = 14^a mod 15 was r=2

Computing GCD[14-1, 15] = 1
Computing GCD[14+1, 15] = 15

Conclusion: No useful factors found.
```

The use of a Fourier transform in conjunction with quantum interference effects provides the key to one class of quantum algorithms. Certainly, a similar idea has been applied to tackle another crucial computational problem: database search. A quantum computer can be used to find the proverbial needle in a haystack. However, we ought not to forget that there are probably lots of other styles of quantum algorithms waiting to be discovered. Quantum algorithms are now known for database search[Grover96], computing Hamiltonian cycles, and solving (but at exponential energy) the traveling salesman problem[Cerny93] and discrete logarithm. The most successful quantum algorithms use some variant of Shor's discrete Fourier transform trick. Shor's quantum algorithm for factoring marks the beginning of a new direction in algorithm design.

Shor's quantum algorithm shows us that factoring is in complexity class P for a quantum computer. But as no one has yet *proved* that factoring is not in P for a classical computer. It is, therefore, still possible that someone will find an efficient (polynomial-time) classical algorithm for factoring too. Moreover, if scientists can find an efficient way to simulate a quantum computer on a classical computer then this would immediately admit an efficient classical factoring algorithm merely by simulating a quantum computer running Shor's algorithm. However, as yet it is not known how to simulate a quantum computer efficiently on a classical computer. This may indeed be an impossibility.

Currently, researchers are attempting to build a prototype quantum computer that will be capable of factoring a small integer. It is likely that we will see a quantum computer specialized to factor the number 15 by around the year 2000. We describe some of the efforts to build real quantum computers in Chapter 11.

This chapter has shown that quantum computers can be more efficient than classical computers for at least one significant application, code-breaking. In the next chapter we describe an even more spectacular advantage. We show how quantum computers can perform a task that cannot be performed, even in principle, by any classical machine. This capability arises because classical computers can *only* compute functions. Certain computational tasks, such as picking a random number, cannot be accomplished by computing any function. Classical computers therefore have to fake such tasks. Quantum computers, however, can perform these tasks directly because they can do more than mere function calculation.

CHAPTER 7

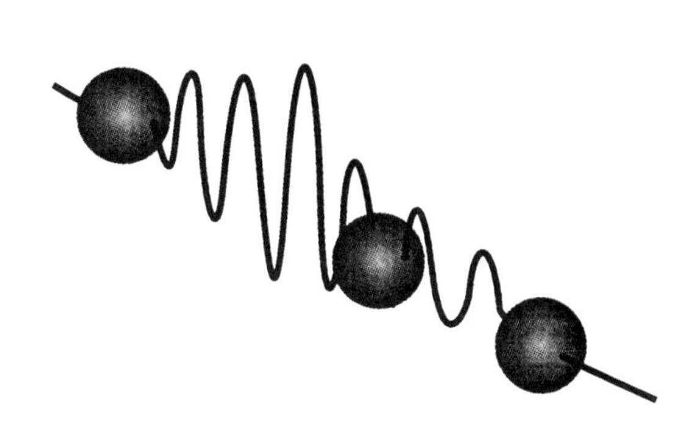

True Randomness

"If you want to build a robust universe, one that will never go wrong, then you don't want to build it like a clock, for the smallest bit of grit could cause it to go awry. However, if things at the base are utterly random, nothing can make them more disordered. Complete randomness at the heart of things is the most stable situation imaginable — a divinely clever way to build a sturdy universe."

— Heinz Pagels

Today our best description of Nature is in terms of quantum physics. Quantum physics, as we currently understand it, relies heavily on the concept of randomness. In particular, when a physical system that is in a superposition of states is observed, it is as if it collapses, spontaneously, into one of its eigenstates. We cannot predict which eigenstate will be selected. All we can do is give the probability of obtaining the various possible outcomes. The inability to predict the state into which a system will collapse upon being observed adds an element of randomness to quantum theory. In this sense, randomness is, indeed, at the heart of things.

In this chapter we show you how to harness randomness in the service of computation. Whereas the last chapter illustrated that a quantum computer can be more *efficient* than a classical computer in performing certain computations, in this chapter we show that a quantum computer is more *proficient* than a classical computer, by describing a computational *task* that a quantum computer can do, but which a classical computer cannot. In particular, we are going to explore the concept of randomness and show that a classical computer can only *pretend* to generate a random number whereas a quantum computer can *actually* generate a random number. This may seem like a minor point but, as we show, it has profound implications for the future of computer-based simulations.

7.1 The Concept of Randomness

Randomness is an elusive concept. To many people "randomness" jumbles together the notions of unpredictability, disorder, chaos, and unintentionality. Mathematicians, however, have made the concept of randomness quite precise. For a start, there is no such thing as *a* random number. It only makes sense to talk about processes for generating sequences of random numbers. However, in most cases, we are ignorant of the underlying generator of a sequence of numbers and so we cannot directly assess whether the sequence of numbers is truly random. Instead, we are faced with the question of whether a particular sequence of numbers has the *appearance* of being random.

Fortunately, there are statistical tests that can answer this question. To pass the test for randomness, a given sequence of numbers must provide a plausible set of samples from some desired distribution and there should be no discernible correlations in the sequence of numbers generated. For example, if you were simulating repeated tosses of a fair coin and you obtained the sequence of outcomes H,T,H,T,H,T,H,T,H,T,H,T,H,T,H,T,H,T,H,T,H,T..., this sequence would certainly pass the distribution check (the results are uniformly distributed between the two values H and T) but it would not pass the correlation check (because it seems as if the outcomes alternate systematically between H and T).

The art in creating computer programs that simulate the generation of true random numbers is to devise algorithmic methods that generate sequences of numbers that pass both the distribution and correlation checks. As we show, even when a random number generator passes such statistical tests, the sequence of numbers it generates may still not be random enough to serve as an approximation to a true random process.

7.2 Does Randomness Exist in Nature?

At certain times, Nature can appear to behave in quite random ways. For example, even today, despite an arsenal of supercomputers and remote sensing satellites, meteorologists are still unable to predict the path of a hurricane accurately. Is our inability to predict such events due to ignorance, or is there some inescapable randomness in the process?

Surprisingly, classical physics places the blame squarely on ignorance and not on inherent randomness. This is because classical physics is a perfectly deterministic theory. If a classical physical system, such as a hurricane, is started off in some definite configuration, then the equations of classical physics allow us to predict, in principle, exactly how it will evolve in the future. There is no randomness in classical physics. In principle, it ought to be possible to predict the path of a hurricane exactly. So why is it so hard to do so in practice?

The problem lies in the fact that the equations governing the motion of the atmosphere are nonlinear. Tiny errors, or uncertainties, in specifying the initial configuration can quickly compound to significant uncertainties in exactly how a system will evolve. In many respects the resulting evolution may look quite random even though it is actually evolving in accordance with some simple deterministic equations. Such behavior is given the paradoxical name "deterministic chaos."

A particularly simple system that exhibits deterministic chaos is the logistic map. This is an iterative process defined by the rule:

$$x_{i+1} = r\,x_i\,(1 - x_i)$$

where r is some arbitrary constant and the system is started off in the state $x_0 = c$ where c is a constant. So, starting with some particular choice for the constants c and r, this rule allows us to generate a sequence of values x_1, x_2, x_3 .... For certain values of r the sequence settles down quickly into a regular periodic sequence. For example, with $x_0 = 0.1$, $r = 3.2$ the evolution converges to a periodic pattern as shown in Fig. 7.1.

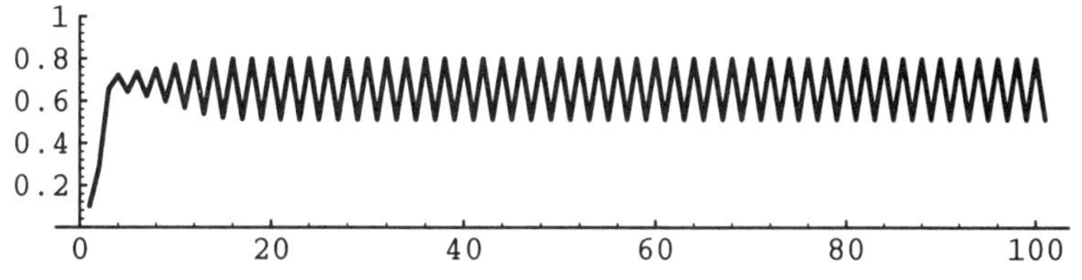

Fig. 7.1 For certain values of r the iterates of a logistic map fall into a periodic pattern.

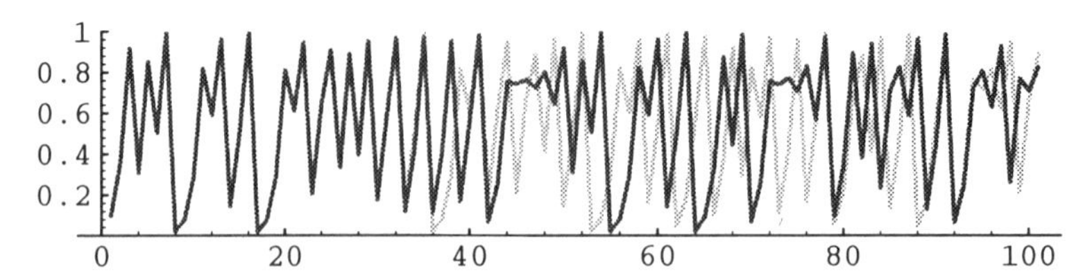

Fig. 7.2 The iterates of a logistic map appear random but are actually determined by a simple equation. They can be acutely sensitive to the initial conditions.

However, with $x_0 = 0.1$, $r = 3.98$ the sequences of values of the x_i never converge to any definite pattern; see Fig. 7.2. Moreover, a minute change in the initial condition can quickly be amplified to a gross change in the behavior. The light gray curve in Fig. 7.2 is obtained by changing the initial condition from $x_0 = 0.1$ to $x_0 = 0.100000001$—a change of one part in 10^9. Yet after just 3 6 iterations a significant difference in the evolution of the system is apparent.

This example shows that, although there is no randomness in the equations of classical physics, it is still possible to generate evolutions that are, effectively, unpredictable, due to our inability to know the exact initial condition in which some dynamical system is started. Could such processes be used as generators for sequences of numbers that would appear random? Unfortunately, in many applications that utilize random numbers, so many random numbers are needed that it would be too slow and too cumbersome to use a (classical) physical process directly as the means to generate random numbers. Moreover, chaotic sequences, such as the one previously shown, harbor subtle correlations that a truly random sequence would not possess. So although chaotic sequences might *appear* random, in fact they are not random at all.

In the 1970s, a new approach to characterizing randomness emerged[Chaitin77]. This connects the idea of randomness to that of compressibility. The field of algorithmic compressibility attempts to characterize the smallest algorithm sufficient to generate a given sequence of numbers. Sequences that have small algorithms are deemed less random than those requiring more elaborate algorithms. By this measure, the sequences of numbers generated by the preceding logistic map (or indeed by the evaluation of any mathematical function) are not random since there is a very compact rule for describing the sequence of numbers that is generated. A truly random sequence would be incompressible as the smallest algorithm sufficient to generate the sequence is the sequence itself. Unfortunately, the answer to the question of whether a particular sequence is incompressible is uncomputable. Thus the concept of randomness is related to that of uncomputability[Gardner79]. Any attempt to generate a true random sequence using a classical algorithm is therefore doomed to failure.

7.3 Uses of Random Numbers

Even if we do know the exact equations governing some dynamical system and the exact initial condition in which the system is started off, it can be a daunting, if not infeasible, computational task to solve such equations exactly. Here again random numbers find an application. Provided the system is sufficiently complicated, for example, by involving millions of interacting components, a statistical description is often more manageable, and more informative, than an exact description.

To create such statistical models of complicated systems, one must stipulate the probability with which its component parts are found in their various possible states. When enough effects occur, it becomes effectively impossible to write down a mathematical model that tracks each of them individually. Instead, statistical effects emerge that allow us to push our computations much further than is possible directly. For example, if you add together many independent, identically distributed random variables, X_i, X_2, ..., X_n with finite means μ and nonzero variances σ^2, then the random variable $(S_n - n\mu)/\sqrt{n\sigma^2}$ involving their sum $S_n = X_1 + X_2 + \text{K } X_n$ converges to a normal distribution as $n \to \infty$ regardless of the exact form of the initial distribution of the X_i (except for the finite mean and nonzero variance requirement)[Grimmett92]. It is the appearance of such regularities from a myriad of simultaneous effects that makes the statistical viewpoint so effective.

Random Numbers Used in Code Making

We encountered another use for random numbers in the last chapter. Recall the one-time pad cryptosystem. This cryptosystem relies upon the sender and receiver each having a copy of a secret pad of "random" numbers, or "keys," by which they encrypt and decrypt secret messages. The one-time pad cryptosystem is extremely secure provided the key pads are only known to the communicating parties.

In fact, during the Second World War, the Soviets routinely used a one-time pad cryptosystem, called the Vernam cipher, for their diplomatic communications. Unfortunately, however, they sent so many secret messages that they consumed all the keys they had precomputed in their key pads. Consequently, instead of deleting used pages of keys from their key pads, they began to reuse them. This enabled cryptanalysts in the West to break the Soviet codes. Eventually, the information gleaned from the intercepted messages uncovered the Rosenberg spy ring and exposed the atomic spy Klaus Fuchs[Hughes95].

Randomized Algorithms

Modern (classical) computer science also exploits the idea of random numbers. There are many kinds of computational problems that appear to be extremely difficult to solve using a classical computer executing a systematic algorithm. However, recently it has been discovered that you can create much more effective algorithms by adding a little randomness in the decision making of an algorithm[Traub94]. Typically, these more haphazard algorithms, called "randomized algorithms," cannot guarantee that they will solve the problem (given that a solution exists) or that they will complete within a definite time. In contrast, the systematic algorithms can usually guarantee that they will eventually find a solution, if one exists, and that they will terminate after a definite (although extremely long) time. Unfortunately, the systematic algorithms are often extremely slow.

Randomized algorithms give up the requirement that an algorithm *must* solve a problem but, nevertheless, keep the probability of success high enough that the chances of error can be safely ignored. Remarkably, such "almost correct" algorithms can be very effective in practice. Two types of algorithm that illustrate some of the tradeoffs have become popular. "Monte Carlo algorithms" are always fast and probably correct and "Las Vegas algorithms" are always correct and probably fast. Depending on the application, speed may be more critical than correctness or vice versa.

Random Numbers Used in Stock Market Prediction

In modern financial markets, many people trade not only in tangible stocks and commodities but also on intangible expectations of how the prices of the stocks and commodities will change in the future. Often they buy and sell the right to buy or sell stock at some future date for some agreed-upon price. In effect, people are making bets on how prices will change in the future. With a prudent investment strategy it is possible to use these kinds of financial instruments to create a "hedge," that is, a portfolio of assets such that you are somewhat immunized from adverse fluctuations in the market.

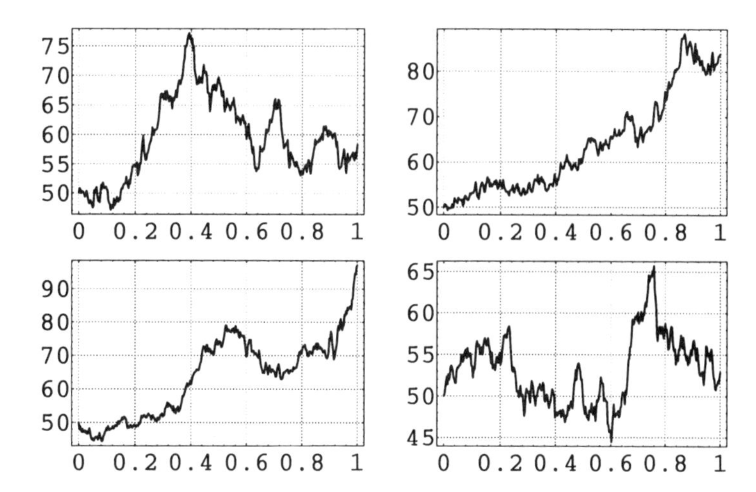

Fig. 7.3 Four realizations of the same underlying random process that describes the behavior of a stock with a mean return of 15% per annum and a volatility of 30%. The vertical axis is the stock price and the horizontal axis the time, measured in years.

There is now a small industry of experts who peer at financial time series charting the ups and down of the market in an attempt to discern some kind of pattern. In reality, the best model of the stock market that is currently known is to treat it as a special kind of random process, known as geometric Brownian motion. In Fig. 7.3 you see four different forecasts of how the same underlying random process might unfold over the next 12 months. As you can see, even though the process is fixed, the possible realizations of the process (i.e., the possible trajectories followed by the stock price) can be quite varied. Nevertheless, if you generate a few hundred thousand sample forecasts, you can gather enlightening statistics on the expected behavior of the asset and hence make more informed investment decisions.

Random Numbers Used in Simulations

Simulations of natural phenomena are, by far, the most voracious consumers of random numbers. Ever since the first "Monte Carlo" (named after the famed casinos in that city) programs of the 1950s[Metropolis53] the need for faithful simulations has grown. In particular, the burgeoning field of molecular biology consumes vast quantities of random numbers in order to simulate the motion of molecules interacting with cells and other molecules.

A particularly difficult computation involves determining how two kinds of molecules are likely to bind to one another. The exact manner in which the two molecules twist, stretch, and vibrate as they are being bombarded by billions of other molecules is an horrendously complex problem. It is quite infeasible to solve it exactly. However, with the help of supercomputers it is becoming possible to *simulate* the motion of a molecule, in a realistic chemical and biological environment, in order to determine its most likely configuration.

A detailed simulation would take into account every atomic vibration of a macromolecule. These occur at around 10^{15} Hz. Consequently, running the simulation of a macromolecule for 100 picoseconds takes about one day of supercomputer time[Hille92]!

The exact shape that a molecule adopts, in a particular environment, determines its medicinal and biochemical uses and, indeed, its potential toxicity. Consequently, faithful simulations of biochemical processes are providing an exciting new way to tackle drug design. Moreover, these simulations may provide new insight into the molecular-level mechanisms of diseases such as cancer and AIDS.

Can People Generate Random Numbers?

You might be wondering how good people are at picking numbers at random. In a famous experiment performed by D. W. Hagelbarger, and described by Claude Shannon, a human subject is asked to create a random sequence of +'s and -'s[Shannon53]. A computer program analyzes each player's past choices in an attempt to identify systematic biases amongst their selections. The program then predicts what symbol the player will pick next. If the player acts truly randomly, then the program should win only 50% of the games. In fact, when the experiment is done, the program wins 55–60% of the games. This means that people are not very good at making unbiased binary choices.

Similar biases are also seen in people's selection of lottery numbers. In fact, it is possible to write a computer program that, although not increasing your odds of winning the lottery, can increase your odds of walking away with 100% of the jackpot if you do win by avoiding the most "popular" lottery numbers!

Given people's poor performance, we might wonder whether computers can do much better. But how would you make a computer program pick a random number?

7.4 Randomness and Classical Computers

"Anyone who considers arithmetical methods of producing random digits is, of course, in a state of sin."
—John von Neumann

Classical computers are very good at following clear precise instructions but they cannot be programmed to do something unpredictable, such as pick a random number. The only thing classical computers can do is evaluate functions. No matter what computational task they are asked to perform they are always constrained to bend the problem into some function evaluation. This is all well and good if the task happens to *be* a function evaluation, but there are plenty of important computational tasks that do not require, or worse, cannot be accomplished by, function evaluation.

Generating a random number happens to be one of the tasks that cannot be accomplished by evaluating any function. Although most modern programming languages contain some kind of command for generating a "random" number, in reality, they can only generate "pseudorandom" numbers. These are sequences of numbers that pass many of the tests that a sequence of random numbers would also pass. But they are not true random numbers, because they are merely the completely predictable output from a definite function evaluation. Thus a classical computer can only feign randomness and thinking otherwise places us, in John Von Neumann words, "in a state of sin!"

Linear-Congruential Generators

One method for generating a sequence of pseudorandom integers is to start with some integer N_0, such that $0 \leq N_0 < n$, and repeatedly apply the rule:

$$N_{k+1} = (\mathtt{l} N_k + m) \bmod n,$$

where $\mathtt{l}$, m, and n are fixed integers and k = 1, 2, 3, ... The resulting sequence of numbers N_1, N_2, N_3, . . ., appears, superficially, to generate a set of random samples from a uniform distribution that lie in the range *0* to n - 1 inclusive.

The problem with linear congruential generators is that their outputs are periodic; that is, they start to repeat, with a period that is at most n. Worse still, for many poor choices of parameters, the period can be considerably less than n. To see this, consider the particular instance of such a generator defined by

$$N_{k+1} = (6 N_k + 7) \bmod 5$$

with N_0 = 2. This generator produces the periodic sequence 4,1,3,0,2,4,1,3,0,2,4,1, . . . which is a sequence with period 5. This is the greatest possible period for any linear congruential generator having $n = 5$. However, merely making n large does not ensure a sequence with a large period. For example, the generator defined by

$$N_{k+1} = (27 N_k + 11) \bmod 54$$

with N_0 = 2 produces the sequence 11,38,11,38,11,38,11,... which has a period of only 2.

In fact, if the prime factors of n are p_i, i = 1,2,3, ..., then the sequence induced by the linear congruential generator, previously given, can be shown to have period n if and only if[Tuckwell95]

1. the only common factor of m and n is 1,
2. $l-1$ is a multiple of p_i for all i, and
3. $l-1$ is a multiple of 4 if n is a multiple of 4.

Despite these limitations, the sequence of numbers output from a linear congruential generator does possess many properties that you would expect of a true random sequence.

RANDU, the random number generator common on IBM mainframe computers in the 1960s, is based on a linear congruential generator having the parameter values $l = 65539$, m = 0 and $n = 2^{31}$. RANDU is adequate for many kinds of simulations that call for random numbers. However, in the 1960s an insidious kind of correlation was found to be lurking in the sequence of pseudorandom numbers spewed out by RANDU[Abbott95]. If successive triples of the numbers RANDU generates are used as a set of coordinates in a three-dimensional space, a remarkable pattern emerges. From most viewpoints the points appear to be randomly distributed in the space but from a special orientation you can actually see that these coordinates all lie in a set of planes[Marsaglia68, Abbott95]. For triples of numbers, in three-dimensional space, the number of planes is usually less than $n^{1/3}$. For groups of d numbers, defining points in a d-dimensional space, the number of hyperplanes is generally less than $(d!n)^{1/d}$. This means that, for a given n, the problem becomes more acute as you move to higher dimensions. For groups of 6 numbers, with $n = 2^{32}$ the number of hyperplanes would be at most 120. For groups of 10 numbers with $n = 2^{32}$ the number of hyperplanes is even less, 41. This effect is illustrated quite graphically in Fig. 7.4.

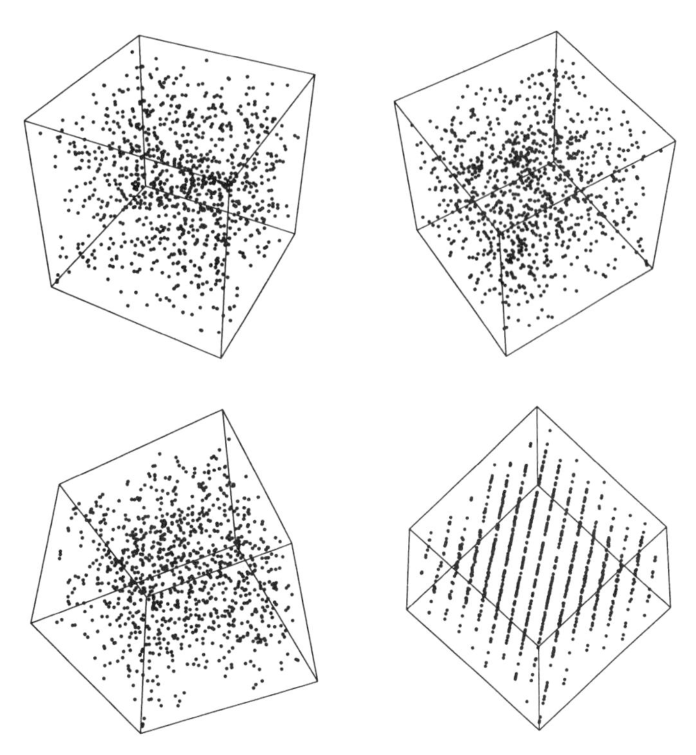

Fig. 7.4 The same cube of points seen from four different viewpoints. The coordinates of the points are given by successive triples of integers output from a linear congruential generator. Notice that from most viewpoints the points appear to be randomly distributed in the cube but from certain special viewpoints, it is evident that they all actually lie in a set of planes.

Shift-Register Generators

To avoid the pitfalls of linear congruential generators researchers have turned to a new class of pseudorandom number generators called "shift-register generators"[Marsaglia90]. In these generators, each successive number depends upon many preceding values (rather than just the last value as in a linear congruential generator). For example, you could make a shift register generator whose kth output is the sum, modulo n, of the (k - i)th and (k - j)th outputs:

$$N_k = \left(a N_{k-i} + b N_{k-j} + c\right) \bmod n .$$

Such a generator again produces a sequence of pseudorandom numbers but this time with a maximum possible period of 2^{qp} where q is the number of binary digits in each pseudorandom number the shift register generates and p is the number of past values that are retained at each iteration[Hayes93]. It is easy to imagine more elaborate schemes, for example, by including more than two past terms in the sum at each iteration step or by changing the way the past values are summed together, for example, by replacing modular addition by a bitwise XOR of the binary representation of the past values:

$$N_k = N_{k-i} \text{ XOR } N_{k-j}.$$

Shift register methods are believed to be amongst the best known[Park88].

7.5 The Plague of Correlations

You might think that the infamous RANDU bug is a thing of the past. Certainly the demand for more ambitious Monte Carlo simulations catalyzed the development of supposedly "better" pseudorandom number generators. However, *Numerical Recipes in C*, the "bible" for numerical algorithms that is widely used by engineers, reports: *"If all scientific papers, whose results are in doubt, because of bad random number generators, were removed from library shelves, there would be a gap on each shelf as big as your fist."*

This warning proved to be prophetic when, in 1992, Alan Ferrenberg and David Landau of the University of Georgia, and Joanna Wong of IBM discovered a subtle bug in several supposedly high quality random number generators[Ferrenberg92].

Ferrenberg, Landau, and Wong were interested in simulating the behavior of a three-dimensional system of interacting spins. In preparation for the full simulation, they wrote a simulator for the simpler two-dimensional case whose exact behavior had already been deduced theoretically. Thus this experiment was merely a test of the simulator against a known benchmark. Amongst the generators they tested they used two versions of a shift register generator:

$$N_k = N_{k-103} \text{ XOR } N_{k-250}$$
$$N_k = N_{k-1063} \text{ XOR } N_{k-1279}$$

To the researchers' surprise, in both cases the simulator gave the wrong predictions. After checking and rechecking their program, in desperation, the researchers replaced the sophisticated pseudorandom number generator with a linear congruential generator — a generator with known deficiencies:

$$N_k = \left(16807\, N_{k-1}\right) \bmod \left(2^{31} - 1\right).$$

To their amazement the more "naive" simulator gave results that were much closer to the known answer. The conclusion was quite striking. Given current (classical) computer technology, any simulation that uses a random number generator must be tested using different generators regardless of how many tests the generator has already passed.

This illustrates the impossibility of generating true random numbers using a classical computer[James90]. However, better and better approximations are appearing all the time[Berdnikov96]. Given the trouble classical pseudorandom number generators have caused, it is prudent to revisit the question of whether they are really necessary. Perhaps if we looked to quantum physics we could find a physical process that is inherently random and which could, therefore, be used as the basis for building a true random number generator.

7.6 Randomness and Quantum Computers

Quantum mechanical indeterminism, the inability to predict into which state a superposition will appear to collapse upon being measured, does provide exactly the right solution. Unlike classical physics, quantum physics *does* contain inherently random phenomena which can be used directly to generate random numbers. In particular if a physical system, representing a qubit, is placed in a state that is an equally weighted superposition of a $|0\rangle$ and a $|1\rangle$ and then measured, quantum physics says that there is exactly a 50:50 chance of obtaining a 0 and a 50:50 chance of obtaining a 1. Here is how it works.

Recall that the state vector describing a qubit in the $|0\rangle$ state is defined by $|0\rangle \equiv \binom{1}{0}$ and the state vector describing a qubit in the $|1\rangle$ state is defined by $|1\rangle \equiv \binom{0}{1}$. Consider the unitary operator defined by

$$U(\theta) = \begin{pmatrix} \cos\theta & -\sin\theta \\ \sin\theta & \cos\theta \end{pmatrix}.$$

This operator has the effect of rotating a state vector $\omega_0|0\rangle + \omega_1|1\rangle$ through an angle θ.

$$U(\theta)\cdot\begin{pmatrix} \omega_0 \\ \omega_1 \end{pmatrix} = \begin{pmatrix} \cos\theta & -\sin\theta \\ \sin\theta & \cos\theta \end{pmatrix}\cdot\begin{pmatrix} \omega_0 \\ \omega_1 \end{pmatrix} = \begin{pmatrix} \omega_0\cos\theta - \omega_1\sin\theta \\ \omega_0\sin\theta + \omega_1\cos\theta \end{pmatrix}.$$

In particular, the operator $U(\frac{\pi}{4})$ transforms the state $|0\rangle$ (i.e., a state which is representing the binary value 0) into the superposition

$\frac{1}{\sqrt{2}}|0\rangle + \frac{1}{\sqrt{2}}|1\rangle$. The latter represents an equally weighted superposition of the binary values 0 *and* 1.

$$U\left(\frac{\pi}{4}\right) \cdot \begin{pmatrix} 1 \\ 0 \end{pmatrix} = \begin{pmatrix} \frac{1}{\sqrt{2}} & -\frac{1}{\sqrt{2}} \\ \frac{1}{\sqrt{2}} & \frac{1}{\sqrt{2}} \end{pmatrix} \cdot \begin{pmatrix} 1 \\ 0 \end{pmatrix}$$
$$= \begin{pmatrix} \frac{1}{\sqrt{2}} \\ \frac{1}{\sqrt{2}} \end{pmatrix}$$
$$= \frac{1}{\sqrt{2}}|0\rangle + \frac{1}{\sqrt{2}}|1\rangle.$$

In order to make a device that outputs a truly random bit, you simply place a simple 2-state system in the $|0\rangle$ state, apply the operator $U(\frac{\pi}{4})$ to rotate the state into the superposition $\frac{1}{\sqrt{2}}|0\rangle + \frac{1}{\sqrt{2}}|1\rangle$ and then observe the superposition. The act of observation causes the superposition to collapse into either the $|0\rangle$ or the $|1\rangle$ state with equal probability. Hence you can exploit quantum mechanical superposition and indeterminism to simulate a perfectly fair coin toss.

Once you have a method for generating fair coin tosses it is easy to combine them to create more elaborate random number generators. For example, suppose you wanted to generate a random integer uniformly in the range 0 to $2^n - 1$ for some positive integer n as in the next section.

7.7 Simulation of a Quantum Computer Generating a True Random Number

Here is a simulation of a quantum computer performing such a computation:

In[]:=

```
RandomQC[Integer, {0,7}]
```

Out[]=

```
Initially, create 3 qubits all in state |0>. The
state of each qubit is {ket[0], ket[0], ket[0]}.

Place each particle in a superposition of |0> and
|1> by rotating each |0> through Pi/4 using the op-
erator U(Pi/4). The state of each particle becomes:
```

$$\left\{\frac{\text{ket[0]}}{\sqrt{2}} + \frac{\text{ket[1]}}{\sqrt{2}}, \frac{\text{ket[0]}}{\sqrt{2}} + \frac{\text{ket[1]}}{\sqrt{2}}, \frac{\text{ket[0]}}{\sqrt{2}} + \frac{\text{ket[1]}}{\sqrt{2}}\right\}$$

```
Create the state of the memory register as the JOINT
state of ALL 3 qubits (using the direct product).
Hence the state of the memory register is:
```

$$\frac{1}{2\sqrt{2}}\left(\text{ket}[0,0,0] + \text{ket}[0,0,1] + \text{ket}[0,1,0] + \text{ket}[0,1,1] + \text{ket}[1,0,0] + \text{ket}[1,0,1] + \text{ket}[1,1,0] + \text{ket}[1,1,1]\right)$$

```
Read the memory register to obtain {0,1,1}.
Convert the bit string {0, 1, 1} to the integer 3.
Random integer in range [0,7] inclusive is 3.
```

Thus, by exploiting superposition and indeterminism, it is possible for a quantum computer to generate a true random number. Sequences of random numbers from quantum random number generators will be free of the subtle correlations that bedevil classical pseudorandom number generators.

This shows that there is at least one important computational task that a quantum computer can perform but which a classical computer cannot.

Indeed, quantum computers may offer a more direct way of simulating many important physical systems than is possible classically. In 1982 Richard Feynman showed that no classical computer could possibly simulate a quantum physical process without incurring an exponential slowdown[Feynman82]. A supercomputer, simulating the motion of a macromolecule in a cellular environment, runs about 100,000 times slower than Nature. However, a general-purpose quantum computer would, in effect, be a perfect quantum simulator[Deutsch85]. Quantum computers might therefore be useful not only for computation, but as a way of testing various predictions of quantum physics as well[Wiesner96].

In the next chapter we investigate yet another feat that a quantum computer can perform but which a classical computer cannot: the ability to communicate with messages that are almost guaranteed to reveal the presence of any eavesdropping.

CHAPTER 8

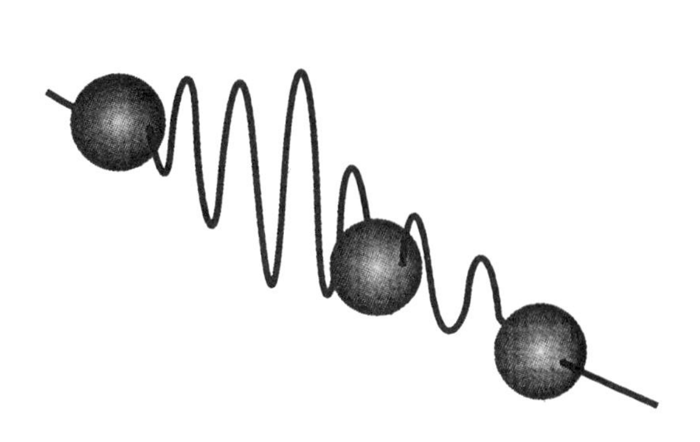

Quantum Cryptography

"In Nature's infinite book of secrecy
A little can I read"
— William Shakespeare

Modern cryptographic schemes such as one-time pads and public key cryptosystems rely on the existence of secret "keys." Such cryptosystems have the property that once the keys are known, any encrypted messages are easily unscrambled, but without the keys it is computationally intractable, at least using any classical computer, to crack a coded message. Consequently, the security of these cryptosystems relies upon the keys remaining secret.

The problem, however, is that the keys can never be *guaranteed* to be secure. One-time pads are vulnerable to attack because, prior to secure messages being exchanged, the sender and legitimate recipient of a message must exchange keys by some physical means and subsequently store them in a secure location. If the keys *could* be guaranteed to be secret, then the one-time pad would be a highly secure cryptosystem. However, potentially, an adversary could intercept and duplicate the keys at the moment they are being exchanged or copy one of the key pads in either party's possession.

Worse still, in the public key cryptosystems, the person wishing to receive messages must broadcast a "public key" that contains a number, which if factored, would reveal the "private key" too. Once the private key is known, then any messages encrypted using the matching public key would be compromised. As we saw in Chapter 5, a quantum computer appears to be able to perform exactly this factoring step very efficiently. So, as of today, the security of public key cryptosystems rests on the presumption that it will be technologically difficult for anyone to build a real quantum computer: a risky assumption indeed given the pace of technological progress.

There is a need, therefore, for a communication scheme that is invulnerable to attack by an adversary who might potentially possess a quantum computer.

In 1984, such a cryptographic protocol was devised by Charles Bennett and Gilles Brassard, that itself rests, fundamentally, on two quantum phenomena. In particular, this cryptographic scheme exploits the impossibility of cloning quantum information and the impossibility of measuring certain pairs of observables simultaneously[Bennett92]. The cloning question relates to the measurement issue, so we will focus on measurement first.

8.1 Heisenberg's Uncertainty Principle

In 1927, the German physicist Werner Heisenberg discovered a fundamental bound on the accuracies with which certain pairs of observables can be measured simultaneously. Knowing the value of one observable more accurately necessarily makes the value of another observable more uncertain.

In Chapter 4 we described how each observable, in quantum theory, is represented mathematically by an "observable operator," that is an Hermitian matrix. The possible values of the observable are the eigenvalues of this matrix.

Now if we measure a particular observable, represented by the operator $\hat{A}$, say, of a quantum system that is in a state $|\psi\rangle$ and we always obtain the same answer, then $|\psi\rangle$ must be an eigenstate of $\hat{A}$. In other words, the column vector $|\psi\rangle$ is an eigenvector of the Hermitian matrix $\hat{A}$, so that $\hat{A}\cdot|\psi\rangle = a|\psi\rangle$ where a is some real number.

Now let us consider measuring a different observable, represented by the operator $\hat{B}$, say, for which $|\psi\rangle$ is not an eigenstate of $\hat{B}$. Consequently, if we were to measure the observable $\hat{B}$, of the quantum system in state $|\psi\rangle$, we would typically obtain different answers each time we made the measurement.

Physicists have a mathematical way of characterizing the spread of the different answers that would be obtained from measuring a collection of identically prepared quantum systems. They look at the "root mean square deviation" of the answers they obtain. For the observable operator $\hat{A}$, the notation $\langle \hat{A} \rangle$ refers to the average (or "expected") value that would be obtained if you repeatedly placed a physical system in a certain state and then measured the observable $\hat{A}$. The root mean square deviation in the expected value of the observable is given by

$$\Delta\hat{A} = \sqrt{\langle \hat{A}^2 \rangle - \langle \hat{A} \rangle^2},$$

where $\hat{A}^2 = \hat{A} \cdot \hat{A}$. Similarly, the root mean square value of the observable $\hat{B}$ is given by

$$\Delta\hat{B} = \sqrt{\langle \hat{B}^2 \rangle - \langle \hat{B} \rangle^2}.$$

The quantities $\Delta\hat{A}$ and $\Delta\hat{B}$ quantify the uncertainty with which the values of the observables $\hat{A}$ and $\hat{B}$ are known. As $|\psi\rangle$ is an eigenstate of the observable $\hat{A}$, then $\Delta\hat{A} = 0$ because the same answer is obtained each time the observable $\hat{A}$ is measured and so there is no uncertainty in the value. However, as $|\psi\rangle$ is not an eigenstate of the observable $\hat{B}$, then measurements on the observable $\hat{B}$ yield any of the possible eigenvalues of $\hat{B}$ with varying degrees of probability. Consequently, $\Delta\hat{B} \neq 0$.

The question then arises as to what happens if we try to measure both observables.

In the case we are considering, the answer depends on the *order* in which we make the measurements. If we measured observable $\hat{A}$ first, then the act of measurement would not perturb the state since $|\psi\rangle$ is already an eigenstate of $\hat{A}$. Then the subsequent measurement of observable $\hat{B}$ would result in one of the possible eigenvalues for $\hat{B}$. However, if we measured observable $\hat{B}$ first then, as $|\psi\rangle$ is not an eigenstate of $\hat{B}$, the act of measuring $\hat{B}$ will perturb the state of the system. After the measurement of $\hat{B}$ the system will be in a state that is some eigenstate of observable $\hat{B}$. Then the subsequent measurement of observable $\hat{A}$ will again perturb the state into some eigenstate of $\hat{A}$. It is therefore natural to look at the difference between the measurements performed in either order. To do so, we construct the "commutator" $[\hat{A}, \hat{B}]$ defined by

$$\left[\hat{A},\hat{B}\right]=\hat{A}\hat{B}-\hat{B}\hat{A},$$

which is simply some new operator.

It can be shown that if two observables are measured simultaneously, then the uncertainty in their joint values must always obey the inequality:

$$\Delta\hat{A}\,\Delta\hat{B}\geq\tfrac{1}{2}\left|\left\langle\left[\hat{A},\hat{B}\right]\right\rangle\right|.$$

This is the most general form of the Heisenberg Uncertainty Principle. The only way that both observables can be measured exactly, simultaneously, is when the commutator $\left[\hat{A},\hat{B}\right]$ reduces to a matrix of all zeroes. In such a situation the operators $\hat{A}$ and $\hat{B}$ are said to "commute."

Many specific instances of the Heisenberg Uncertainty Principle are known. The most famous concerns a measurement of a particle's position and momentum simultaneously. In order to determine the position accurately, it is necessary to use light with a very short wavelength. This is because the ability to provide position information is comparable to the wavelength of the object providing the

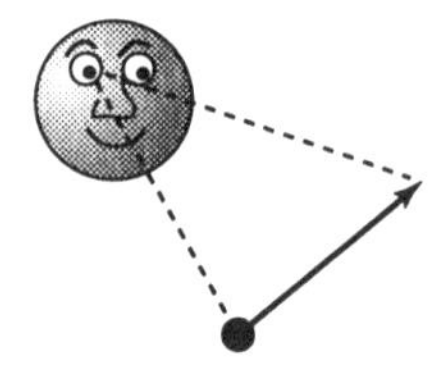

Classically you can measure both position and momentum exactly without any disturbance to either

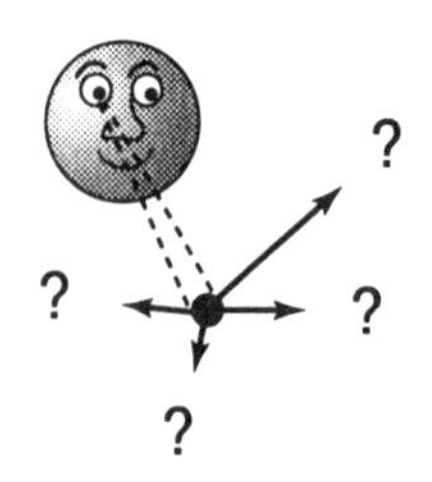

Quantum mechanically, measuring position exactly randomizes momentum

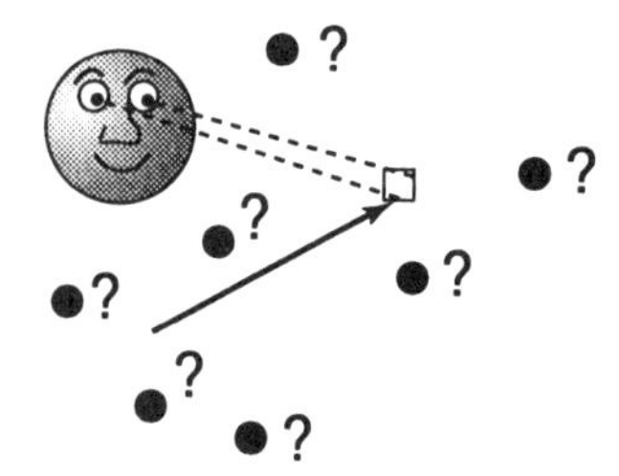

Quantum mechanically, measuring momentum exactly randomizes position

Fig. 8.1 The Heisenberg Uncertainty Principle precludes exact measurements of certain pairs of observables at the same time.

position information. The problem is that short wavelength light imparts a large momentum kick to the electron when it scatters off it to provide the position information. Conversely, if we want an accurate momentum measurement, then we have to use very long wavelengths which gives us very poor position information.

Similar tradeoffs arise when measuring any pair of conjugate variables. This unavoidable disturbance in the value of a conjugate variable when its partner variable is measured provides the key to quantum cryptography. In particular, as we show in the next section, it is possible to encode a bit in the direction of polarization of a photon.

8.2 Polarization

By now you will be used to the idea that any message text can be reduced, ultimately, to some sequence of 0s and 1s. An example of such an encoding was given in Chapter 6 in the discussion of one-time pad cryptosystems. For quantum cryptography, the idea is to reduce a message to a sequence of bits and then to create a stream of photons placed in certain quantum states corresponding to those bits. By choosing a clever encoding of message bits in photon states, and by exploiting the Heisenberg Uncertainty Principle applied to measurements of photon states, it is possible to transmit a message that is almost certain to reveal the presence of any eavesdropper.

Photons are electromagnetic waves that have many interesting properties. One of the properties of a photon that can be used to encode a bit is the photon's polarization state.

Photons are "transverse electromagnetic waves." This means that the electric and the magnetic fields are perpendicular to the direction in which they propagate. Moreover, the electric and magnetic fields are perpendicular to each other. Thus, in the usual three-dimensional coordinate system with mutually perpendicular x-, y-, and z-axes, if a photon is propagating in the positive z-direction, the electric and magnetic fields will oscillate in the x-z plane and the y-z plane, respectively.

The photon property we are interested in is called *polarization* and refers to the bias of the electric field in the electromagnetic field of the photon. *Linear* polarization means that as the photon propagates the electric field stays in the same plane. In *circularly* polarized

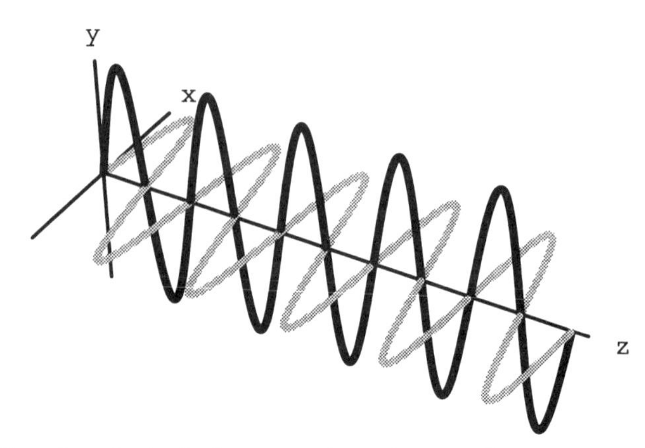

Fig. 8.2 A linearly polarized photon consists of an oscillating electric field and an oscillating magnetic field that are perpendicular to each other and to the direction of propagation.

light the electric field rotates at a certain frequency as the photon propagates. Quantum cryptography can be implemented with linearly polarized light, circularly polarized light, or a combination of the two. However, we restrict our discussions to implementations using linearly polarized light only as this is a little simpler to explain.

In order to encode a bit in the direction of polarization of a photon, it is necessary to place a photon in a particular polarization state. This amounts to creating a photon whose electric field is oscillating in a desired plane. One way to do this is simply to pass the photon through a polarizer whose polarization axis is set at the desired angle.

The development of modern polarizers began with Edwin Land, the founder of Polaroid Corporation, in 1928, when he was an undergraduate at Harvard College. His interest in polarization was piqued when he read about the strange properties of crystals that were formed when iodine was dropped into the urine of a dog that had been fed quinine. The crystals turned out to be made of the material that Land later used to make large-scale polarizers[Hecht74].

According to quantum theory, one of two things can happen to a single photon passing through a polarizer: either it will emerge with its electric field oscillating in the desired plane or else it will not emerge at all. In the latter case, the photon is absorbed by the polarizer and its energy re-emitted later in the form of heat.

If the axis of the polarizer makes an angle of θ with the plane of the electric field of the photon fed into the polarizer, there is a probability of $\cos^2\theta$ that the photon will emerge with its polarization set at the desired angle and a probability of $1 - \cos^2\theta$ that it will be absorbed.

Two polarizers with their polarization axes set at $90°$ from each other will not pass any light. This is consistent with quantum theory which gives the probability of a photon emerging from the first polarizer then passing through the second polarizer as $\cos^2\pi/2 = 0$, meaning that the photon is certainly absorbed.

8.3 Using Polarized Photons to Encode a Message

A slightly more sophisticated method for placing a photon in a definite polarization state uses a device known as a Pockels cell.

By using a Pockels cell, it is possible to create a photon with its electric field oscillating in any desired plane. We can therefore (arbitrarily) call polarized photons whose electric fields oscillate in a plane at either $0°$ or $90°$ to some reference line "rectilinear" and those whose electric fields oscillate in a plane at $45°$ or $135°$ "diagonal". Furthermore, we can stipulate that photons polarized at angles of $0°$ and $45°$ are to represent the binary value 0 and those polarized at angles of $90°$ and $135°$ represent the binary value 1. Once this correspondence has been made, a sequence of bits can be used to control the bias in a Pockels cell and hence determine the polarization orientations from the stream of photons emerging from the cell. This allows a sequence of bits to be converted into a sequence of polarized photons. These may then be fed into some communication channel, such as an optical fiber.

8.4 Measuring the Polarization of a Photon

In order to recover the bits encoded in the polarization orientation of a stream of photons, it is necessary for the recipient to measure the polarizations. Fortunately, Nature has provided us with a material beautifully suited for just this purpose.

A calcite, or calcium carbonate ($CaCO_3$), crystal has the property of birefringence. This means that the electrons in the crystal are not bound with equal strength in each direction. Consequently, a photon passing through the crystal will feel a different electromagnetic force depending on the orientation of its electric field relative to the polarization axis in the crystal. For example, suppose the calcite's

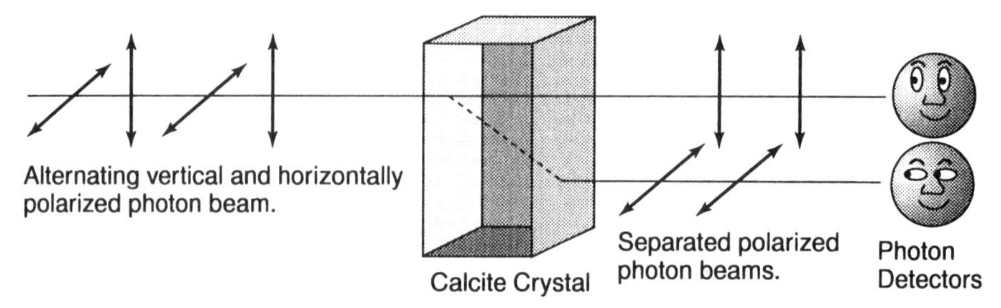

Fig. 8.3 How a birefringent crystal can be used to separate photons based on their polarization.

polarization axis is aligned so that vertically polarized photons pass straight through it. A photon with a horizontal polarization will also pass through the crystal but it will emerge from the crystal shifted from its original trajectory as shown in Fig. 8.3.

This allows us to use a calcite crystal to determine whether a given photon has a vertical or horizontal polarization. However, what happens when diagonally polarized light passes through a vertically oriented polarizer? The Uncertainty Principle says that the polarizer provides no information about the original polarization so that some of the photons will not be shifted and the rest will be shifted depending on the angle of the photon's electric field axis to the calcite's axis.

Thus we can use the location at which a photon emerges from a calcite crystal as a way of signaling (i.e., measuring) that photon's polarization.

8.5 Uncertainty Principle for Polarized Photons

To read a message "written" in a stream of polarized photons, we would have to measure (observe) the direction of polarization of each of the photons. What are the observable operators for measuring photon polarization?

In the rectilinear basis, the observable operator is:

$$\hat{P}_r = \begin{pmatrix} 1 & 0 \\ 0 & -1 \end{pmatrix}_r$$

This has eigenvalues ± 1 (corresponding to observing the photon emerge from the upper or lower portions of a vertically oriented calcite crystal respectively) and eigenstates $|0\rangle_r = \begin{pmatrix} 1 \\ 0 \end{pmatrix}_r$ and $|1\rangle_r = \begin{pmatrix} 0 \\ 1 \end{pmatrix}_r$.

The subscripts r just remind us that we are in the rectilinear (vertical/horizontal) basis.

Likewise, in the diagonal (slant-up/slant-down) basis, the observable operator for polarization orientation is:

$$\hat{P}_d = \begin{pmatrix} 1 & 0 \\ 0 & -1 \end{pmatrix}_d$$

with eigenvalues ±1 (corresponding to observing the photon emerge from the upper or lower portions of a diagonally oriented calcite crystal, respectively) and eigenstates $|0\rangle_d = \begin{pmatrix} 1 \\ 0 \end{pmatrix}_d$ and $|1\rangle_d = \begin{pmatrix} 0 \\ 1 \end{pmatrix}_d$. The subscripts d indicate that these operators and state vectors are in the diagonal basis.

Now the question is, can you measure both the rectilinear and diagonal polarizations simultaneously? To answer this question, we turn to Heisenberg's Uncertainty Principle. This states that we can only measure two observables exactly, simultaneously, if their commutator vanishes. However, to construct the commutator, both observables must be couched in the same basis.

In general, a given state vector can be expressed with respect to many possible bases. In our case, we are interested in how to transform state vectors and operators in (say) the diagonal polarization basis into state vectors and operators in the rectilinear polarization basis. If U is the transformation matrix that carries each basis vector in space A into a basis vector in space B, then arbitrary state vectors $|\psi\rangle_A$ (in space A) and arbitrary operators $\hat{\Omega}_A$ (in space A) can be transformed into equivalent state vectors and operators, respectively (in space B), using the rules:

$$|\psi\rangle_B = U^{\dagger} |\psi\rangle_A$$
$$\hat{\Omega}_B = U^{\dagger}\, \hat{\Omega}_A\, U$$

where $U^{\dagger}$ is the conjugate transpose of U. Incidentally, such transformation operators are always unitary.

We can apply these rules to transform the observable operator for polarization orientation in the diagonal basis into the equivalent operator in the rectilinear basis. The transformation, $U(\theta)$, that connects the two bases is simply a rotation through an angle θ where

$$U(\theta) = \begin{pmatrix} \cos\theta & -\sin\theta \\ \sin\theta & \cos\theta \end{pmatrix}.$$

To rotate the diagonal basis into the rectilinear basis, we rotate by $-\frac{\pi}{4}$. Consequently, the observable for diagonal polarization, in the rectilinear basis, is given by:

$$\hat{P}_r' = U\left(-\frac{\pi}{4}\right)\cdot\begin{pmatrix}1 & 0\\ 0 & -1\end{pmatrix}_d = \begin{pmatrix}\frac{1}{\sqrt{2}} & -\frac{1}{\sqrt{2}}\\ -\frac{1}{\sqrt{2}} & -\frac{1}{\sqrt{2}}\end{pmatrix}_r.$$

The commutator of $\hat{P}_r$ and $\hat{P}_r'$ does not vanish. Hence it is impossible to measure both rectilinear and diagonal polarizations exactly, simultaneously. Any attempt to measure the rectilinear polarization orientation necessarily perturbs (in fact, randomizes) the diagonal polarization orientation and vice versa. This is the crucial physical principle that can be exploited to thwart eavesdropping during the exchange of a secret cryptographic key.

8.6 Quantum Cryptography Using Polarized Photons

We now have all the ingredients needed to design a protocol for exchanging a secret key, that is, a sequence of bits known only to Alice and Bob.

Alice chooses a set of bits out of which she and Bob will construct a key. Initially, neither Alice nor Bob have a particular key in mind. The key they will use will emerge out of the communication protocol that they will follow. Consequently, the exact sequence of bits that Alice sends to Bob is not that important. All that matters is that they, and only they, come to learn the identity of a common subset of the bits. These common, but private, bits are used as the key.

8.7 Simulation of Quantum Cryptography in the Absence of Eavesdropping

Let us begin by considering the case in which there is no eavesdropping.

To illustrate the steps involved in secure key distribution between Alice and Bob, we have written a simulator that generates a sequence of graphics summarizing the operations that the various parties must perform. All of the following diagrams are generated automatically when you run the quantum cryptography simulator that is available in the code supplement.

The top level command is `DistributeKey`. It takes three arguments; the probability with which you want to detect eavesdropping; the number of bits you want to use in your key; and what is essentially a switch that allows you to turn eavesdropping on or off. You have to pretend that you do not really know whether eavesdropping is taking place to make it an honest simulation, even though you, as the controller of the simulator, can turn eavesdropping on and off at will.

To run the simulator (with eavesdropping turned off) such that you would have a 75% chance of detecting any eavesdropping and resulting in a secure key based on 4 bits, you would issue the following command.

In[]:=

```
DistributeKey[0.75, 4, Eavesdropping->False]
```

Note that in practice, you would really want the probability of detecting eavesdropping to be much higher (e.g., 99%), and you would want to have more bits in your key. We picked these smaller values so that the graphics that the simulator generates would fit on a page in this book! If you run the code yourself, you can use more realistic figures.

We annotate the output to help you interpret what is going on. To begin with, Fig. 8.4 shows the diagram illustrating the sequence of steps Alice makes to encode bits as polarized photons. Alice chooses some set of bits (first row). Then, for each bit, she chooses to encode it in either the rectilinear polarization (+) or in the diagonal polarization (×) of a photon (second row). This choice of polarization orientation must be made randomly. Alice then sends the photons she has created to Bob over an open communication channel (third row).

Next consider Bob's actions upon receipt of the photons, as shown in Fig. 8.5. Upon receipt of the photons (first row), Bob chooses a polarizer orientation (second row) with which he measures the direction of polarization of the incoming photons. Hence Bob reconstructs a set of bits (third row).

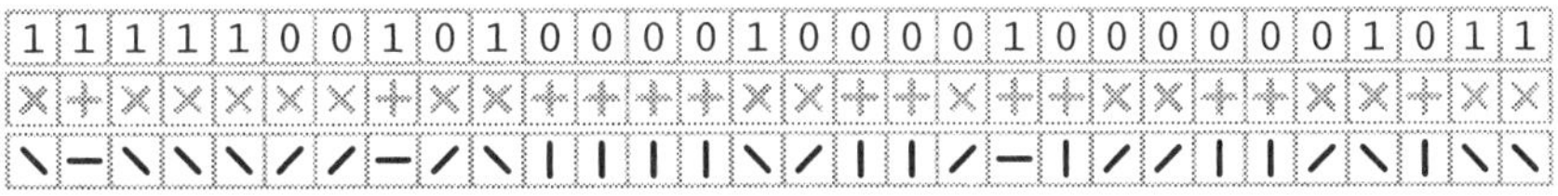

<table>
<tr><td>1</td><td>1</td><td>1</td><td>1</td><td>1</td><td>0</td><td>0</td><td>1</td><td>0</td><td>1</td><td>0</td><td>0</td><td>0</td><td>0</td><td>1</td><td>0</td><td>0</td><td>0</td><td>0</td><td>1</td><td>0</td><td>0</td><td>0</td><td>0</td><td>0</td><td>0</td><td>1</td><td>0</td><td>1</td><td>1</td></tr>
<tr><td>×</td><td>+</td><td>×</td><td>×</td><td>×</td><td>×</td><td>×</td><td>+</td><td>×</td><td>×</td><td>+</td><td>+</td><td>+</td><td>+</td><td>×</td><td>×</td><td>+</td><td>+</td><td>×</td><td>+</td><td>+</td><td>×</td><td>×</td><td>+</td><td>+</td><td>×</td><td>×</td><td>+</td><td>×</td><td>×</td></tr>
<tr><td>\</td><td>—</td><td>\</td><td>\</td><td>\</td><td>/</td><td>/</td><td>—</td><td>/</td><td>\</td><td>|</td><td>|</td><td>|</td><td>|</td><td>\</td><td>/</td><td>|</td><td>|</td><td>/</td><td>—</td><td>|</td><td>/</td><td>/</td><td>|</td><td>|</td><td>/</td><td>\</td><td>|</td><td>\</td><td>\</td></tr>
</table>

Fig. 8.4 Alice encodes bits as polarized photons.

<table>
<tr><td>\</td><td>—</td><td>\</td><td>\</td><td>\</td><td>/</td><td>/</td><td>—</td><td>/</td><td>\</td><td>|</td><td>|</td><td>|</td><td>|</td><td>\</td><td>/</td><td>|</td><td>|</td><td>/</td><td>—</td><td>|</td><td>/</td><td>/</td><td>|</td><td>|</td><td>/</td><td>\</td><td>|</td><td>\</td><td>\</td></tr>
<tr><td>+</td><td>+</td><td>×</td><td>+</td><td>×</td><td>×</td><td>+</td><td>×</td><td>+</td><td>×</td><td>+</td><td>+</td><td>×</td><td>×</td><td>+</td><td>×</td><td>×</td><td>+</td><td>×</td><td>+</td><td>×</td><td>+</td><td>+</td><td>+</td><td>×</td><td>+</td><td>×</td><td>+</td><td>×</td><td>+</td></tr>
<tr><td>0</td><td>1</td><td>1</td><td>1</td><td>1</td><td>0</td><td>0</td><td>0</td><td>1</td><td>1</td><td>0</td><td>0</td><td>0</td><td>1</td><td>0</td><td>0</td><td>0</td><td>0</td><td>0</td><td>1</td><td>1</td><td>1</td><td>1</td><td>0</td><td>1</td><td>1</td><td>1</td><td>0</td><td>1</td><td>0</td></tr>
</table>

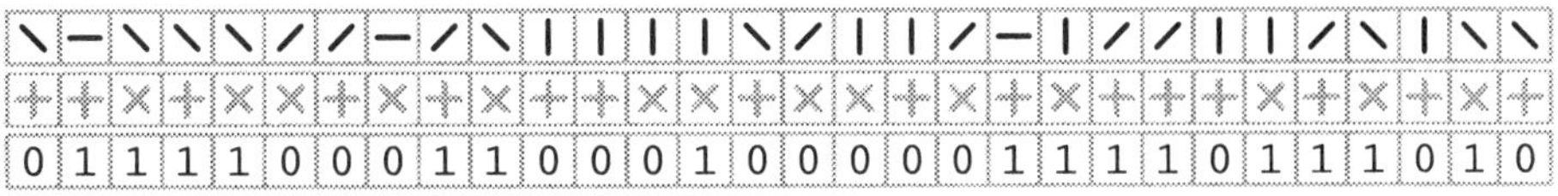

Fig. 8.5 Bob decodes polarized photons as bits.

Now Alice and Bob enter into a public (insecure) communication in which Alice divulges to Bob the types of polarizers that she used to encode a subset of the bits. Likewise Bob divulges to Alice the types of polarizers he used to decode the same subset of bits, as shown in Fig. 8.6. For those cases in which they used the same type of polarizers, Alice tells Bob what bits he ought to have measured. Assuming that the encoding, decoding, and transmission steps are error free, and provided there is no eavesdropping, Bob's test bits ought to agree with Alice's test bits 100%.

The more bits that are tested, the more likely it is that a potential eavesdropper is detected. In fact, for each bit tested by Alice and Bob, the probability of that test revealing the presence of an eavesdropper (given that an eavesdropper is indeed present) is $\frac{1}{4}$. Thus, if N bits are tested, the probability of detecting an eavesdropper (given that one is present) is $1-\left(\frac{3}{4}\right)^N$. A sketch of this function is shown in Fig. 8.7. As you can see, the probability of detecting eavesdropping approaches 1 asymptotically as the number of bits tested tends to infinity. Thus we can make the probability of detecting eavesdropping as close to certainty as we please simply by testing more bits.

	1	1			0				1		0				0		0		1							1	0		
	+	×			×				×		+				×		+		+							×	+		
	+	×			×				×		+				×		+		+							×	+		
	1	1			0				1		0				0		0		1							1	0		

Fig. 8.6 Alice and Bob compare a subset of the bits to test for the presence of eavesdropping.

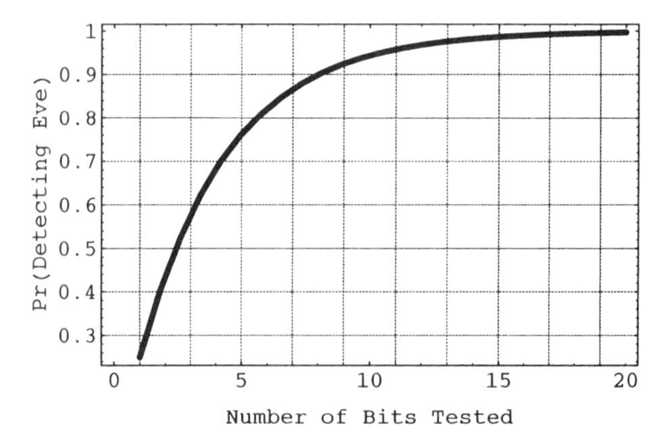

Fig. 8.7 Probability of detecting eavesdropping as a function of the number of bits tested by Alice and Bob. For a bit to be tested, Alice and Bob must have used the same polarizer orientation to encode and decode that bit, respectively.

×			×	×		×	+	×		+		+	+	×		+		×		+	×	×	+	+	×			×	×
+			+	×		+	×	+		+		×	×	+		×		×		×	+	+	+	×	+			×	+
0			1	1		0	0	1		0		0	1	0		0		0		1	1	1	0	1	1			1	0
☹			☹	☺		☹	☹	☹		☺		☹	☹	☹		☹		☺		☹	☹	☹	☺	☹	☹			☺	☹
				1						0								0					0					1	

Fig. 8.8 Key exchange step.

Once Alice and Bob have decided that the channel is secure, Alice then tells Bob what polarization orientations she used for each of her bits but not what those bits were (first row of Fig. 8.8). Bob then compares his polarizer orientations with those of Alice (second row) and also records his own answers (third row). Bob then, categorizes each bit in terms of whether he used the same polarizer orientation as Alice (fourth row). Bob then looks at the bits he read for the matching cases (fifth row).

This sequence of actions allows Bob to deduce a set of bits known only to Alice and himself. To see this, compare the top line of Fig. 8.4 with the bottom line of Fig. 8.8. You will find that Alice and Bob do agree on the bits for those cases in which they used the same polarizer orientations.

Once Alice and Bob deduce a common sequence of bits, this sequence can be used as the basis for a key in a provably secure classical cryptosystem such as a one time pad.

8.8 Simulation of Quantum Cryptography in the Presence of Eavesdropping

Now consider what happens if there is an eavesdropper, Eve, present. To turn eavesdropping on in the simulator, we set the switch `Eavesdropping->True`. So with the options we had before we issue the command:

In[]:=

```
DistributeKey[0.75, 4, Eavesdropping->True]
```

(actually by seeding the random number generator in the simulator, we have fixed things so that we are using the same bit sequence that we used before). Now although you know eavesdropping is taking place, Alice and Bob do not. So the first step proceeds as before with Alice encoding her bits in polarized photons, as in Fig. 8.9.

1	1	1	1	1	0	0	1	0	1	0	0	0	0	1	0	0	0	0	1	0	0	0	0	0	0	1	0	1	1
×	+	×	×	×	×	×	+	×	×	+	+	+	+	×	×	+	+	×	+	+	×	×	+	+	×	×	+	×	×
\	—	\	\	\	/	/	—	/	\	|	|	|	|	\	/	|	|	/	—	|	/	/	|	|	/	\	|	\	\

Fig. 8.9 Alice encodes her bits as polarized photons.

\	—	\	\	\	/	/	—	/	\	|	|	|	|	\	/	|	|	/	—	|	/	/	|	|	/	\	|	\	\
+	+	×	+	×	+	+	×	×	+	+	+	×	+	+	×	+	×	+	×	×	×	+	×	×	+	×	+	×	+
0	1	1	1	1	0	0	1	0	0	0	0	1	0	0	0	0	0	1	1	1	0	1	1	0	0	1	0	1	0

Fig. 8.10 Eve intercepts the photons Alice sent to Bob and tries to decode them. Eve then sends the photons she decoded on to Bob using whatever polarizer orientations Eve had picked.

This time, however, there is an eavesdropper, Eve, who is intercepting Alice's photons and making her own measurements of their polarizations in an effort to see what bits Alice is sending to Bob. Eve goes through the operations that Bob would have done: she intercepts the photons (first row), picks polarizer orientations (second row), and decodes the polarized photons as bits (Fig. 8.10).

In an effort to cover her tracks Eve then retransmits the photons she measured to Bob. Eve is free to do a complete recoding of her measured bits into photons polarized in whatever orientation she chooses. But the simplest situation has Eve using the same sequence of orientations that she used during her decoding step.

From Bob's perspective, at this moment he is unaware of Eve's presence, so Bob proceeds to decode the photons he thinks are coming from Alice, but which are actually coming from Eve (see Fig. 8.11). Bob intercepts the photons (first row), picks polarizer orientations (second row), and decodes the photons as a sequence of bits.

Now Alice and Bob compare both their polarizer orientations and their bits for a subset of the bits Alice sent. On those cases where they agree on polarizer orientation they should also agree on the bit sent and received. Here there is an error in the third bit tested that reveals the presence of Eve, the eavesdropper, as shown in Fig. 8.12.

|	—	\	—	\	|	|	\	/	|	|	|	\	|	|	/	|	/	—	\	\	/	—	\	/	|	\	|	\	|
+	+	×	+	×	×	+	×	+	×	+	+	×	×	+	×	×	+	×	+	×	+	+	+	×	+	×	+	×	+
0	1	1	1	1	1	0	1	1	1	0	0	1	0	0	0	0	0	1	1	1	1	1	0	0	0	1	0	1	0

Fig. 8.11 Bob decodes the photons unaware of Eve's presence.

		1		1	0				1	0	0				0				1				0				0		
		×		×	×				×	+	+				×				+				+				+		
		×		×	×				×	+	+				×				+				+				+		
		1		1	1				1	0	0				0				1				0				0		

Fig. 8.12 Alice and Bob detect the presence of Eve.

Consequently, Alice and Bob decide to abort their communications.

8.9 The Working Prototype

It is rare for exclamation marks to appear in the titles of scientific papers, yet in 1989 Charles Bennett and four colleagues were so exuberant in their achievement of building a prototype quantum cryptography machine that they wrote a paper[Bennett89] entitled "The dawn of a new era for quantum cryptography: The experimental prototype is working!" David Deutsch has described this as the first device whose capabilities exceed those of a Turing Machine[Deutsch89].

The original scheme for quantum cryptography was similar to that previously described except that it used four types of polarizations rather than just two. The prototype was built at IBM in 1989. Figure 8.13 shows a schematic[Bennett92a]. The original machine could only send a secure key over a distance of 30 cm. However, technological progress has been so rapid that it is now possible to transmit qubits, securely, a distance of over 30 km[Marand95].

The source of photons is provided by a green light emitting diode (LED). The light from the LED passes through a pinhole, a lens, and an aperture to produce a collimated beam of photons. The light then passes through a color filter to narrow the frequency spread of the green light. The light coming out of the LED is unpolarized and the next step is to polarize the light with a linear polarizer. To generate the necessary rectilinear or diagonal polarization a pair of Pockels cells are used. The Pockels cell was invented in 1893 by German

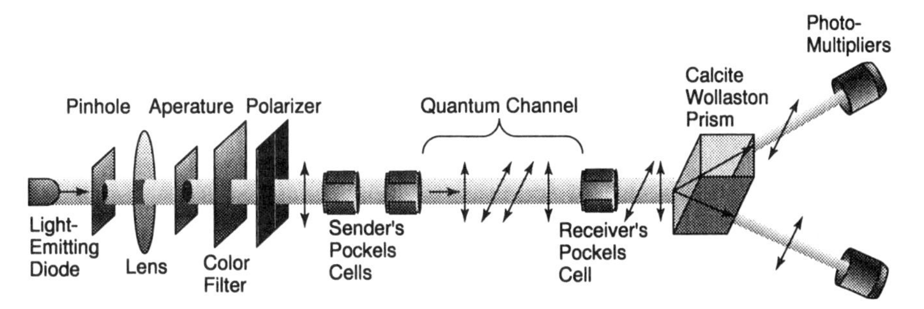

Fig. 8.13 Schematic of the first quantum cryptography prototype. The actual prototype is about a meter long.

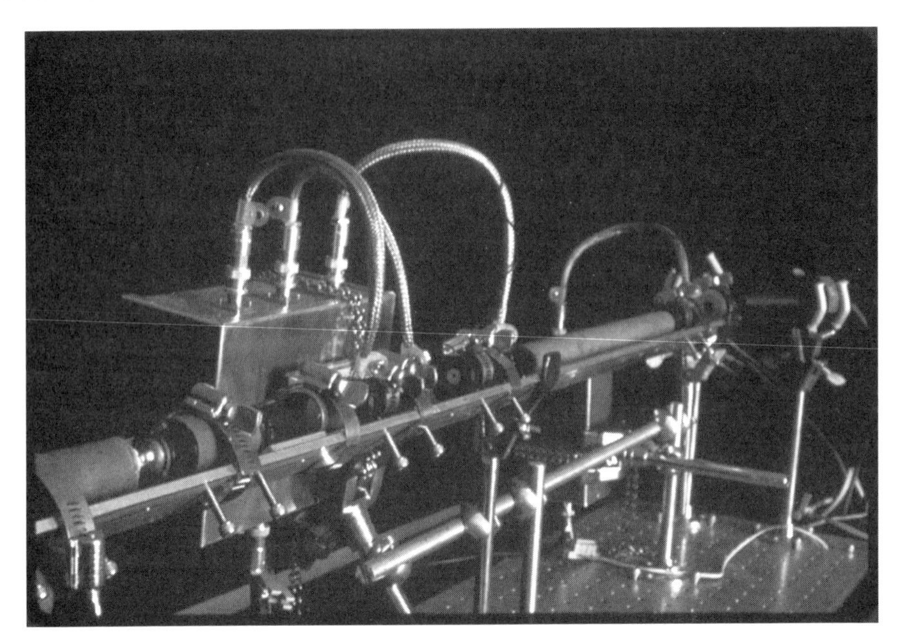

First quantum cryptography machine built by Charles Bennett, Gilles Brassard, and colleagues at the IBM T.J. Watson Research Center in 1989. (Photograph provided by Robert L. Prochnow.)

physicist Friedrich Pockels and it basically acts as a birefringent switch. The birefringence is induced by an externally applied electric field. The first Pockels cell can rotate a polarization by 45° if it is switched on. The second cell can rotate a polarization by 90° when it is switched on. Note that the order of the Pockels cells does not matter. Depending on which cell or cells are turned four possible polarization states can be generated, 0°, 45°, 90°, and 135°.

In an after-dinner speech at the 1994 Physics of Computation Conference, Gilles Brassard, who was one of the creators of the prototype, recalled whimsically that the original machine was not that secure after all, as the devices used to place photons in particular polarization states made noticeably different noises depending on the type of polarization selected! Fortunately, such technological quirks have not impeded progress in quantum cryptography.

8.10 Other Approaches to Quantum Cryptography

Since 1984, when the original quantum cryptography protocol was invented, there have been many proposals for alternative schemes for achieving secure key distribution. Although these all rely, in some manner, on uniquely quantum effects, they do not all use the polarization properties of photons.

Phase-Coding Protocols

For example, in 1992 a new quantum cryptographic scheme was devised that uses phase modulation (rather than polarization)[Ekert92]. A photon is an oscillating electromagnetic wave. The phase of a photon simply represents where this wave is in its oscillation. We can choose a particular phase to represent the binary value 0 and a different phase to represent the binary value 1. Thus the bits that will comprise a cryptographic key can be encoded in the phases of a sequence of photons.

In the phase-coding method of quantum cryptography, short wavelength light is split into two longer wavelength beams by passing it through a beamsplitter (see Fig. 8.14). As the two longer wavelength photons are created simultaneously, their phases are in step. One of these photons is sent to Alice and the other to Bob.

Alice and Bob each possess an interferometer consisting of a long arm (the upper path of each interferometer, consisting of three legs) and a short arm (the lower path of each interferometer, consisting of one leg). Each long arm contains a device that can, if activated, shift

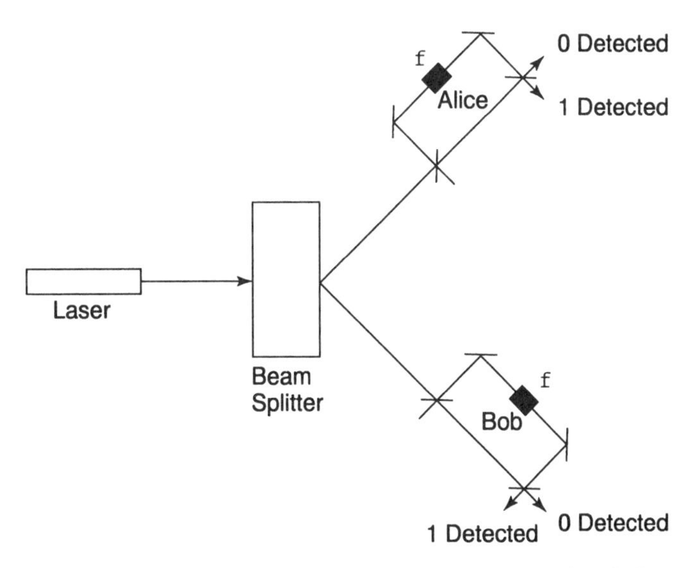

Fig. 8.14 Schematic showing quantum cryptography with correlated phases.

the phase of the photon passing through it. As Alice and Bob receive one member of the pair of phase-correlated photons they choose, randomly and independently, whether to activate their phase shifter. On those occasions when Alice and Bob both make the same decision (both "on" or both "off") they both observe identical results in their detectors; that is, they both observe a "0" or a "1." By comparing notes on whether they did or did not have their phase shifters on for each photon that passes through their respective interferometers, they can quickly establish a common subset of matching answers. They may then use these bits as the key in a cryptosystem such as a one-time pad. This phase-coding technique for secure key distribution has been demonstrated successfully over a distance of 10 km of optical fiber[Townsend93], [Townsend93a].

Entangled States and Rejected Data Protocols

In 1991 Arthur Ekert, of Oxford University, devised a radically different approach to quantum cryptography[Ekert91]. Ekert's scheme is based on a property of pairs of correlated quantum systems, known as entanglement. When a pair of particles are created simultaneously, such as when a spherically symmetric atom emits a pair of photons in opposite directions, they behave as a single quantum system. When, subsequently, the particles become separated, the influence of one member of the pair over the other persists. For example, although initially neither photon has a definite value for rectilinear polarization, as soon as the rectilinear polarization of one photon in the pair is measured, that of the second photon is determined instantaneously, regardless of the distance between the photons or the nature of the intervening medium. This effect, dubbed the EPR effect, can be used as the basis for establishing a secret shared key between two parties, Alice and Bob.

Rather than using polarized photons, Ekert's scheme is described in terms of angular momentum measurements on correlated particles. Here is how it works. We suppose that between Alice and Bob there sits a device that generates pairs of entangled particles. As each pair is generated, one particle in the pair is sent to Alice and the other to Bob. As the particles arrive, Alice measures their components of angular momentum at angles of 0¯, 45¯, or 90¯ in a plane perpendicular to the axis connecting her to Bob. Likewise Bob measures the components of angular momentum of his particles at angles of 45¯, 90¯, and 135¯. By counting the number of particles received, Alice and Bob can keep a tally of the answers they measure for the angular momenta of the *i*th particle pair they received measured at an angle $\theta_1 \in \{0^{\circ}, 45^{\circ}, 90^{\circ}\}$ for Alice and $\theta_2 \in \{45^{\circ}, 90^{\circ}, 135^{\circ}\}$ for Bob.

Individually, the sequence of measurements Alice and Bob obtain will each appear completely random. However, whenever Alice and Bob choose the same orientation for their detectors then, provided that no particles go astray and there is no eavesdropping, Alice's measurement will be perfectly anticorrelated with Bob's measurement.

Thus to exchange a key, Alice and Bob announce publicly which orientations they used for their detectors. For the cases in which they used the same orientation, they make a secret note of the result that they measured. For the cases in which they used different orientations, they announce publicly what result they obtained. Provided Alice and Bob picked the orientations of their detectors randomly and independently, the correlation between their results is exactly calculable and should come out as $-2\sqrt{2}$. A significant departure from this value would indicate the presence of an eavesdropper. However, if the correlation works out as expected, Alice and Bob can then trust the perfectly anticorrelated results they obtained for their secret bits. That is, if Alice measures a sequence of spins that encode the bits 010110101, then Bob will measure a sequence of spins that encode the bits 101001010. Bob simply has to flip these bits to determine Alice's secret bit sequence and hence to establish a shared key.

An experimental demonstration of this approach, which uses entangled pairs, was accomplished in 1994[Rarity94].

The use of entangled states is not limited to secure key distribution, however. In the next chapter, we describe quantum teleportation, a technique for transferring a state from one particle to another without that state having to move through the intervening space. This may sound like science fiction but it is squarely rooted in science fact.

CHAPTER 9

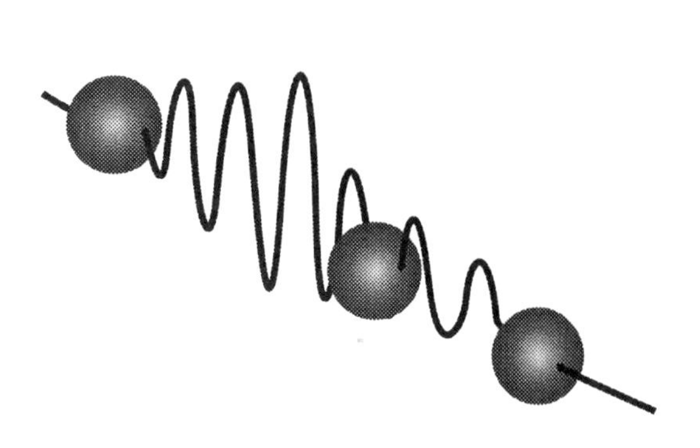

Quantum Teleportation

"Gentlemen, beam me aboard."
— Captain Kirk

9.1 What Is Teleportation?

Teleportation is commonly understood as a fictional method for transferring an object between two locations by a process of dissociation, information transmission, and reconstitution. The net effect is the destruction of the original object at the source and the creation of an exact replica at the intended destination. A key ingredient of teleportation is that the actual object does not traverse the intervening distance. Instead, the object is scanned to extract sufficient information to recreate the original, the information is transmitted, and an exact replica is re-assembled at the destination out of the material that is locally available, apparently without the need for any fancy machinery. Teleportation is therefore distinct from fax transmission, which leaves the original intact and only creates an approximate replica and (hypothetically perfect) cloning which would result in two identical versions of the original.

Although teleportation is a popular artifice in science fiction stories for beaming action heroes around the Universe and introducing paradoxes of identity into a storyline, until recently no serious attention had been paid to the physical principles on which true teleportation might be based. The presumption of most scientists, if they had any, was that teleportation was impossible because, as the Heisenberg Uncertainty Principle attests, it is impossible to measure *all* the attributes of a quantum state exactly, simultaneously[1]. For example, it is impossible to measure both the position and momentum of a particle exactly, simultaneously. Consequently, it appeared that the scanning step in teleportation was doomed to failure because it would never yield *complete* information about the original.

The situation changed in 1993 when a team of physicists and computer scientists pooled their talents to come up with the first scientifically plausible account of how to teleport an unknown quantum state. In a paper whose author list reads like a "Who's Who?" of quantum information theory, Charles Bennett, Gilles Brassard, Claude Crepeau, Richard Jozsa, Asher Peres, and William Wootters showed that it was possible to exploit yet another aspect of quantum theory, the notion of entangled states and nonlocal influences, to circumvent the limitations of the Heisenberg Uncertainty Principle and hence create an exact replica of an arbitrary quantum state, but only if the original state was destroyed in the process.

Entangled states can be created whenever quantum systems interact with one another. The net effect is often to produce a pair of particles that have highly correlated values for some attribute. For example, you could have an entangled state of a pair of spin $-\frac{1}{2}$ particles such that if one is measured to have a spin of $+\frac{1}{2}$, then the other must have a spin of $-\frac{1}{2}$. Likewise, you could have a pair of photons in an entangled state such that if one is measured to have a polarization of 0^{-}, then the other will have a polarization of 90^{-}.

In the context of quantum teleportation, the EPR entangled pair of particles serve as two ends of a quantum communication channel: one particle being retained by the person wishing to teleport the quantum state and the other by the person wishing to receive it. Thus, in order to teleport a quantum state, the sender and receiver must each already possess one member of a pair of entangled particles.

Bennett and his colleagues showed that the information needed to recreate the quantum state of a simple 2-state quantum system could be separated into a classical component and a quantum component, the classical part transmitted through a classical communication channel and the quantum part transmitted via (not through) a quantum communication channel, and then recombined at the in-

[1] Although you can, of course, know the state vector itself exactly if you know how the state it describes was constructed.

tended destination to reincarnate the original state. The basic idea is for the sender to make a measurement of the *joint* state of the particle whose state is to be teleported and one of the entangled particles to send the result of the measurement (a classical message) over a conventional communication channel (such as a radio) and for the receiver to use the information in this classical message to determine which of four possible operations to apply to his member of the pair of entangled particles. As we show in this chapter, this allows the receiver to place his member of the entangled pair of particles in a state that is an exact replica of the state that the sender wished to teleport.

Notice that this scheme teleports the "quantum state" of an object, not the object itself. This is slightly different from the usual science fiction view of teleporting an object. Consequently, we cannot use this scheme to teleport an electron in its entirety from one place to another, but we can teleport the *spin* orientation of one electron at a particular location to another electron at a different location. The net effect, however, is similar: a particle in a specific state at the source location has its state destroyed and reincarnated on another particle at the destination without the original particle traversing the intervening distance.

Why, you might ask, would you want to teleport a quantum state? In the context of quantum computing the answer lies in the use of quantum states to encode qubits (superpositions of conventional bits, i.e., 0s and 1s). If we can teleport a quantum state between two locations, and we use quantum states to encode qubits, then, in principle, we can teleport a qubit between two locations. So teleportation might provide an alternative way to shunt quantum information around inside a quantum computer or indeed between quantum computers. This might be especially useful if some qubit needs to be kept secret. Using quantum teleportation, a qubit could be passed around without ever being transmitted over an insecure (public) channel.

In this chapter we take a look at the physics underpinning quantum teleportation and quantum information theory in general. This takes us into the realm of some of the most astonishing scientific discoveries that have ever been made.

9.2 Physics Behind Teleportation

"That one body may act upon another at a distance through a vacuum without the mediation of anything else ... is to me so great an absurdity, that I believe no man, who has in philosophical matters a competent faculty for thinking, can ever fall into."
— Isaac Newton

Teleportation, as it is meant here, relies upon a quintessential quantum phenomenon known as the EPR effect, named after its discoverers Albert Einstein, Boris Podolsky, and Nathan Rosen [Einstein35]. The EPR effect describes the enduring interconnection between pairs of quantum systems that had, at some earlier time, interacted with one another. Surprisingly, many interpretations of quantum theory predict that such an interaction is instantaneous, unmediated, and immune to the nature of the intervening medium. Judging from the preceding quote, this prediction would not have sat well with Isaac Newton. It certainly did not sit well with Einstein, Podolsky, and Rosen who tried to use the apparent absurdity of the prediction to prove that quantum mechanics gave an incomplete description of physical reality. As quantum teleportation relies crucially on such an "action at a distance" effect, it is important to take a minor diversion to convince you that the effect is real and that reality is, in fact, nonlocal.

9.3 Local Versus Nonlocal Interactions

First we had better clarify the difference between local and nonlocal interactions. A local interaction is one that involves direct contact, or employs an intermediary that is in direct contact. The forces with which we are familiar in everyday life, such as friction and gravity, are *local interactions*. In the case of friction, the physical contact between two bodies is really mediated by an electromagnetic field, which in turn comes about by the action of an intermediary, the carrier of the electromagnetic force, called the photon. Photons travel at the speed of light, which although fast is still finite. Consequently, electromagnetic influences cannot propagate faster than the speed of light in a vacuum. Moreover, electromagnetic forces tend to weaken the farther you go from the source.

Locality does not necessarily imply "nearby," however. Gravity, for example, is a force that exerts its influence over astronomically large distances. Nevertheless, gravity is still regarded as a local interaction because it is mediated by particles, gravitons, which travel between gravitating objects. It too drops off in strength as the distance between the gravitating objects increases and cannot travel faster than the speed of light.

An important corollary of local interactions is the following. If two events occur in regions of spacetime such that no signal, not even one traveling at the speed of light, could ever reach one region from the other, these two events ought to be completely independent of one another. Why? Because if no signal could ever travel from one region to the other, how could what happens in one region ever be communicated to the other? In fact, special relativity has a special name for two such regions: it says that they are "spacelike separated."

In short, local interactions can be characterized by three criteria: they are mediated by another entity, such as a particle or field; they propagate no faster than the speed of light; and their strength drops off with distance. And locality predicts that events in spacelike separated regions ought to be independent of one another.

Scientists have shown that all the known forces in the Universe, the electromagnetic, the gravitational, the strong nuclear,[2] and the weak nuclear forces are all *local*, in this sense. One might think, therefore, that that is an end to it, and that reality must be local. After all, if *all* the known forces are local, what is left to be nonlocal?

Well, what is left is the "collapse of the state vector." State vectors, you will recall, provide the mathematical description of quantum systems. When we make measurements, the state vectors collapse into eigenstates. At least this is the account of measurement according to one interpretation of quantum theory. Now the intriguing point is that there is nothing in quantum theory that explains, mediates, or determines the exact mechanism of the collapse. In particular, the collapse of a state vector involves no *forces* of any kind. This provides quantum theory with an "out"; a devious way to evade the censorship of locality.

What exactly would a nonlocal influence be? We can just flip each criterion for a local interaction to say that a nonlocal interaction is *not* mediated by anything, *not* limited to acting at the speed of light, and does *not* drop off in strength with distance.

Thus nonlocal interactions would appear to be magic! Many scientists have an instinctive distaste for nonlocal interactions. Cer-

---

[2] Actually the strong force is weak at *very* small distances and becomes much stronger until about the size of a nucleon (10^{-13}cm). For distances of separation greater than that of a nucleon the strong force exerts no influence. The strong force is carried by the gluon.

tainly, they would seem to be in direct conflict with Einstein's Theory of Special Relativity which says that nothing can travel faster than the speed of light. Indeed, it was the discrepancy between the predictions of relativity and quantum theory concerning the correlations between events in spacelike separated regions that led Einstein, Podolsky, and Rosen to the EPR effect concerning the behavior of entangled quantum systems.

Remarkably, the scientific evidence that arose in the mid-1980s proved, empirically, that reality is nonlocal. It *is* this remarkable nonlocal type of interaction that is exploited in quantum teleportation.

9.4 Entanglement

The issue of nonlocal influences enters into quantum theory by way of the outcomes of measurements on pairs of entangled states whose component particles are spacelike separated.

Entangled states arise, naturally, as a result of interactions between particles, such as when a pair of particles are created simultaneously under the requirement that some attribute, such as spin or polarization, be conserved. According to quantum theory, although the exact state of each particle is not determined until a measurement is made, their states must, nonetheless, be correlated. Thus whenever a measurement is made on one member of the entangled pair, the states of both particles become definite, usually opposite (in some sense), and the entanglement ceases.

Regardless of the exact mechanism for generating entangled particles, if we write the state of particle 1 as $|\psi_1\rangle = \frac{1}{\sqrt{2}}(|0\rangle_1 + |1\rangle_1)$, and the state of particle 2 as $|\psi_2\rangle = \frac{1}{\sqrt{2}}(|0\rangle_2 + |1\rangle_2)$, the joint state of the entangled pair is often found to be

$$|\Psi_{12}\rangle = \frac{1}{\sqrt{2}}(|0\rangle_1 \otimes |1\rangle_2 - |1\rangle_1 \otimes |0\rangle_2).$$

Notice that this is *not* simply the direct product of the individual states of the component particles. Instead, it represents a true entanglement.

States can also become entangled during the natural unitary evolution of a composite quantum system. Consider two separate quantum systems that are initially in states $|\Psi\rangle = \frac{1}{\sqrt{2}}(|\psi_a\rangle + |\psi_b\rangle)$ (a superposition) and $|\phi_0\rangle$, respectively. As the two systems are initially independent, with no interaction between them, we can write their joint state as a direct product:

$$|\Phi_0\rangle = |\Psi\rangle \otimes |\phi_0\rangle = \tfrac{1}{\sqrt{2}}\left(|\psi_a\rangle + |\psi_b\rangle\right) \otimes |\phi_0\rangle.$$

Note how the mathematical description of the 2-state system can be factored, or separated, at this initial stage as the direct product of the state of particle 1 with that of particle 2. Now suppose there exists a unitary operator that acts to advance the total state vector in time in the following way,

$$|\Phi(t)\rangle = \hat{U}(t) \cdot |\Phi_0\rangle,$$

$$\hat{U} \cdot |\psi_a\rangle \otimes |\phi_0\rangle = |\psi_a\rangle \otimes |\phi_a\rangle,$$

$$\hat{U} \cdot |\psi_b\rangle \otimes |\phi_0\rangle = |\psi_b\rangle \otimes |\phi_b\rangle$$

Substituting these last two equations into the equation for $|\Phi(t)\rangle$ we have:

$$\begin{aligned} |\Phi(t)\rangle &= \tfrac{1}{\sqrt{2}} \hat{U} \cdot \left(|\psi_a\rangle \otimes |\phi_0\rangle + |\psi_b\rangle \otimes |\phi_0\rangle\right) \\ &= \tfrac{1}{\sqrt{2}}\left(|\psi_a\rangle \otimes |\phi_a\rangle + |\psi_b\rangle \otimes |\phi_b\rangle\right) \end{aligned}$$

Notice what has happened to the structure of the state vector of the joint system. $|\Phi(t)\rangle$ can no longer be written, simply, as the direct product of states of each subsystem. The only way to untangle them would be to apply the inverse unitary operator.

This unfactorizable joint state is now an *entangled state*. If the states of the entangled particles are used to encode bits, then the entangled joint state represents what is called an *ebit*. Unlike a classical bit or qubit, an ebit is, intrinsically, a *shared* resource [Bennett95]. It is always distributed between two particles. The states of these particles are correlated, but undetermined until measured. Thus an ebit provides the basis for a restricted kind of quantum communication channel. It is a communication channel in the sense that once either particle comprising the ebit is measured, the states of both particles become definite (and, in the context of quantum computing, often represent opposite bits). However, the channel is restricted in the sense that you cannot send an intentional message (i.e., a deliberate sequence of 0s and 1s) between two parties simply by having each party measure his or her members of a set of shared ebits, because, individually, the sequence of results (the decoded bits) would appear random.

Nevertheless, Bennett and his colleagues found a way to use the quantum channel afforded by an ebit in conjunction with a classical communication channel to send a deliberate message, albeit a rather simple one, consisting of a single qubit.

9.5 Spooky Action at a Distance

"The most profound discovery of science."
— Henry Stapp

So far we have talked about locality and entanglement separately. Next we explore the implications of combining these concepts. We are hardly the first to do so. Back in 1935, Albert Einstein, Boris Podolsky, and Nathan Rosen recognized that the combination of locality and entanglement implied a contradiction between quantum theory and Einstein's own conception of reality, embodied in his special theory of relativity. The EPR argument has been reformulated many times and we adopt a version that is correct in spirit but different in detail from the original.

Here is the problem. Suppose that a quantum system, such as an atom, emits a pair of photons whose polarization states are entangled. According to quantum theory, neither photon has a definite value for its polarization until its polarization is actually measured. Nevertheless, due to the entanglement, the states must be correlated so that if one photon is found, upon being measured, to be polarized at an angle of 0¯, the other, if measured, would then be found, with absolute certainty, to be polarized at an angle of 90¯. The critical question is, *when* are the polarization states actually determined?

Quantum theory predicts that the polarization of the second photon is determined at the instant, note the *instant*, that the polarization of the first photon is measured. Quantum theory does not provide any account of how the polarization of the photons is determined or communicated. In particular, none of the four known forces in the Universe are involved. In fact, quantum theory predicts that the interaction is unmediated, instantaneous, and unaffected by the nature of the intervening medium. This is the hallmark of a nonlocal influence.

Einstein, Podolsky, and Rosen were uncomfortable with the implication of nonlocal influences, so they set about finding a hole in quantum theory. They asked what would happen if you measured the polarizations of both photons simultaneously, or at least within such a short time interval that no signal, even one traveling at the speed of light, could possibly travel from one photon to the other.

In Einstein's view of reality, no influence can propagate faster than the speed of light. Hence Einstein, Podolsky, and Rosen argued that the events of measuring each photon would be in spacelike separated regions and hence ought to be independent of one another. They believed that the correlations in the measured polarizations of the photons hinted at the existence of "hidden variables" that determined the polarization states of the photons from the outset. In this view, it is our ignorance of these hidden variables that prevents us from knowing the photon polarizations from the outset, not some mythical, nonlocal, "action at a distance." Moreover, as quantum theory makes no mention of these hidden variables, EPR contended that quantum theory must necessarily be incomplete; that is, that some key ingredient was missing.

9.6 Bell's Theorem

Now here comes the twist. It could be argued that it is simply a matter of philosophical *taste* as to whether you believe the quantum account or the hidden variable account of how the two photons come to have correlated polarization states upon being measured. But what if there were some experimentally testable difference between the predictions of the two theories? Could a physical test shed light on a philosophical question?

In the 1960s John Bell showed that there was an empirically testable difference between the predictions of any hidden variable theory and the predictions of quantum mechanics. And, when the tests were actually conducted, they showed that quantum theory was correct and the hidden variable theories were wrong. This provided experimental evidence that reality is nonlocal — a rather remarkable conclusion.

The theoretical basis underlying this landmark conclusion is a deceptively simple inequality known as Bell's Theorem.

In order to fully explain teleportation we need to understand Bell's Theorem or Bell's Inequality, named after physicist John Bell[Bell64]. Bell's Inequality and its subsequent experimental verification showed decisively that reality is nonlocal. Figure 9.1 shows the basic setup we use in our derivation. Alice and Bob each have a polarizer oriented at angles θ_1 and θ_2, respectively. Thus the angle between their polarizers is $\theta_{12} = \theta_2 - \theta_1$. There are four possible states for outcomes of this experiment depending on whether the photon is measured in the vertical (v_1 or v_2) or horizontal (h_1 or h_2)

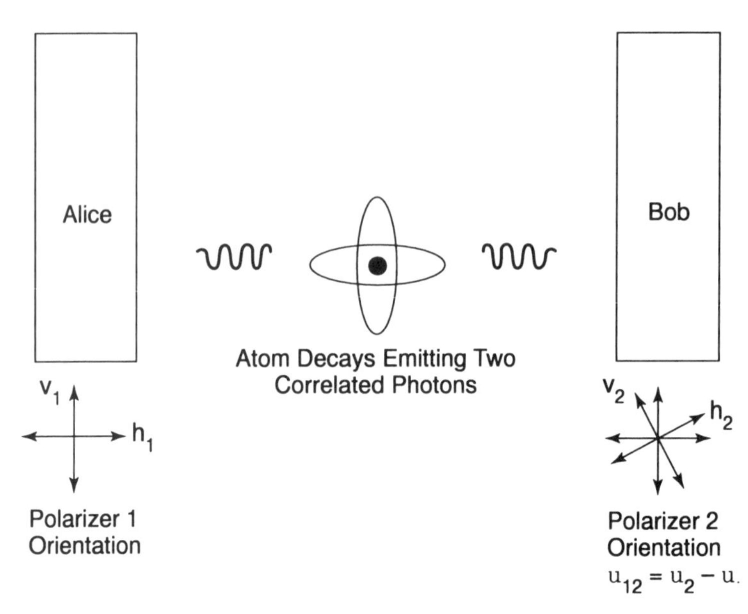

Fig. 9.1 The key point of the experimental setup is that the measurement by one of the polarizers affects the outcome of the other depending on the relative orientation of the two polarizers. What exactly this effect will be is the gist of the Bell's Inequality.

direction in the respective polarizer (1 or 2). The outcomes form a basis and are written as

$$\left|\psi_{v_1v_2}\right\rangle = \left|\psi_{v_1}\right\rangle \otimes \left|\psi_{v_2}\right\rangle$$

$$\left|\psi_{v_1h_2}\right\rangle = \left|\psi_{v_1}\right\rangle \otimes \left|\psi_{h_2}\right\rangle$$

$$\left|\psi_{h_1v_2}\right\rangle = \left|\psi_{h_1}\right\rangle \otimes \left|\psi_{v_2}\right\rangle$$

$$\left|\psi_{h_1h_2}\right\rangle = \left|\psi_{h_1}\right\rangle \otimes \left|\psi_{h_2}\right\rangle$$

We can use heuristic arguments from classical optics to find out the coefficients for the total wavefunction for the rotated system 2 in terms of system 1. Recall from the previous chapter that the strength of the electric field along a particular polarization axis is given by the cosine of the angle between the electric field and the polarizer axis. In terms of ket vectors,

$$\left|h_2\right\rangle = \cos\theta_{12}\left|h_1\right\rangle + \sin\theta_{12}\left|v_1\right\rangle$$

so that the projection of h_2 along the axis of $\left|h_1\right\rangle$ is

$$\left\langle h_1 \middle| h_2\right\rangle = \cos\theta_{12}\left\langle h_1 \middle| h_1\right\rangle + \sin\theta_{12}\left\langle h_1 \middle| v_1\right\rangle = \cos\theta_{12},$$

where the first bra-ket is the unity matrix and the second term drops out by the orthogonality of h_1 and v_1; that is, $\left\langle h_1 \middle| h_1\right\rangle = 1$ and $\left\langle h_1 \middle| v_1\right\rangle = 0$.

In classical optics we measure the intensity of light which is the sum of the amplitude squared of the electric field vectors of many photons interacting with our measuring device which will vary depending on the degree of correlation between the photons and the polarizers as given by

$$\text{Intensity} = \sum_i \left| \vec{E}^i_{\text{photon}} \cdot \hat{n}_{\text{polarizer}} \right|^2 .$$

The intensity is a measure of the probability that a photon will interact in a certain way and thus in this experiment we can tie classical photon intensity to quantum probability. This can be seen using the probability amplitude of the ket vectors,

$$\left| \langle \psi_{h_1} | \psi_{h_2} \rangle \right|^2 = \left| \langle h_1 | h_2 \rangle \right|^2 = \cos^2 \theta_{12} .$$

It follows then that the total wavefunction is given in terms of these bases as in

$$|\Psi\rangle = |\psi_{v_1v_2}\rangle \otimes \langle \psi_{v_1v_2} | \Psi \rangle + |\psi_{v_1h_2}\rangle \otimes \langle \psi_{v_1h_2} | \Psi \rangle + |\psi_{h_1v_2}\rangle \otimes \langle \psi_{h_1v_2} | \Psi \rangle + |\psi_{h_1h_2}\rangle \otimes \langle \psi_{h_1h_2} | \Psi \rangle$$

$$|\Psi\rangle = \frac{1}{\sqrt{2}} \left[|\psi_{v_1v_2}\rangle \cos\theta_{12} + |\psi_{v_1h_2}\rangle \sin\theta_{12} + |\psi_{h_1v_2}\rangle \sin\theta_{12} + |\psi_{h_1h_2}\rangle \cos\theta_{12} \right]$$

where the coefficients on $|\Psi\rangle$ are the projection operators onto the particular basis axis. The probability for a measurement for any of the outcomes of the total wavefunction is

$$P_{v_1v_2} = \left| \langle \psi_{v_1v_2} | \Psi \rangle \right|^2 = \tfrac{1}{2} \cos^2 \theta_{12}$$

$$P_{v_1h_2} = \left| \langle \psi_{v_1h_2} | \Psi \rangle \right|^2 = \tfrac{1}{2} \sin^2 \theta_{12}$$

$$P_{h_1v_2} = \left| \langle \psi_{h_1v_2} | \Psi \rangle \right|^2 = \tfrac{1}{2} \sin^2 \theta_{12}$$

$$P_{h_1h_2} = \left| \langle \psi_{h_1h_2} | \Psi \rangle \right|^2 = \tfrac{1}{2} \cos^2 \theta_{12}$$

where P_{xy} is the probability of detecting a photon along the x and y axes of the two detectors. Note that the probabilities for all the measurements add up to 1 as they should. Each of the probabilities represents the chance that a photon can be detected if the polarizers are set as in Fig. 9.1.

Now we need one more piece to put together our derivation of Bell's Inequality. This comes from a simple probability argument. Suppose we add a third measuring polarizer with axes in v_3 and h_3 and at an angle θ_3 from the same common reference as the first two polarizers. We can write down the following relationships which are from straightforward probability arguments,

$$P_{v_1h_2} = P_{v_1h_2v_3} + P_{v_1h_2h_3},$$

where the right-hand side takes into account both possible measurements of the third measurement and so the two terms together account for all possibilities on the left-hand side. Similarly for other combinations,

$$P_{v_2h_3} = P_{v_1v_2h_3} + P_{h_1v_2h_3},$$

and

$$P_{v_1h_3} = P_{v_1v_2h_3} + P_{v_1h_2h_3}.$$

Again the probabilities refer to the probability that a photon can be detected in the third polarizer given its relative orientation to the first two polarizers. From these relations it follows that

$$P_{v_1h_2} \geq P_{v_1h_2h_3}$$

and

$$P_{v_2h_3} \geq P_{v_1v_2h_3},$$

from which it follows

$$P_{v_1h_2} + P_{v_2h_3} \geq P_{v_1h_2h_3} + P_{v_1v_2h_3}$$

or more simply

$$P_{v_1h_2} + P_{v_2h_3} \geq P_{v_1h_3}$$

which is Bell's Inequality. Note the similarity to a "triangle inequality" where the two "sides" are longer than the longest "side." Also in analogy to this triangle inequality we should note that it applies to Euclidean space and that Bell's Inequality applies only to worlds with local interactions. The derivation we have given for Bell's Inequality is just one instance and there are more general versions as well[Clauser69]. Bell's Inequality is a statement about the correlations between probabilities. Figure 9.3 shows that this inequality is violated, which implies that there are no local hidden variable theories for quantum mechanics — reality really is spooky!

Another way to put it is to say that our notion of probability is not correct. The key difference between the classical beliefs and the quantum reality is that in the latter we must take into account the

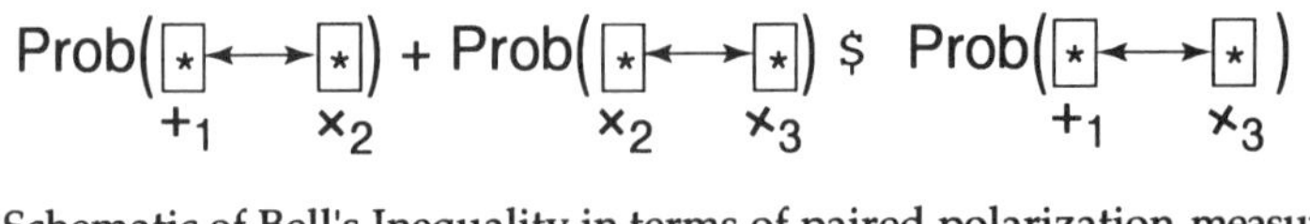

Prob([*]←→[*]) + Prob([*]←→[*]) $ Prob([*]←→[*])

+1 ×2 ×2 ×3 +1 ×3

Fig. 9.2 Schematic of Bell's Inequality in terms of paired polarization measurements. The * in the polarizer box and refers to the detection of a photon. Experiments show that there are cases where the inequality is violated which proves that reality is nonlocal.

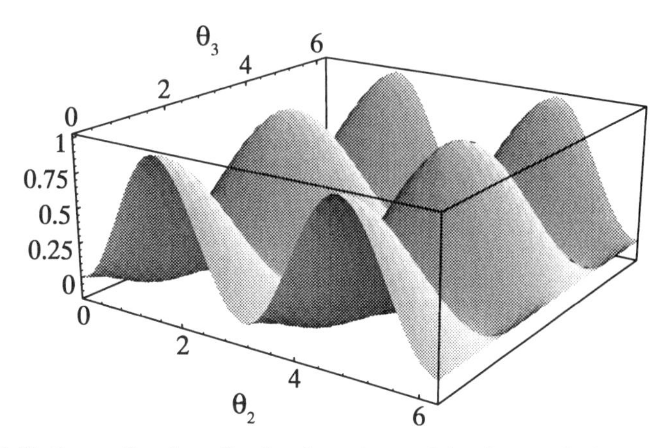

Fig. 9.3 Bell's Inequality for a fixed value of one of the three polarizer angles $\theta_1 = 0$. The horizontal axes are the angles between the polarizers θ_2 and θ_3 in radians and the vertical axis is the function $\sin^2(\theta_2 - \theta_1) + \sin^2(\theta_3 - \theta_2) - \sin^2(\theta_3 - \theta_1)$.

entire experimental apparatus when determining what will eventually happen. Just because two particles become physically separated does not mean that their states cease to be correlated. The EPR correlation is a latent interaction that comes into play when measurements are made. Either we abandon our notion of locality or we abandon our notion of probabilistic correlations.

In terms of the angles of the polarizers, Bell's Inequality can be stated as

$$\frac{1}{2}\sin^2(\theta_2 - \theta_1) + \frac{1}{2}\sin^2(\theta_3 - \theta_2) \geq \frac{1}{2}\sin^2(\theta_3 - \theta_1).$$

If reality is local, this inequality should always be true regardless of the angles at which the polarizers are set. However, if we fix $\theta_1 = 0$ and plot the difference between the left and right-hand sides of Bell's Inequality, we obtain the surface shown in Fig 9.3.

If reality is local, this surface should be *always* greater than or equal to 0. If any value is less than 0, then it shows that physics is not local. We can see that some portions of this surface are indeed below zero by rotating the preceding plot and adding in a plane through the origin, as shown in Fig. 9.4. The parts of the surface that protrude through the horizontal plane indicate that there are angles for the polarizers for which Bell's inequality is violated. Hence local hidden variable theories must be wrong and reality is shown to permit nonlocal influences!

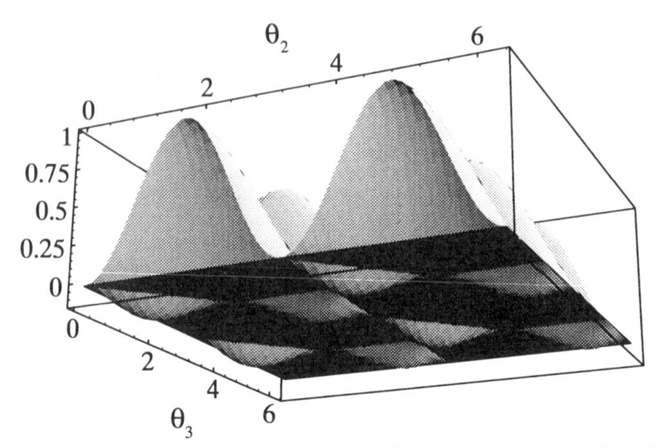

Fig. 9.4 Rotation of Fig. 9.3 showing that Bell's Inequality is violated for certain settings of the polarizers.

Bell's Inequality is much more than a mere theoretical result. It has been demonstrated experimentally using the detection of photons after they had traveled different paths[Aspect82]. The basic idea was to change the path of one of the entangled photons while they were in flight so that they could not "know" which possible path the other had taken. Any "communication" between the photons regarding the setup of the experiment was restricted to the moment of the measurement.

In the next section we discuss a simple form of teleportation, that of teleporting a simple quantum state such as the spin of a particle [Bennett93]. Unlike the science fiction accounts of teleportation, there is nothing speculative about the physical basis for quantum teleportation that Bennett and his colleagues have proposed.

9.7 How to Teleport One Qubit

Next we show how the EPR effect can be used in conjunction with a classical message to teleport a single qubit between two parties. As in the quantum cryptography examples, let Alice be the sender and let Bob be the receiver. It is worth noting that if Alice can teleport a single qubit to Bob, then she can, in fact, teleport arbitrary messages to Bob by decomposing them into sequences of qubits and teleporting each qubit separately. So although the ability to teleport a single qubit may seem inconsequential, it actually provides the foundation for a rich new communication technology.

As a qubit is, in general, a superposition of a 0 and a 1, Alice may not be able to make a measurement on the particle encoding the qubit without disrupting its state and hence corrupting the qubit. So it is best to assume that Alice does not know the details of the qubit (i.e., the state) that she wishes to teleport to Bob. Fortunately, this does not matter. Alice can teleport a known qubit or an unknown qubit just as easily. Her goal is simply to leave Bob in possession of a particle whose state is an exact replica of the state of the particle with which Alice begins.

To make the argument more mathematical, let us call the particle that encodes the qubit Alice wishes to teleport to Bob "particle 1." Physically, a qubit is encoded in the state of some simple 2-state quantum system, such as the spin of an electron or the direction of polarization of a photon. The state of particle 1 can therefore be written as

$$|\phi\rangle = a|\uparrow_1\rangle + b|\downarrow_1\rangle = \begin{pmatrix} a \\ b \end{pmatrix},$$

where $|a|^2 + |b|^2 = 1$. If we were using the spin state of an electron to encode the qubit, the up arrow would represent the spin-up state and the down arrow would represent the spin-down state. The subscripts "1" on the arrows just indicate that we are talking about particle 1.

To teleport the qubit we need two other particles, which we will call particle 2 and particle 3. Particles 2 and 3 must be entangled. Mathematically, this means that their joint state is not factorizable as the direct product of two simpler states. Physically, the entanglement comes about from the two particles being produced jointly during some quantum process that requires some attribute to be conserved. The state for these two entangled particles is written as

$$\tfrac{1}{\sqrt{2}}\left(|\uparrow_2\rangle \otimes |\downarrow_3\rangle - |\downarrow_2\rangle \otimes |\uparrow_3\rangle\right),$$

where, again, the subscripts simply stipulate about which particles we are speaking.

At this point particle 1 is still not correlated with particles 2 and 3, either classically or quantum mechanically, so we can still write the combined wave function for the 3-particle system as a direct product state:

$$|\Psi\rangle = |\phi\rangle \otimes \tfrac{1}{\sqrt{2}}\left(|\uparrow_2\rangle \otimes |\downarrow_3\rangle - |\downarrow_2\rangle \otimes |\uparrow_3\rangle\right),$$

or, equivalently,

$$|\Psi\rangle = \tfrac{a}{\sqrt{2}}\left(|\uparrow_1\rangle \otimes |\uparrow_2\rangle \otimes |\downarrow_3\rangle - |\uparrow_1\rangle \otimes |\downarrow_2\rangle \otimes |\uparrow_3\rangle\right) +$$
$$\tfrac{b}{\sqrt{2}}\left(|\downarrow_1\rangle \otimes |\uparrow_2\rangle \otimes |\downarrow_3\rangle - |\downarrow_1\rangle \otimes |\downarrow_2\rangle \otimes |\uparrow_3\rangle\right)$$

Alice keeps particle 2, which is one member of the pair of entangled particles, and sends the other member of the entangled pair, particle 3, to Bob; see Fig. 9.5. To accomplish teleportation Alice must make a measurement that couples the state of particle 1 with the state of particle 2. Such a measurement might be a total spin measurement of the two-particle (1 and 2) system—for instance, by sending them through a magnetic field and measuring the deflection of particles 1 and 2. A side effect of such a measurement is the selection of a definite state for particle 3.

Since this measurement is with respect to the combined particle 1-2 system it is convenient to represent the total system with respect to the basis of the two-particle system. The complete orthonormal basis vectors are given by the so-called Bell operator basis[Braunstein92]

$$\left|\Psi^A\right\rangle = \tfrac{1}{\sqrt{2}}\left(\left|\uparrow_1\right\rangle\left|\downarrow_2\right\rangle - \left|\downarrow_1\right\rangle\left|\uparrow_2\right\rangle\right),$$
$$\left|\Psi^B\right\rangle = \tfrac{1}{\sqrt{2}}\left(\left|\uparrow_1\right\rangle\left|\downarrow_2\right\rangle + \left|\downarrow_1\right\rangle\left|\uparrow_2\right\rangle\right),$$
$$\left|\Psi^C\right\rangle = \tfrac{1}{\sqrt{2}}\left(\left|\uparrow_1\right\rangle\left|\uparrow_2\right\rangle - \left|\downarrow_1\right\rangle\left|\downarrow_2\right\rangle\right),$$
$$\left|\Psi^D\right\rangle = \tfrac{1}{\sqrt{2}}\left(\left|\uparrow_1\right\rangle\left|\uparrow_2\right\rangle + \left|\downarrow_1\right\rangle\left|\downarrow_2\right\rangle\right).$$

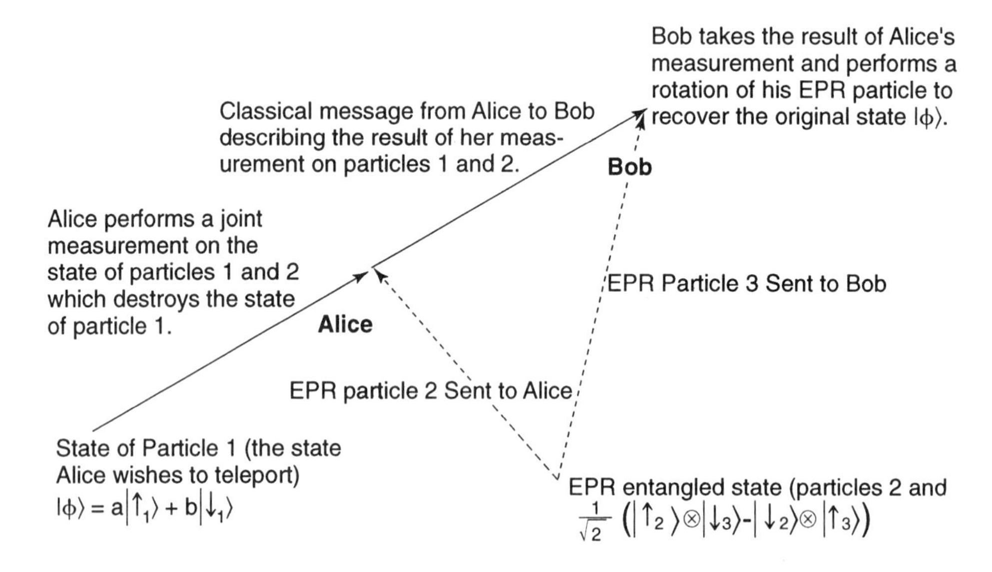

Fig. 9.5 Schematic view of quantum teleportation using EPR.

We can then rewrite the total wave function for all three particles involved in the teleportation in terms of the basis vectors,

$$|\Psi\rangle = \frac{1}{2}\left[\begin{array}{c}|\Psi^A\rangle\left(-a|\uparrow_3\rangle - b|\downarrow_3\rangle\right) + |\Psi^B\rangle\left(-a|\uparrow_3\rangle + b|\downarrow_3\rangle\right) + \\ |\Psi^C\rangle\left(a|\downarrow_3\rangle + b|\uparrow_3\rangle\right) + |\Psi^D\rangle\left(a|\downarrow_3\rangle - b|\uparrow_3\rangle\right)\end{array}\right]$$

or, remembering that ket vectors can be just a shorthand notation for column vectors,

$$|\Psi\rangle = \frac{1}{2}\left[|\Psi^A\rangle\begin{pmatrix}-a\\-b\end{pmatrix}_3 + |\Psi^B\rangle\begin{pmatrix}-a\\b\end{pmatrix}_3 + |\Psi^C\rangle\begin{pmatrix}b\\a\end{pmatrix}_3 + |\Psi^D\rangle\begin{pmatrix}-b\\a\end{pmatrix}_3\right].$$

From this equation it is easy to see the eigenvectors of Alice's measurements. After Alice makes her measurement on particles 1 and 2, they will be in a joint state described by one of the eigenvectors $|\Psi^A\rangle, |\Psi^B\rangle, |\Psi^C\rangle$, or $|\Psi^D\rangle$ and particle 3 will be in one of the states described by $\begin{pmatrix}-a\\-b\end{pmatrix}_3, \begin{pmatrix}-a\\b\end{pmatrix}_3, \begin{pmatrix}b\\a\end{pmatrix}_3$, or $\begin{pmatrix}-b\\a\end{pmatrix}_3$, respectively. Each pair of outcomes has a 1/4 chance of being the result of Alice's measurement. Note that the first term, $|\Psi^A\rangle\begin{pmatrix}-a\\-b\end{pmatrix}_3$ would leave particle 3 in exactly the initial state that Alice had originally (i.e., $|\phi\rangle = \begin{pmatrix}a\\b\end{pmatrix}_1$) except for a change of sign which is due to an unimportant phase factor. The other possible outcomes can recover Alice's initial state by the use of an appropriate rotation. If Alice informs Bob which of the four possible measurements *she* observed when she measured particles 1 and 2 (i.e., one of $|\Psi^A\rangle, |\Psi^B\rangle, |\Psi^C\rangle, |\Psi^D\rangle$), then Bob can perform the appropriate rotation to recover the initial state and the teleportation operation will be complete. The rotations needed by Bob to complete the teleportation process are shown in Table 9.1.

Table 9.1 For each possible outcome that Alice can obtain for the joint state of particles 1 and 2 there is a different rotation (unitary evolution) that Bob must perform on his particle 3 in order to recover the original state of particle 1. This final rotation completes the teleportation of the state of particle 1 (in Alice's possession) to the state of particle 3 (in Bob's possession).

Alice's result for the state of particles 1 and 2.	The rotation Bob must perform on particle 3.
$\lvert\Psi^A\rangle$	$\begin{pmatrix}1 & 0\\0 & 1\end{pmatrix}$
$\lvert\Psi^B\rangle$	$\begin{pmatrix}-1 & 0\\0 & 1\end{pmatrix}$
$\lvert\Psi^C\rangle$	$\begin{pmatrix}0 & 1\\1 & 0\end{pmatrix}$
$\lvert\Psi^D\rangle$	$\begin{pmatrix}0 & -1\\1 & 0\end{pmatrix}$

9.8 Teleportation Circuit for a Quantum Computer

How might teleportation be accomplished inside a quantum computer? At the very least, there would have to be a quantum circuit that is capable of the performing the four basic stages of teleportation. First, a quantum channel must be established between the source location and the destination location. Second, the unknown state to be teleported must become entangled with one end of this quantum channel. After that, the results of a measurement made on the unknown state and the state at the source end of the quantum communication channel must be communicated (via a classical message) to the destination location. Finally, this classical message must be combined with the state at the destination end of the quantum communication channel in such a way as to re-incarnate the orginal (unknown) state at the destination location.

Now this sounds like a very difficult problem. Certainly, one could imagine that it will take a very elaborate apparatus to accomplish this feat! It turns out, however, that this is not the case. Canadian computer scientist Gilles Brassard has devised a hypothetical quantum circuit that can accomplish all of the necessary tasks using a handful of quantum logic gates that act, at most, on pairs of qubits at a time[Brassard96].

Fig. 9.6 shows a schematic of Brassard's teleportation circuit. The circuit can be thought of as consisting as two parts, one mimicking the actions of Alice (the person teleporting the state) and the other mimicking the actions of Bob (the person re-assembling the state from the classical and quantum messages generated by Alice).

Let's take a look at the teleportation circuit in detail. Notice that there are three inputs and three outputs. The gates labelled L and R perform rotations on single qubits. The gates labelled S and T perform phase shifts on single qubits. The remaining type of gate, drawn as a large circle at one end and a black dot at the other, represents an XOR gate, which is sometimes called a "controlled-NOT" gate. Each XOR gate acts on a pair of qubits.

In the language of unitary matrices, the rotation operators are:

$$L = \frac{1}{\sqrt{2}}\begin{pmatrix} 1 & -1 \\ 1 & 1 \end{pmatrix},$$

which implies $L|0\rangle = \frac{1}{\sqrt{2}}(|0\rangle + |1\rangle)$ and $L|1\rangle = \frac{1}{\sqrt{2}}(-|0\rangle + |1\rangle)$ and

$$R = \frac{1}{\sqrt{2}}\begin{pmatrix} 1 & 1 \\ -1 & 1 \end{pmatrix},$$

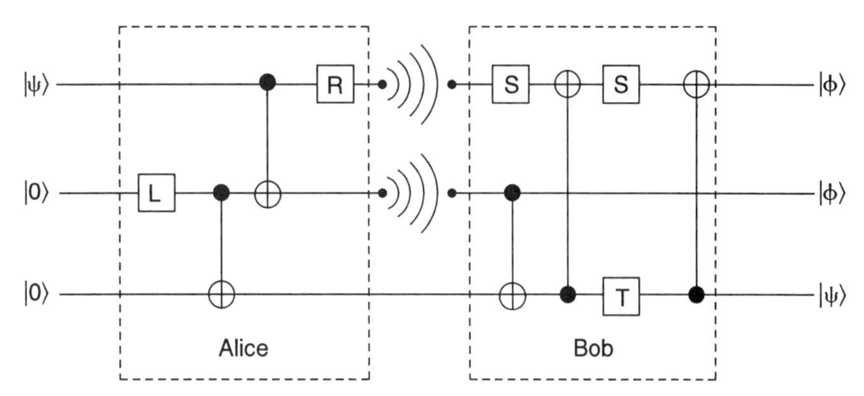

Fig. 9.6 A quantum circuit for teleporting an unknown state between two parts of a quantum computer. Time advances from left to right. Hence, operations are performed from left to right too. The operations that occur at the same time (vertically aligned) commute so they may be performed in either order. When all the operations have been completed the original quantum state on the top line of the input will have been transferred to the bottom line of the output.

which implies $R|0\rangle = \frac{1}{\sqrt{2}}(|0\rangle - |1\rangle)$ and $R|1\rangle = \frac{1}{\sqrt{2}}(|0\rangle + |1\rangle)$. Notice that L and R both create superpositions from eigenstates. If a qubit is represented, physically, in the polarization state of a photon, these rotations may be achieved through the use of wave-shifting plates in a birefringent crystal. If the qubits are really spins, the rotations can be accomplished by the use of a pulsed magnetic field. Alternatively, if the qubits are really atomic energy states, the rotations can be achieved by the use of laser pulses. So it does not matter what embodiment you pick for qubits as the same computational operations can be accomplished one way or another.

Similarly, the phase shift operators are defined by:

$$S = \begin{pmatrix} i & 0 \\ 0 & 1 \end{pmatrix},$$

which implies $S|0\rangle = i|0\rangle$ and $S|1\rangle = |1\rangle$ corresponding to 90° and 0° phase shifts, respectively, and

$$T = \begin{pmatrix} -1 & 0 \\ 0 & -i \end{pmatrix},$$

which implies $T|0\rangle = -|0\rangle$ and $T|1\rangle = -i|1\rangle$ corresponding to 180° and -90° phase shifts, respectively.

Finally, the XOR operator is described by the matrix:

$$XOR = \begin{pmatrix} 1 & 0 & 0 & 0 \\ 0 & 1 & 0 & 0 \\ 0 & 0 & 0 & 1 \\ 0 & 0 & 1 & 0 \end{pmatrix}.$$

What makes the teleportation circuit unusual, in comparison to other quantum circuits, is that the top and middle lines are *severed* at the mid-point of the circuit. Instead of the usual values passing along these lines, the states on these lines are measured, at the outputs of Alice's part of the circuit, to produce two answers, each consisting of a single classical bit. These answers are then communicated to the top and middle inputs of Bob's circuit in the form of a classical 2-bit message.

Let's walk through an example to illustrate the operation of each part of the teleportation circuit.

Step 1: Alice Creates a Quantum Channel

For Alice to teleport an arbitrary, unknown, quantum state to Bob, she begins by creating a pair of particles whose quantum states are entangled with one another. Alice keeps one of these particles and sends the other to Bob (via the bottom line of the teleportation circuit). Although the two particles become physically remote from one another, the correlation between their states persists so long as neither particle is measured nor interacts with its environment in any way. Such correlated particles are referred to as "ebits" in the jargon of quantum computing.

To create the ebits, Alice pushes two standard states (two particles each in state $|0\rangle$) through the circuit depicted in Figure 9.7.

The circuit has two inputs and two outputs and consists of an L gate and an XOR gate. Algebraically, the circuit can be described as the dot product of the matrices representing the action of each gate. If LOP[*i,m*] is an L gate that acts on the *i*th of *m* qubits and XORGate[*i,j,m*] is an XOR gate that acts on the *i*th and *j*th of *m* qubits, then the overall operation of the circuit is described, mathematically, by XORGate[1,2,2] • LOP[1,2].

The pair of inputs, initially in the states $|0\rangle$ and $|0\rangle$, evolve under the action of this circuit into the state $\frac{1}{\sqrt{2}}(|00\rangle + |11\rangle)$. This state is an entangled state in the sense that, if you measured one of the outputs and obtained the answer "0", then a subsequent measurement on the other output would also yield the answer "0". Likewise, if you measured one of the outputs and obtained the answer "1", a subsequent measurement on the other output would also yield a "1".

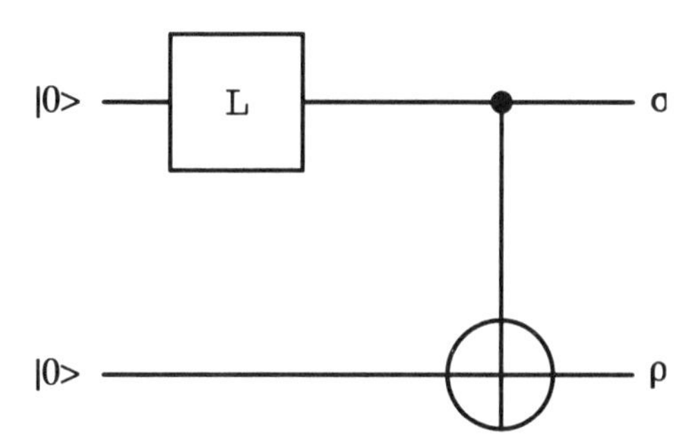

Fig. 9.7 Alice creates a quantum communication channel by creating a pair of entangled qubits.

Hence, although each output is in a superposition of 0 and 1, the joint state of the outputs are strongly correlated with one another. By retaining one of these output states and sending the other to Bob, Alice establishes a quantum communication channel between herself and Bob via the enduring correlation between the states labelled σ and ρ.

Step 2: Entanglement (also Alice's job)

Next, suppose Alice wants to teleport a particular quantum state, $|\psi\rangle$, to Bob. Alice need not know what this state is in order to teleport it successfully, so without loss of generality we can say that the state is "unknown". To teleport the state, $|\psi\rangle$, Alice entangles it with her end of the quantum communication channel that she established in Step 1. Again the entangling is done by pushing certain states through a particular quantum circuit, in this case the circuit shown in Figure 9.8.

Step 3: Measurement (Alice's last job)

Next, Alice measures the bits on the top and the middle outputs from her circuit. This will yield a pair of classical bits. Alice tells Bob what bits she read by sending him a message over a conventional (classical) communication channel such as a radio broadcast, a telephone, or a letter. Curiously, if Alice uses a radio broadcast, she need not even know Bob's exact location in order to teleport the state to him successfully!

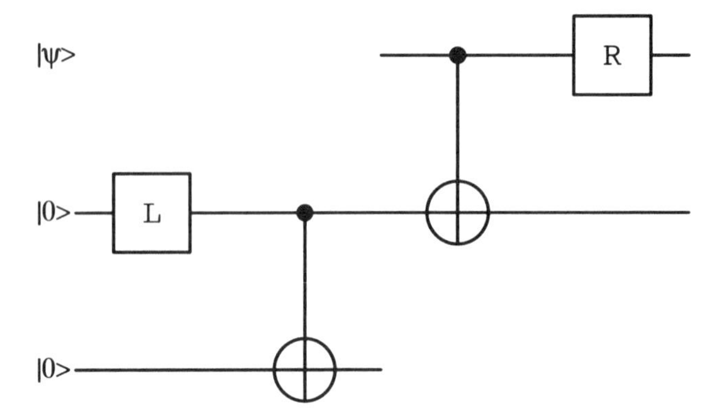

Fig. 9.8 Alice entangles the unknown state $|\psi\rangle$ with the state σ that comprises one end of the quantum communication channel.

Step 4 (Bob's job)

Upon receipt of Alice's 2-bit classical message, Bob converts the classical bits into the corresponding qubits i.e. a 0 is converted to a $|0\rangle$ and a 1 is converted to a $|1\rangle$. He then feeds these states into the corresponding inputs to his circuit, i.e., the upper and middle lines on the left hand side of Figure 9.9.

Notice that, although Alice does not measure the output state on the bottom line of her circuit explicitly, its state is affected by her measurements on the top and middle outputs. This is because the state on the bottom line is entangled with the state on the middle line due to the operations Alice performed in Step 1. It is precisely this side effect of Alice's measurements that is the means by which partial information about the unknown state is conveyed to Bob via the quantum communication channel, i.e., the entangled qubits Alice created at the outset.

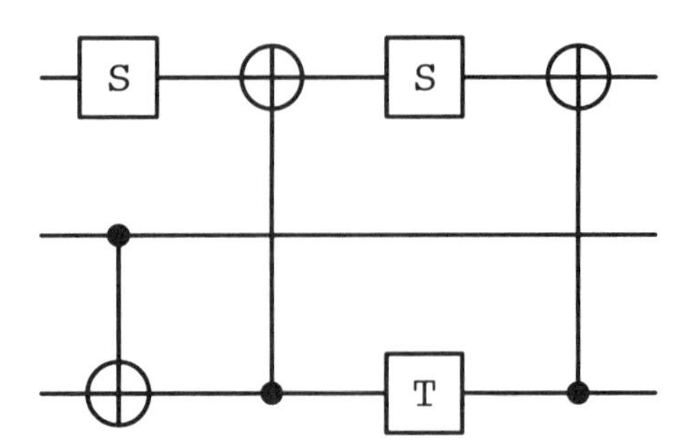

Fig. 9.9 Bob's disentangling circuit. The unknown state will re-appear on the bottom output.

Thus the states entering Bob's portion of the circuit contain the complete information about the unknown state $|\psi\rangle$ partitioned between a classical channel (the 2-bit message) and a quantum channel (the entanglement Alice created in Step 1 and modified in Step 3). Bob's circuit can be thought of as applying a different rotation to the state on the bottom input to his circuit depending on the answer Alice sent him in the classical message.

Hence the entire circuit, consisting of Alice, Bob and the intermediate measurement operations, constitutes one possible design for a quantum teleportation device.

9.9 Simulation of Quantum Teleportation

The CD-ROM contains a simulator for Brassard's quantum teleportation circuit. You can use the simulator to step through the various stages of the teleportation circuit and to generate a self-documenting trace of the teleportation process as a state is pushed through the device.

To run the simulator, begin by creating a random superpostion of $|0\rangle$ and $|1\rangle$ using the function `CreateUnknownState[]`. This state will serve as the supposedly unknown state that Alice wishes to teleport to Bob.

In[]:=
```
unknownState = CreateUnknownState[]
```
Out[]=
```
(0.0288733 + 0.0848611 I) ket[0] + 0.995974 ket[1]
```

Be aware that the amplitudes is our "unknown" state contain an arbitrary phase factor. This makes it clear to see that the teleportation circuit can indeed teleport an *arbitrary* quantum state rather than just special states having purely real amplitudes.

Having created an "unknown" state, you can generate a self-documenting trace of the teleportation process by entering the command `Teleport[unknownState]`. The code will then create a description of the entire teleportation process for the "unknown" state you created. Here is a sample run of the simulator:

In[]:=
```
Teleport[unknownState]
```
Out[]=
```
Unknown  State:
The state to be teleported is:
  (0.0288733 + 0.0848611 I) ket[0] + 0.995974 ket[1]
This state is unknown to both Alice and Bob.
```

```
Step 1:
Alice creates a pair of ebits in the joint state:
   1/Sqrt[2] (ket[0,0]+ket[1,1])
Alice keeps one ebit and sends the other to Bob.

Step 2:
Alice entangles the unknown state with the ebit she
retained in Step 1.
To do this, Alice pushes the states:
   (0.0288733 + 0.0848611 I) ket[0] + 0.995974 ket[1]
   |0>
   |0>
through her circuit. The resulting state (at the
mid-point of the circuit) is:
    (0.0144367 + 0.0424305 I) ket[0, 0, 0] +
                     0.497987 ket[0, 0, 1] +
                     0.497987 ket[0, 1, 0] +
    (0.0144367 + 0.0424305 I) ket[0, 1, 1] +
   (-0.0144367 - 0.0424305 I) ket[1, 0, 0] +
                     0.497987 ket[1, 0, 1] +
                     0.497987 ket[1, 1, 0] +
   (-0.0144367 - 0.0424305 I) ket[1, 1, 1]

Step 3:
Alice measures the state on lines 1 and 2 to obtain
the bits {1, 0} and conveys these results to Bob via
a classical 2-bit message.

Step 4:
Bob receives Alices classical 2-bit message and
creates the inputs |1> and |0> for input to line 1
and line 2 of his circuit. Because of the ebit Alice
created, and the measurement she made, the state on
line 3 entering Bob's circuit is:
   (-0.0288733 - 0.0848611 I) ket[0] + 0.995974 ket[1]

Step 5:
Bob pushes the three states:
   |1>,
   |0> and
   (-0.0288733 - 0.0848611 I) ket[0] + 0.995974 ket[1]
through his circuit to obtain the output:
   (0.0288733 + 0.0848611 I) ket[1, 0, 0] +
                    0.995974 ket[1, 0, 1]
Thus the state on line 3 of the output is:
   (0.0288733 + 0.0848611 I) ket[0] + 0.995974 ket[1]
```

```
Compare this with the "unknown" state on line 1 of
Alice's input:
   (0.0288733 + 0.0848611 I) ket[0] + 0.995974 ket[1]
They are the same! Thus the unknown state on line 1
of the input has been teleported to line 3 of the
output.
```

Further examples of simulations of the teleportation circuit are given in the code supplement to this chapter.

Teleportation inside a quantum computer could be used as a security feature wherein only one version of certain sensitive data is ensured to exist at any one time in the machine. This is a better way to transmit secure data than the quantum cryptographic methods we have discussed earlier because only one version of the message can ever exist in the universe at one time[Brassard96]. We need not worry about the original message being stolen after it has been teleported because it no longer exists at the source location! Furthermore, any eavesdropper would have to steal both the entangled particle and the classical particle in order to have any chance of capturing the information.

9.10 Experimental Status of Quantum Teleportation

The technology needed to teleport a qubit is surprisingly simple. In fact, it is arguably less complicated than that required for quantum cryptography, provided you perform the teleportation without delay after creating the entanglement.

We are hardly at the point of being able to teleport an entire person, however, even though there seems to be no physical restriction on such a possibility. Samuel Braunstein has estimated how much information you would need to transmit in order to teleport a human being. Starting from the observation that the visible human project, sponsored by the American National Institute of Health, requires about 10 Gigabytes of bits (about 10 CD-ROMs) to hold the information needed to describe the full three-dimensional structure of a human to a 1 mm^3 resolution, Braunstein estimates that an entire human could be described, down to the atomic level, using roughly 10^{32} bits. With current communication channel capacities, Braunstein estimates that it would take about a hundred million centuries to transmit this information down a single channel.

Prospects for a small-scale teleportation prototype, capable of teleporting a single qubit, are much better. Gilles Brassard ranks quantum teleportation as significantly easier than quantum factorization or more general quantum computations. In Brassard's words, "... a working demonstration of quantum teleportation is likely to be seen before the quantum factorization of even a very small integer."

However, some significant technological hurdles need to be overcome before quantum teleportation will be a useful means of secure communication. The most glaring problem is how to maintain the state of the entangled particles for the time necessary to transport the classical message. A larger-scale deployment of quantum teleportation would require the maintenance of stockpiles of entangled particles indefinitely, so that we could have them ready whenever we had the need to teleport a message.

9.11 Other Uses of Entangled Bits

There are ways of using the remarkable properties entangled bits (ebits) other than in quantum teleportation. The ability to ship information (sequences of bits) efficiently between a source and a receiver is of crucial importance in many scientific and commercial endeavors. To date, communication channels have been assumed to be conduits of classical information, i.e., sequences of 0s and 1s. However, with the advent of quantum information theory a new possibility has arisen, based on exploiting the fundamental quantum properties of qubits (quantum bits) and ebits (entangled bits) rather than classical bits, to send information down a quantum communication channel.

The classical version of information theory was pioneered by Claude Shannon[Shannon49]. Since this time, several important theorems have been proved concerning the capabilities of classical communication channels such as their capability to carry information and their robustness against errors. Now new quantum versions of those theorems are appearing[Adami96, Lloyd96a]. Perhaps not surprisingly, the capabilities of quantum communication channels are found to be superior to classical communication channels.

Table 9.2 shows the relationship between classical and quantum notions of information[Bennett95]. The basic idea is that different types of bits can be traded to achieve the information-carrying capacities other types of bits.

Table 9.2 The information-carrying capacities of bits, qubits, and ebits can be implemented by consuming combinations of other types of bits.

Information	Implemented by Consuming	Meaning
1 bit	1 qubit	A classical bit can be sent if you can send a qubit.
1 ebit	1 qubit	An ebit (i.e., two entangled bits) can be created if you can send a qubit. (Alice creates an EPR pair and sends one to Bob.)
1 qubit	1 ebit + 2 bits	A qubit can be sent if you can create an ebit and send 2 classical bits (i.e., teleportation).
2 bits	1 ebit + 1 qubit	2 classical bits can be sent if you can create an ebit and send a qubit (superdense coding).

Superdense Codes

An interesting application of ebits and qubits, that illustrates one of the trades summarized in Table 9.2, concerns the use of an ebit in conjunction with the transmission of a qubit to send two classical bits of information.

Prior to any communication taking place, the source and receiver, conspire to collect and store corresponding pairs of EPR particles. To send two classical bits, one of 4 unitary operations are applied at the source on one member of a shared EPR pair (an ebit), thereby placing the pair as a whole into one of the Bell states. Subsequently, this particle is sent to the receiver. Upon receipt, the original 2 classical bits can be recovered by measuring the transmitted particle and the other (twin) EPR particle already in the receiver's possession. Of course, to be practicable, it will be necessary to understand how to distribute and preserve EPR pairs until they are needed.

Note that the net effect of this process is that two bits of information are transmitted even though only one particle (the treated EPR particle) is sent at the moment of communication. Hence the name "superdense coding" for this technique.

Superdense coding is a way of putting more information into a state and is closely related to teleportation [Bennett92b]. A schematic view of superdense coding is shown in Fig. 9.10.

In Fig. 9.10 the sender Alice wants to transmit two classical bits. An ebit in the form of an EPR pair is given to the sender Alice and

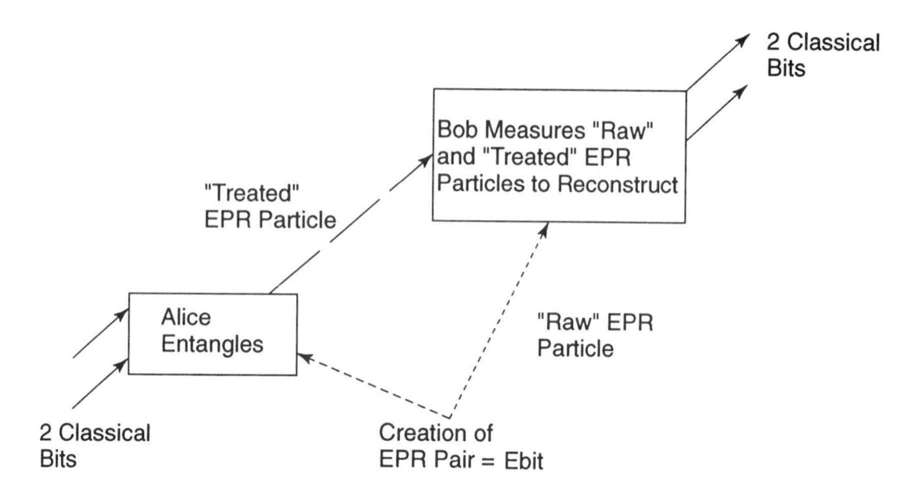

Fig. 9.10 Schematic diagram of superdense coding.

the receiver Bob. Alice entangles the ebit with the two classical particles and transmits the "treated" particle which is now in a Bell state to Bob. Bob then measures the "raw" half of the EPR pair as well as the treated half to reconstruct the two classical bits.

Quantum Data Compression

Another application of entanglement is in the domain of data compression[Schumacher95]. One approach to quantum data compression works by exploiting the redundancy in a message, composed of a sequence of qubits in non-orthogonal states, to allow certain qubits to be dropped (at the source) from the quantum message, yet have their information content re-constituted later (at the reciever) by augmenting the received quantum message with qubits prepared in a certain "standard" state. The net effect is that a quantum message can be sent from the source to a receiver using fewer qubits than "ought" to be necessary. Thus if one state is used to represent 0 and the other (non-orthogonal) state is used to represent 1 at the source, it is therefore possible to send an n-bit message using fewer than n bits. It remains to be seen how efficient this process is in comparison to (or augmented with) classical compression schemes.

One possible application of these ideas is in the area of communication over optical fibers. It already seems quite feasible to squeeze about 5% to 10% more information (more bits) down a conventional optical fiber by using qubits and ebits[Fuchs].

One of the biggest problems in realizing quantum teleportation, superdense coding and quantum data compression lies in the difficulty of maintaining the delicate quantum coherences necessary to achieve a perfectly functioning device. The inevitable coupling between a quantum system and its environment, as well as imperfections in the design and operation, create several potential sources of errors. However, it is now known how to use several entangled qubits to represent one logical qubit (i.e., a computationally active qubit). The entangling essentially spreads out the state of the qubit in such a way that errors in any "part" of the entangled qubit can be detected, diagnosed, and corrected. Thus entangling qubits with the environment may introduce errors, but entangling qubits with themselves might immunize them from such errors. We take up these ideas in the next chapter.

CHAPTER 10

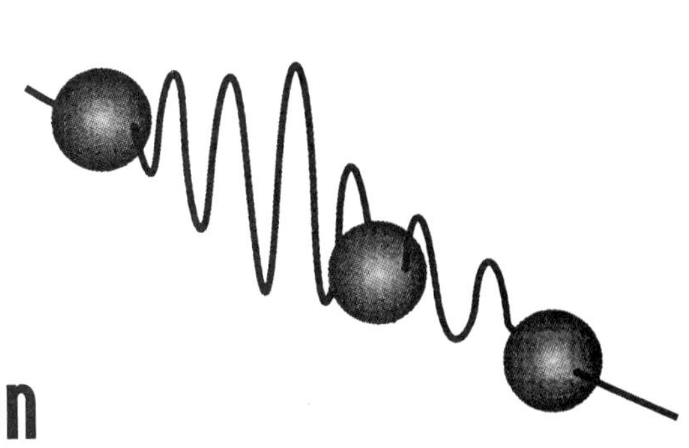

Quantum Error Correction

"I wish to God these calculations had been executed by steam!"
— Charles Babbage

In Chapter 5, we examined some of the errors that can arise during the three phases of operation of a quantum computer, namely, preparation, evolution, and measurement. We saw that software errors (i.e., computational mistakes due to imprecision in the initial state) do not grow over time and that hardware errors (i.e., computational mistakes due to imperfections in the Hamiltonian) grow only quadratically[Zurek84]. This sounds like good news for quantum computers. Unfortunately, these conclusions are based on the assumption that a quantum computer is perfectly isolated from its environment. Although this assumption simplifies the mathematical analysis of error propagation, it is not very realistic. Any real quantum computer is going to incur other kinds of errors caused by myriad physical processes such as decoherence, cosmic radiation, and spontaneous emission. These latter processes mean that even maintaining the state of a quantum memory register of an idle quantum computer is fraught with difficulties. Similarly, preserving a pair of EPR entangled particles until they are needed for quantum teleportation or quantum cryptography is tricky. These kinds of problems have led some to question the feasibility of quantum computers [Landauer95], [Unruh95].

In this chapter we describe the effects of error processes, such as decoherence and spontaneous emission, on the state of a quantum memory register, and examine some of the proposals for correcting the resulting errors. Remarkably, the new error-correction techniques appear to be good enough to envisage nontrivial quantum computations. In particular, the recent developments in fault-tolerant and concatenated error-correction schemes (explained later), which are robust against noise, bring a practical quantum computer significantly closer to reality.

Before we can discuss error correction, we had better be clear about the error processes that are at work.

In the idealized models of quantum computers that we studied in Chapters 3 and 4, the qubits representing the computational state of the computer were assumed to be perfectly isolated from their environment. In other words, from the moment the quantum computer is prepared in some initial state to start the computation off, to the moment it is measured to extract an answer, the logical qubits of ideal quantum computers are supposed to evolve unitarily in accordance with Schrödinger's equation.

Unfortunately, such an idealization is, strictly speaking, unattainable. Any real quantum system tends to couple to its environment over time. In the process, information leaks out of the logical state of the qubits in the quantum memory register. If we did not model the effect of the environment explicitly, it would look as if the logical qubits were no longer evolving in accordance with Schrödinger's equation. Indeed, this coupling between a quantum system and its environment, and the resulting loss of coherence, is what prevents quantum effects from being evident at the macroscopic level.

10.1 Decoherence and Dissipation

Technically, these "external errors" are the biggest impediments to realizing a true quantum computer. Certainly, it is extremely difficult to isolate a quantum memory register from its environment. For one thing, a memory register has to be built out of *something* so there must be some supporting infrastructure (i.e., "scaffolding") in the vicinity of the computationally active components. Thus there is a chance that the infrastructure will couple to the computational components. In addition, there can be a coupling between the memory register and an ambient thermal heat bath. Also, stray particles such as cosmic rays or gas molecules can interact with the memory register. So, as we can see, there are many physical processes that can perturb the state of a quantum register. Broadly speaking, however, these fall under the headings of dissipation and decoherence.

Dissipation is a process by which a qubit loses energy to its environment. Thus, for example, if an excited state is used to represent a $|1\rangle$ and a lower energy state is used to represent a $|0\rangle$, a qubit might transition, spontaneously, from the $|1\rangle$ state to the $|0\rangle$ state emitting a photon in the process. In computational terms, a bit in the quantum memory register would have "flipped."

Decoherence is more insidious. Rather than a gross bit flip, decoherence is a coupling between two initially isolated quantum systems (the qubits and its environment) that tends to randomize the relative phases of the possible states of the memory register. As a result, the interference effects, needed in any true quantum computation, are destroyed and the state of the qubit becomes entangled with the state of the environment.

Mathematically speaking, decoherence is best modeled using a formalism called "density matrices." All the equations of quantum physics can be rewritten using density matrices instead of state vectors. However, density matrices provide a more compact description of so-called "mixed states" of composite quantum systems. Mixed states are characterized by not being expressible as a direct product of states of their component parts. For example, if a quantum memory register is described by the pure state $|\psi\rangle$, its corresponding density matrix ρ is defined by $\rho = |\psi\rangle \otimes \langle\psi|$ (which is always a square matrix). For the case of an isolated qubit, $|\psi\rangle = \omega_0|0\rangle + \omega_1|1\rangle$, and its density matrix is given by

$$\rho = |\psi\rangle \otimes \langle\psi| = \begin{pmatrix} |\omega_0|^2 & \omega_0\omega_1^* \\ \omega_0^*\omega_1 & |\omega_1|^2 \end{pmatrix},$$

where the asterisk denotes taking the complex conjugate. The effect of decoherence is to damp out the off-diagonal elements in the density matrix, in this case $\omega_0\omega_1^*$ and $\omega_0^*\omega_1$. In reality this is a time-dependent process, and the density matrix changes over time:

$$\rho_{\text{decohering}} = \begin{pmatrix} |\omega_0|^2 & e^{-t/\tau}\omega_0\omega_1^* \\ e^{-t/\tau}\omega_0^*\omega_1 & |\omega_1|^2 \end{pmatrix},$$

where τ, called the "decoherence time," sets the characteristic timescale of the decoherence process.

However, in the simple model of decoherence that we adopt later, we model the end effect only. Thus after the state has decohered appreciably, the off-diagonal terms will die away (see Fig. 10.1).

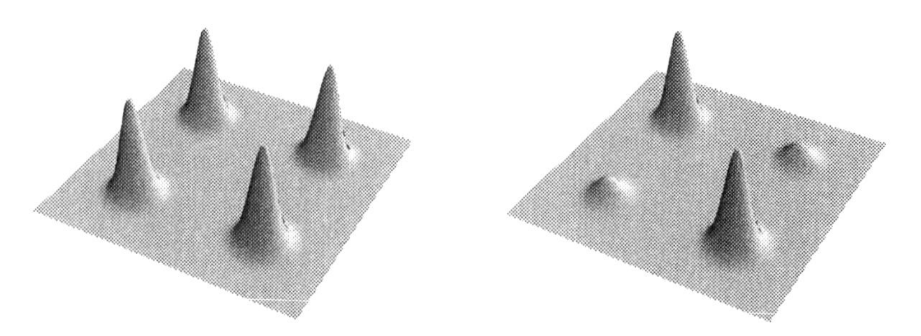

Fig. 10.1 A decohering density matrix. On the left we have the initial density matrix and on the right the density matrix after time t.

and its density matrix will have become:

$$\rho_{\text{decohered}} = \begin{pmatrix} |\omega_0|^2 & 0 \\ 0 & |\omega_1|^2 \end{pmatrix}.$$

Usually, decoherence occurs on a faster timescale than dissipation. The time it takes a memory register to decohere depends, principally, upon what kind of quantum systems it is made from, the size of the register, the temperature of the thermal environment, and the rate of collisions with ambient gas molecules. A crude estimate of decoherence times in various settings can be obtained from the Heisenberg Uncertainty Principle, in energy and time:

$$\Delta t \approx \frac{\mathrm{h}}{\Delta E} = \frac{\mathrm{h}}{kT},$$

where k is Boltzmann's constant (approximately 1.38 10^{-23} Joules per K) and T is the absolute temperature of the environment. In this estimate we have taken the uncertainty in the energy to be of the order of the energy of a typical particle at the ambient temperature. At room temperature, this gives a typical decoherence time of about 10^{-14} s. At lower temperatures, systems take longer to decohere. For example, at the temperature of liquid helium, it takes about 100 times as long for a system to decohere as it does at room temperature. Consequently, the simplest way to try to combat decoherence is simply to operate the computer at a lower temperature. Table 10.1 summarizes some characteristic decoherence times, under various physical scenarios. These estimates were derived using a more sophisticated analysis[Joos85].

Table 10.1 Approximate decoherence times (in seconds) for various sized systems in different thermal and gaseous environments.

	—Temperature Effects—			—Bombardment Effects—	
System Size (cm)	Cosmic Background Radiation	Room Temperature (300 K)	Sunlight on the Earth	Vacuum (10^{6} particles / cm^3)	Air
10^{-3}	10^{-7}	10^{-14}	10^{-16}	10^{-18}	10^{-35}
10^{-5}	10^{15}	10^{-3}	10^{-8}	10^{-10}	10^{-23}
10^{-6}	10^{24}	10^{5}	10^{-2}	10^{-6}	10^{-19}

Once we have chilled our quantum computer and sealed it in as best a vacuum as we can, what else can we do to slow down decoherence? Well, we could try building the quantum memory register out of different types of quantum systems. Certain quantum systems are much more resilient to decoherence than others. For example, David DiVincenzo has collected some statistics on the decoherence properties of various materials[DiVincenzo95], as shown in Table 10.2. The figures in Table 10.2 suggest that it might be possible to build a quantum memory register that can support a significant number of computational steps. More detailed results for ion trap quantum computers are reported in the next chapter and are actually quite encouraging, at least with respect to the possibility of factoring small integers.

Table 10.2 The maximum number of computational steps that can be accomplished without losing coherence for various quantum systems.

Quantum System	Time per Gate Operation (sec)	Coherence Duration (sec)	Maximum No. Coherent Steps
Mössbauer nucleus	10^{-19}	10^{-10}	10^{9}
GaAs electrons	10^{-13}	10^{-10}	10^{3}
Gold electrons	10^{-14}	10^{-8}	10^{6}
Trapped indium ions	10^{-14}	10^{-1}	10^{13}
Optical microcavity	10^{-14}	10^{-5}	10^{9}
Electron spin	10^{-7}	10^{-3}	10^{4}
Electron quantum dot	10^{-6}	10^{-3}	10^{3}
Nuclear spin	10^{-3}	10^{4}	10^{7}

Recent analyses of decoherence in quantum memory registers, in which all the qubits are assumed to couple to the same environment, suggest that there are certain kinds of entangled states of the register for which no decoherence occurs at all, even when the coupling to the environment is not weak[Duan96]. It might someday be possible to exploit these entangled states to immunize a register against decoherence. Unfortunately, for the present it seems that we can only postpone decoherence rather than prevent it.

However, if decoherence can only be postponed for a *finite* length of time, this poses a severe problem for anyone wanting to build a *universal* quantum computer. That is not to say that quantum computers that can only run for a certain length of time coherently will be useless: there might still be many useful computations such machines could do within their time constraints. Moreover, there might be ways to partition certain computations across several quantum computers in such a way that each subcomputation could be completed within the coherence time bound. Nevertheless, from a theoretical perspective, we would like to do better. We would like, ideally, to design a quantum computer that could, in principle, maintain coherent quantum computations indefinitely. In such a setting, since we can not prevent decoherence, we need to think about ways of undoing its effects. Thus there needs to be a way of doing quantum error correction.

10.2 Models of Errors

How can we describe the effect of errors on the state of a quantum memory register mathematically? This might seem somewhat arbitrary. After all, how can we circumscribe what can go wrong? Surprisingly, it can be shown that any possible kind of error in the state of a single qubit, including the "no error" possibility, can be described in terms of the application of one of four basic operators to the state representing the qubit. These operators are called the Pauli spin matrices, named after their discoverer Wolfgang Pauli, who found them during the early days of quantum mechanics long before quantum computers, or error models, had ever been imagined. Any 2x2 matrix can be written as a sum involving simple multiples of one or more of the Pauli matrices:

$$\sigma_x = \begin{pmatrix} 0 & 1 \\ 1 & 0 \end{pmatrix}, \sigma_y = \begin{pmatrix} 0 & -i \\ i & 0 \end{pmatrix}, \sigma_z = \begin{pmatrix} 1 & 0 \\ 0 & -1 \end{pmatrix}, I = \begin{pmatrix} 1 & 0 \\ 0 & 1 \end{pmatrix}.$$

Hence the mathematical effect of any kind of error on a qubit can also be expressed as a sum of the Pauli matrices.

Let us begin by looking at the action of these matrices on a single qubit. These operations are supposed to mimic the introduction of an error into the state of a single qubit. We are quite general and assume that the qubit starts off in an arbitrary superposed state given by

$$|\psi\rangle = \alpha|0\rangle + \beta|1\rangle = \begin{pmatrix}\alpha\\ \beta\end{pmatrix},$$

where $|\alpha|^2 + |\beta|^2 = 1$. Applying σ_z to this qubit results in the following,

$$\sigma_z \cdot (\alpha|0\rangle + \beta|1\rangle) = \sigma_z \cdot \begin{pmatrix}\alpha\\ \beta\end{pmatrix} = \begin{pmatrix}\alpha\\ -\beta\end{pmatrix} = \alpha|0\rangle - \beta|1\rangle.$$

That is, σ_z causes the "correct" state to evolve according to the rule $\alpha|0\rangle + \beta|1\rangle \to \alpha|0\rangle - \beta|1\rangle$. This operation has changed the phase of the qubit. Consequently, we call such an operation a "phase shift error."

Next let us look at the affect of σ_x on the state of a qubit.

$$\sigma_x \cdot (\alpha|0\rangle + \beta|1\rangle) = \sigma_x \cdot \begin{pmatrix}\alpha\\ \beta\end{pmatrix} = \begin{pmatrix}\beta\\ \alpha\end{pmatrix} = \alpha|1\rangle + \beta|0\rangle.$$

This time the action of the operator has caused the bits to flip. That is, σ_x causes the transformation $\alpha|0\rangle + \beta|1\rangle \to \alpha|1\rangle + \beta|0\rangle$. So let us call such an operation a "bit flip error."

Now the action of the identity matrix on a state is to leave the state unchanged. So I must represent the "no error" possibility. That only leaves us to consider what happens when we apply σ_y to the state of the qubit:

$$\sigma_y \cdot (\alpha|0\rangle + \beta|1\rangle) = \sigma_y \cdot \begin{pmatrix}\alpha\\ \beta\end{pmatrix} = \begin{pmatrix}-i\beta\\ i\alpha\end{pmatrix} = i\alpha|1\rangle - i\beta|0\rangle.$$

This operation corresponds to *both* a phase shift and a bit flip. That is, σ_y causes the transformation $\alpha|0\rangle + \beta|1\rangle \to i\alpha|1\rangle - i\beta|0\rangle$. Thus any error in a single qubit can be described by the action of a linear combination of the σ_x, σ_y, σ_z, and I operators.

Okay, so far so good. But the memory register of a quantum computer is going to be composed of lots of qubits, not just one. How do we describe an error in the ith of m qubits?

To describe the action of these matrices on the ith of m qubits, we embed each matrix in the $2^m \times 2^m$ dimensional identity operator. For example, the operator that would apply σ_z to the second of three qubits, to create a 1 qubit error on the middle of three qubits, would be given by:

$$I \otimes \sigma_z \otimes I = \begin{pmatrix} 1 & 0 \\ 0 & 1 \end{pmatrix} \otimes \begin{pmatrix} 1 & 0 \\ 0 & -1 \end{pmatrix} \otimes \begin{pmatrix} 1 & 0 \\ 0 & 1 \end{pmatrix}.$$

We do not need to do this by hand because the code in the electronic supplement can do this for us. Here, by way of example, is the operator corresponding to the application of σ_z acting on the second of three qubits.

In[]:=

```
PauliZ[2,3] // MatrixForm
```

Out[]//MatrixForm=

$$\begin{pmatrix} 1 & 0 & 0 & 0 & 0 & 0 & 0 & 0 \\ 0 & 1 & 0 & 0 & 0 & 0 & 0 & 0 \\ 0 & 0 & -1 & 0 & 0 & 0 & 0 & 0 \\ 0 & 0 & 0 & -1 & 0 & 0 & 0 & 0 \\ 0 & 0 & 0 & 0 & 1 & 0 & 0 & 0 \\ 0 & 0 & 0 & 0 & 0 & 1 & 0 & 0 \\ 0 & 0 & 0 & 0 & 0 & 0 & -1 & 0 \\ 0 & 0 & 0 & 0 & 0 & 0 & 0 & -1 \end{pmatrix}$$

`PauliX[i,m]` and `PauliY[i,m]` are defined similarly. Given the embeddable Pauli matrices we can now define three functions for introducing an error into the *i*th qubit of an *m*-qubit quantum memory register:

`BitFlipError[`*state, i, m*`]`

`PhaseShiftError[`*state, i, m*`]`

`BitFlipAndPhaseShiftError[`*state, i, m*`]`

Here are some examples of various types of errors being introduced into an arbitrary state of a 3-qubit memory register. We obtain an arbitrary symbolic state description using the function `ArbitraryState[`*m*`]`. Initially, the state of the register is described by

In[]:=

```
state = ArbitraryState[3]
```

Out[]=

```
c[0] ket[0, 0, 0] + c[1] ket[0, 0, 1] +
c[2] ket[0, 1, 0] + c[3] ket[0, 1, 1] +
c[4] ket[1, 0, 0] + c[5] ket[1, 0, 1] +
c[6] ket[1, 1, 0] + c[7] ket[1, 1, 1]
```

where ket[0,0,0] represents the ket $|000\rangle$, ket[0,0,1] represents the ket $|001\rangle$, and so on, and the coefficients c[0], c[1], and so on are probability amplitudes. The output is just an arbitrary superposition of the eight possible states of a 3-qubit memory register. We can simulate a bit flip error on the first qubit by calling:

In[]:=
```
BitFlipError[state,1,3]
```
Out[]=
```
c[4] ket[0, 0, 0] + c[5] ket[0, 0, 1] +
c[6] ket[0, 1, 0] + c[7] ket[0, 1, 1] +
c[0] ket[1, 0, 0] + c[1] ket[1, 0, 1] +
c[2] ket[1, 1, 0] + c[3] ket[1, 1, 1]
```

Notice that the amplitude that was multiplying the ket $|0\psi\phi\rangle$ becomes swapped with the amplitude multiplying the ket $|1\psi\phi\rangle$.

Similarly, a phase shift error on the first qubit is given by

In[]:=
```
PhaseShiftError[state,1,3]
```
Out[]=
```
c[0] ket[0, 0, 0] + c[1] ket[0, 0, 1] +
c[2] ket[0, 1, 0] + c[3] ket[0, 1, 1] -
c[4] ket[1, 0, 0] - c[5] ket[1, 0, 1] -
c[6] ket[1, 1, 0] - c[7] ket[1, 1, 1]
```

Finally, a bit flip and phase shift error on the first qubit is given by

In[]:=
```
BitFlipAndPhaseShiftError[state,1,3]
```
Out[]=
```
-I c[4] ket[0, 0, 0] - I c[5] ket[0, 0, 1] -
 I c[6] ket[0, 1, 0] - I c[7] ket[0, 1, 1] +
 I c[0] ket[1, 0, 0] + I c[1] ket[1, 0, 1] +
 I c[2] ket[1, 1, 0] + I c[3] ket[1, 1, 1]
```

Errors in other qubits can be obtained in a similar way. Likewise, multiple errors can be simulated by nesting calls to `BitFlipError`, `PhaseShiftError`, and `BitFlipAndPhaseShiftError`.

So now we know how to describe the four possible types of error that can afflict a single qubit due to interaction with its environment. The next question is how to correct such an error.

10.3 Classical Versus Quantum Error Correction

Nowadays, error correction is commonplace in conventional computers and communication devices. Error correction is so fast and reliable that we do not hesitate to trust the outputs from our classical computations.

It is not as easy, however, to detect an error in a quantum computation as it is in a classical computation. With classical computers, it is possible to measure the state of the physical system used to encode a bit without disrupting the bit. Thus, if a voltage were used to represent a classical bit, we could, in principle, detect a slight drop from its nominal value and then give the voltage a little nudge to restore it to the correct level. In a quantum computation, the qubits may exist in a whole continuum of possible superposed states, any one of which might, in fact, be the intended state. Moreover, even if we knew what state a qubit should be in, we cannot merely inspect the qubit to see if it is correct because the act of inspecting it would necessarily disturb it unless, of course, the qubit happened to be in an eigenstate of the operator corresponding to the observable that we measured.

Such considerations make quantum error correction an extremely challenging problem. Indeed, currently, quantum error correction is by far the most active area of research in quantum computing. Progress is so rapid that it is hard to do justice to all the contributions. Instead, we focus on some of the principles that are emerging and sketch some schemes that seem particularly promising.

10.4 Elementary Error Correction Using Redundancy

Classical computers can be made more reliable through the use of redundancy. Instead of a single computer being used to perform a given computation, several computers are used to perform the same computation simultaneously. If the computers are all following the same deterministic program, they should all produce identical results at each stage of the computation. However, if an error occurs in one of the computers, its computational state will begin to diverge from that of the others. If we periodically poll all the computers and reset their computational states to the majority opinion, we will typically be able to correct errors that arose in a few of the computers since the last poll was taken. This type of majority voting scheme is currently used in the Space Shuttle to improve the reliability of the on-board decision making.

For majority voting to be effective, however, a number of assumptions must hold. For a start, the individual chances of any one computer obtaining the "correct" result must be greater than 50%. If this were not true then the majority opinion is more likely to be wrong than it is to be right! Moreover, the duplicated computations need to be independent of one another so that the errors incurred by the different computers are uncorrelated. This can be difficult to guarantee if the separate computers use the same type of hardware to run the same program. Finally, to guarantee a majority opinion, we

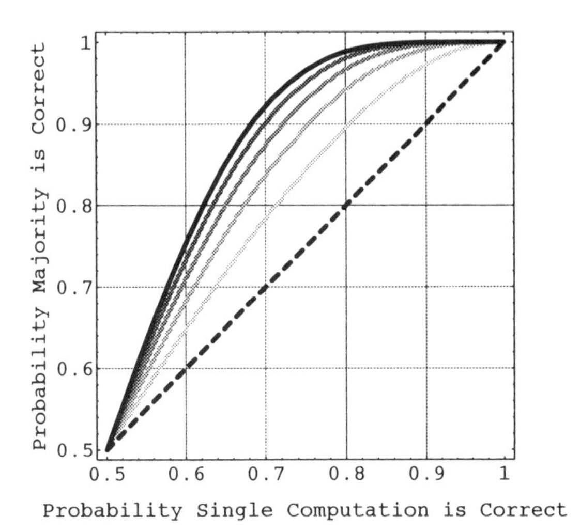

Fig.10.2 Probability that the majority vote is correct as a function of the probability that each independent computation is correct.

must use an odd number of computers. The more we use, the better our chances of fixing potential errors. In fact, if there are $2n - 1$ computers (for $n = 1, 2, 3, \ldots$) and the individual probability of each computer obtaining the correct answer is p then the probability that the majority opinion is correct is given by:

$$\Pr(\text{Majority Correct}) = \sum_{i=n}^{2n-1} \binom{2n-1}{i} p^i (1-p)^{2n-1-i}.$$

Fig. 10.2 is a plot that shows how the probability of the majority vote being correct increases as the probability of success of the individual computations increases for various numbers of duplicated computations.

If there is only one computer (i.e., $n = 1$) the probability of the majority vote being correct is identical to the probability of the individual computations being correct (dashed line). However, as we increase the number of duplicated computations (lightest curve has $n = 2$, darkest curve has $n = 6$), the probability of the majority being correct can exceed the probability of the individual being correct.

10.5 The Problem With a Quantum Version of Majority Voting

Unfortunately, a quantum version of majority voting is tricky. In general, at any intermediate point in a quantum computation the state of the memory register will be a superposition of possible states of the register weighted according to various amplitudes. If we replicated the same quantum computation across different quantum computers, then, in the absence of any errors, the memory registers would be in identical states at any given point of the computation. This is because the quantum mechanical evolution of each computer is governed by the same Schrödinger equation, which is a deterministic differential equation. Unfortunately, although the states of all the memory registers would be identical, if we measured the memory registers we would not, of course, obtain identical answers. Observing the memory register of a particular machine would appear to collapse the state of the register into some eigenstate of the measurement operator in an unpredictable way. Consequently, you cannot observe the state of a quantum computer as it is working without disrupting the state, irrevocably, in an unpredictable fashion. Thus a direct use of majority voting is not possible because the very acting of determining the majority opinion would alter the majority opinion.

In certain special cases, you might be able to do a little better. For example, if we could guarantee that, at certain moments the memory register ought to be in some unknown eigenstate of the register-measurement operator, then measuring the state of the register at those moments would enable the state (i.e., the eigenstate) to be determined without introducing any error. However, such measurements would have to be timed perfectly to avoid problems.

Thus, in general, a majority voting scheme is not appropriate for correcting the state of a quantum memory register *during* a quantum computation. Nevertheless, it might be useful for correcting the final answer from a quantum computation, provided that it could be guaranteed that the answer would be unique. Such a scheme has been devised by Seth Lloyd for obtaining the majority opinion as to the value of a bit by replicating the computation of the bit on three separate quantum computers[Lloyd93].

In Lloyd's scheme, we suppose that three bits representing answers to three independent, identical, computations are stored in the states of three particles A, B and C. Let us represent these three bits as the triple $|ABC\rangle$. If there are more 0s than there are 1s amongst A, B, and C then the "majority vote" is a 0. Conversely, if there are more 1s than 0s, the majority vote is a 1.

The majority vote is determined in three steps, each involving a pairwise comparison of the value of one of the bits against the values of the other two bits. The bit being tested is flipped (a 0 becomes a 1 or vice versa) if the other two bits agree with each other

Table 10.3 Majority voting in pairs.

Initial State	Bit 1 Compared With Bits 2 & 3	Bit 2 Compared With Bits 1 & 3	Bit 3 Compared With Bits 1 & 2
$\|000\rangle$	$\|000\rangle$	$\|000\rangle$	$\|000\rangle$
$\|001\rangle$	$\|001\rangle$	$\|001\rangle$	$\|000\rangle$
$\|010\rangle$	$\|010\rangle$	$\|000\rangle$	$\|000\rangle$
$\|011\rangle$	$\|111\rangle$	$\|111\rangle$	$\|111\rangle$
$\|100\rangle$	$\|000\rangle$	$\|000\rangle$	$\|000\rangle$
$\|101\rangle$	$\|101\rangle$	$\|111\rangle$	$\|111\rangle$
$\|110\rangle$	$\|110\rangle$	$\|110\rangle$	$\|111\rangle$
$\|111\rangle$	$\|111\rangle$	$\|111\rangle$	$\|111\rangle$

but not with the bit being tested. Otherwise, when the two bits used for comparison disagree with each other, then the bit being tested is not flipped. With three bits, three tests are sufficient to guarantee that a majority decision can be reached. That is, after the majority voting the joint state of all three bits will either be $|000\rangle$ or $|111\rangle$, meaning that the correct decision for the bit was 0 or 1, respectively.

The possible realizations of this scheme are outlined in Table 10.3. Notice that the leftmost column lists the eight possible values of the bits initially. The rightmost column lists the corresponding final states after the majority voting has been completed. In all cases, the final state is either $|000\rangle$ or $|111\rangle$ meaning that the overall decision as to the correct answer to the bit computation was a 0 or 1, respectively.

10.6 Error Correction via Symmetrization

There is, however, a more cunning way of exploiting redundancy for error correction during the course of quantum computation. Andre Berthiaume, David Deutsch, and Richard Jozsa have invented a method for coupling a set of R replica quantum computations to an ancilla system in such a way that measurements on the ancilla will correct errors in the state of the computers, even though the errors are never measured explicitly[Bertiaume94a], [Barenco96]. The best way to understand this is to begin with a simpler ancilla system with which we are already familiar.

Recall that, in Feynman's quantum computer (discussed in Chapter 4) there is a set of cursor qubits and a set of program qubits. The cursor bits keep track of the progress of the computation (i.e., how many steps of the computation have been accomplished) and the program bits keep track of the result of the computation. As soon as we find the cursor at its terminal position, we know that the state of the program bits, at that moment, contains an answer to the computation.

Mathematically, the joint state of all the qubits is described by a state vector in the Hilbert space given by the direct product of the Hilbert spaces of the individual qubits (i.e., cursor bits + program bits). The effect of measuring the cursor is to project this state vector into a subspace given by the direct product of the Hilbert spaces of just the program bits. We can visualize this, crudely, in the diagram in Fig. 10.3. This idea of coupling two systems, so that measuring the state of one system projects the state of the other into a specific subspace, can be used to perform error correction. This technique is most appropriate for correcting several qubits that are slightly wrong rather than correcting a single qubit that is terribly wrong[Peres96].

Here is how it works. We imagine that we have R replicas of our quantum computer, all performing exactly the same computation. If all R computers function perfectly, and the correct state of one of the computations at time t is $|\psi(t)\rangle$, their joint state at time t will be described by the direct product:

$$|\Psi(t)\rangle = \underbrace{|\psi(t)\rangle \otimes |\psi(t)\rangle \otimes \ldots |\psi(t)\rangle}_{\text{R identical computations}}.$$

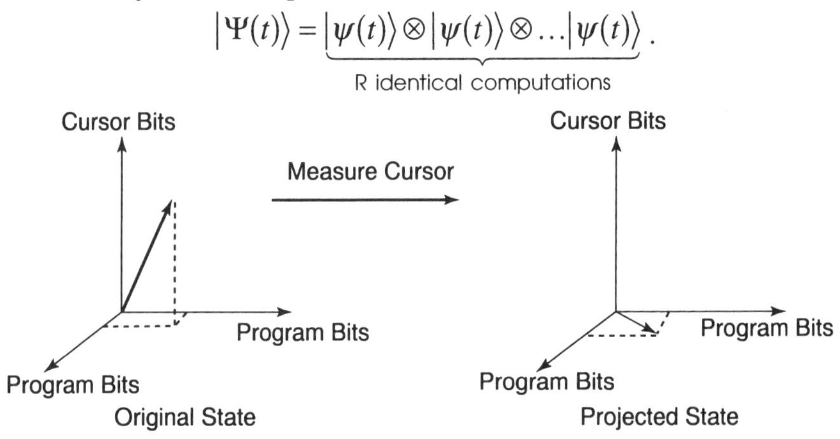

Fig.10.3 Partial measurements project a state into a subspace.

Notice that this state looks symmetric, being simply a direct product of R *identical* kets. However, if small, continuous, random drifts of all the qubits occur, the actual joint state of the R computers will be something like:

$$|\Psi(t)\rangle = \underbrace{|\psi^{(1)}(t)\rangle \otimes |\psi^{(2)}(t)\rangle \otimes \ldots |\psi^{(R)}(t)\rangle}_{\text{R slightly different computations}}$$

where the $|\psi^{(1)}(t)\rangle$, $|\psi^{(2)}(t)\rangle$, and so on, are all slightly different and slightly wrong.

Now, even though we do not know any of the individual states in the direct product we do know that, if all the computations are error-free, we ought to have a nice symmetric form for the joint state of the R computations. If we imagine carving up the entire Hilbert space into a portion that contains such symmetric states and a portion that does not, then we will find that the symmetric states occupy a tiny fraction of the entire Hilbert space. All the erroneous joint states lie somewhere outside this symmetric subspace. Consequently, the idea behind error correction via symmetrization is to devise a measurement that projects the joint state of the R computers back into the symmetric subspace (where it belongs).

This technique is very reminiscent of the use of the cursor and program bits in Feynman's quantum computer. Just as Feynman avoided measuring the program bits directly, by coupling them to the cursor bits, so too do we couple an ancilla to the joint state of the R computers.

First we create an ancilla system that can exist in $R!$ possible states. The $R!$ is significant because there are $R!$ possible permutations of R objects. Initially the ancilla is placed in a state of all zeroes. Next we apply a unitary operator that places the ancilla in an equally weighted superposition of its $R!$ possible states,

$$\hat{U} \cdot |00\text{K } 0\rangle = \frac{1}{\sqrt{R!}} \sum_{i=0}^{R!-1} |i\rangle .$$

Then we apply a unitary operation on the hybrid ancilla/R-computer system, which applies the ith permutation to the R-computers if the ancilla is in state $|i\rangle$. Finally, we perform the inverse of the $\hat{U}$ operation on just the ancilla bits.

If the R computations were error-free, then the permutations would leave the joint state invariant. So applying $\hat{U}$ to $|00\text{K } 0\rangle$, permuting the joint state, and then applying $\hat{U}^{-1}$ to the ancilla bits in the result should simply restore the original state of the ancilla (all zeroes). Thus whenever after this coupling process we find the ancilla in the state $|00\mathrm{K}\ 0\rangle$, we know that the state of the R computers is in the symmetric subspace, whether or not there was an error present amongst the R computers originally.

10.7 Quantum Error-Correcting Codes

The preceding scheme for stabilizing a quantum computation assumes that the errors incurred among the observations that project the state of the computers back into the symmetric subspace are not too large. However, certain error processes, such as spontaneous emission, can result in sudden large errors. These kinds of errors require a different error-correction strategy based on the idea of quantum error-correcting codes.

Classical codes are used routinely to immunize classical computations from errors such as accidental bit flips. The basic idea is to encode each logical bit in a set of codeword bits. The codewords are chosen so that they are maximally distinguishable. If an accidental bit flip occurs in a codeword, the correct codeword can still be regained by replacing the buggy codeword with the nearest neighbor codeword.

A similar strategy is used in quantum codes. Several physical qubits are used to encode one logical qubit. The actual value of the logical qubit is stored in the correlations between several physical qubits so that even if there is disruption to a few qubits in the codewords, there is still sufficient information in the correlations to identify which error occurred and thence correct the qubit.

The first such code was devised by Peter Shor in 1995[Shor95]. Peter Shor's scheme uses nine physical qubits to encode one logical qubit. The information about the state of the logical qubit is distributed amongst the correlations between the states of the nine entangled qubits. Even if one of the nine entangled qubits becomes corrupted, there is still enough information in the states of the qubits to figure out the error and hence, restore the correct logical qubit.

Raymond Laflamme, Cesar Miquel, Juan Paz, and Wojciech Zurek, working at the Los Alamos National Laboratory, improved upon Shor's scheme by showing that the state of a single logical qubit could be protected by entangling it with only five physical qubits. The resulting entanglement is resilient to one error of a general type (i.e., a phase shift and/or a bit flip) [Laflamme96].

10.8 Quantum Circuit for Correcting a Phase Shift and/or Bit Flip Error

Once a quantum coding scheme has been devised, there is still the question of how to implement it. In this section we describe a quantum circuit that implements a particular encoding/decoding scheme. We simulate the operation of this circuit, as well as the introduction of an error into the entangled state in which the logical state of a qubit is stored. Most quantum error correction schemes follow the same general pattern of this circuit although the details will change from scheme to scheme. The simulation illustrates the key features of any quantum error correction circuitry:

1. Encode the state, $|\psi\rangle$, as an entangled state of n-qubits.
2. Simulate the introduction of an error in the encoded state to produce a new buggy state of the n-qubits.
3. Decode the (buggy) state of the entangled qubits.
4. Determine what error occurred by computing the "error syndrome."
5. Given the error syndrome, apply the appropriate unitary operator (rotation) to correct the state.

Operationally, each of these steps can be accomplished by means of a quantum circuit. In the following sections we describe a simulator for such a quantum error-correction circuit. The simulator is based on a simplified version of the Laflamme, Miquel, Paz, and Zurek scheme devised by Samuel Braunstein[Braunstein96].

Quantum Codes

Recall that the idea is to map the quantum state of each logical qubit into the entangled state of several qubits. There is a good deal of latitude as to how this can be done. In the simplified Laflamme, scheme, the eigenstates of the logical qubit $|0_L\rangle$ and $|1_L\rangle$ are encoded as follows.

$$|0_L\rangle \equiv \frac{1}{2\sqrt{2}}\begin{pmatrix} |00000\rangle + |00110\rangle + |01001\rangle - |01111\rangle + \\ |10011\rangle + |10101\rangle + |11010\rangle - |11100\rangle \end{pmatrix}$$

$$|1_L\rangle \equiv \frac{1}{2\sqrt{2}}\begin{pmatrix} |11111\rangle + |11001\rangle + |10110\rangle - |10000\rangle - \\ |01100\rangle - |01010\rangle - |00101\rangle - |00011\rangle \end{pmatrix}$$

Thus the general state of a logical qubit, $\alpha|0\rangle + \beta|1\rangle$, is mapped into the state of the five encoding qubits $\alpha|0_L\rangle + \beta|1_L\rangle$. Subsequently, if there is a phase shift error, a bit flip error, or a combined error amongst any of the five encoding qubits, there is still sufficient information in their entanglement to be able to determine the original (correct) state of the single logical qubit.

Performing the Encoding

How would you achieve such an encoding in practice? One way is to entangle the state $|\psi\rangle$ of the qubit that you wish to preserve with the states of four ancilla particles that are all in the state $|0\rangle$ initially. This can be accomplished using a quantum circuit built from rotation gates L and $L^{\dagger}$, that act on single qubits, and XOR gates (equivalent to the controlled-NOT gate, CNOT, that was introduced in Chapter 4).

The single qubit rotation gates are:

$$L = \begin{pmatrix} \frac{1}{\sqrt{2}} & -\frac{1}{\sqrt{2}} \\ \frac{1}{\sqrt{2}} & \frac{1}{\sqrt{2}} \end{pmatrix}, \quad L^{\dagger} = \begin{pmatrix} \frac{1}{\sqrt{2}} & \frac{1}{\sqrt{2}} \\ -\frac{1}{\sqrt{2}} & \frac{1}{\sqrt{2}} \end{pmatrix}.$$

Their effects can be described via their truth tables:

In[]:=

```
TruthTable[L]
```

Out[]=

$$\text{ket}[0] \to \frac{\text{ket}[0]}{\sqrt{2}} + \frac{\text{ket}[1]}{\sqrt{2}}$$

$$\text{ket}[1] \to -\frac{\text{ket}[0]}{\sqrt{2}} + \frac{\text{ket}[1]}{\sqrt{2}}$$

In[]:=

```
TruthTable[L†]
```

Out[]=

$$\text{ket}[0] \to \frac{\text{ket}[0]}{\sqrt{2}} - \frac{\text{ket}[1]}{\sqrt{2}}$$

$$\text{ket}[1] \to \frac{\text{ket}[0]}{\sqrt{2}} + \frac{\text{ket}[1]}{\sqrt{2}}$$

In quantum circuits these single qubit rotations are usually represented as labeled boxes with one input line and one output line such as shown in Fig. 10.4.

Fig. 10.4 Icons used in a quantum circuit to represent the L and $L^{\dagger}$ gates. These gates both act on a single qubit.

Likewise, the action of the XOR gate, defined by the operator:

$$\text{XOR} = \begin{pmatrix} 1 & 0 & 0 & 0 \\ 0 & 1 & 0 & 0 \\ 0 & 0 & 0 & 1 \\ 0 & 0 & 1 & 0 \end{pmatrix}$$

is summarized in its truth table:

In[]:=

```
TruthTable[XOR]
```

Out[]=

```
ket[0, 0] -> ket[0, 0]
ket[0, 1] -> ket[0, 1]
ket[1, 0] -> ket[1, 1]
ket[1, 1] -> ket[1, 0]
```

Notice that whereas the rotation gates act on single qubits, the XOR gate acts on pairs of qubits. In the truth table for XOR, the pair of bits to the left of each arrow represent the inputs and the pair of bits to the right represent the corresponding outputs. As we can see, the second output bit is always the exclusive OR of the two input bits and the first output bit simply echoes the first input bit. In quantum circuits, the XOR operation is represented, typically, by the icon illustrated in Fig. 10.5.

Armed with the 3 types of gates L, $L^{\dagger}$, and XOR, it is possible to design a quantum circuit that encodes a single qubit, in an arbitrary quantum state, into an entanglement of 5 qubits in accordance with the 5-qubit quantum code that we gave previously.

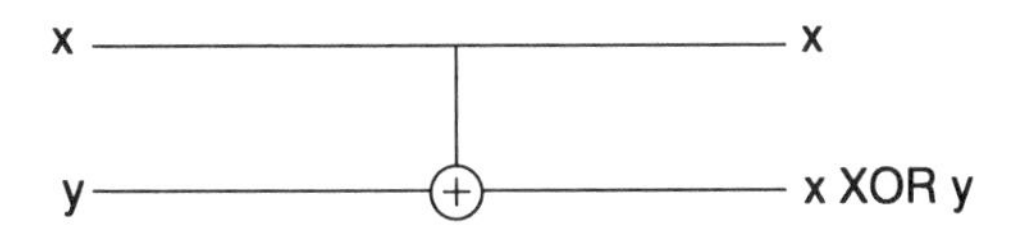

Fig. 10.5 Icon used in a quantum circuit to represent the XOR gate (also called the CNOT gate). XOR acts on two qubits.

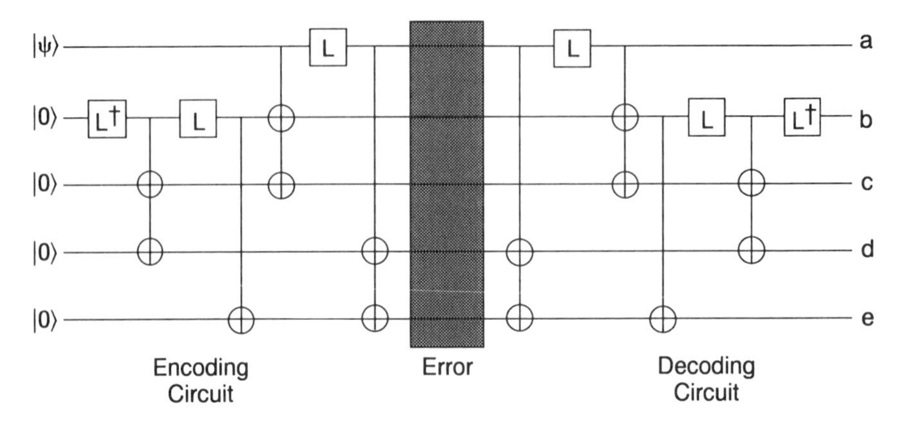

Fig. 10.6 Quantum circuit for encoding and decoding a single qubit.

The encoding step takes place in the left-hand portion of the circuit. The (logical) qubit whose state $|\psi\rangle$ is to be preserved is fed into line 1 of the circuit. Four ancilla particles, each in state $|0\rangle$, are fed into lines 2 through 5. The circuit performs the XOR, L, and $L^{\dagger}$ operations shown so that the state of the complete 5-qubit system entering the gray shaded area corresponds to the encoded qubit.

Let us simulate the action of this circuit as follows. First we create a random quantum state for a single qubit using the function `ArbitraryState` found in the electronic supplement. The first argument of `ArbitraryState` is simply the number of qubits in the (assumed pure) state being generated. The second argument merely sets the option that the result should have numeric amplitudes (rather than symbolic amplitudes, which is the default).

In[]:=
```
state1 = ArbitraryState[1, Numeric->True]
```
Out[]=
```
(0.294135 + 0.328409 I)  ket[0] +
(-0.565992 - 0.696624 I) ket[1]
```

Next we encode the chosen state, `state1`, by entangling it with 4 ancilla particles all in the state $|0\rangle$ initially. This amounts to setting the top line in the preceding circuit to `state1`, setting the other lines to $|0\rangle$, and then driving the inputs through the left-hand portion of the circuit. The resulting (encoded) state that arrives at the center of the circuit is:

In[]:=
```
encodedQubit = EncodeQubit[state1]
```

Out[]=

```
( 0.103992 + 0.11611 I)  ket[0, 0, 0, 0, 0] +
(-0.200108 - 0.246294 I) ket[0, 0, 0, 1, 1] +
( 0.200108 + 0.246294 I) ket[0, 0, 1, 0, 1] +
( 0.103992 + 0.11611 I)  ket[0, 0, 1, 1, 0] +
( 0.103992 + 0.11611 I)  ket[0, 1, 0, 0, 1] +
( 0.200108 + 0.246294 I) ket[0, 1, 0, 1, 0] +
( 0.200108 + 0.246294 I) ket[0, 1, 1, 0, 0] +
(-0.103992 - 0.11611 I)  ket[0, 1, 1, 1, 1] +
( 0.200108 + 0.246294 I) ket[1, 0, 0, 0, 0] +
( 0.103992 + 0.11611 I)  ket[1, 0, 0, 1, 1] +
( 0.103992 + 0.11611 I)  ket[1, 0, 1, 0, 1] +
(-0.200108 - 0.246294 I) ket[1, 0, 1, 1, 0] +
(-0.200108 - 0.246294 I) ket[1, 1, 0, 0, 1] +
( 0.103992 + 0.11611 I)  ket[1, 1, 0, 1, 0] +
(-0.103992 - 0.11611 I)  ket[1, 1, 1, 0, 0] +
(-0.200108 - 0.246294 I) ket[1, 1, 1, 1, 1]
```

Simulating an Error in the Encoded Qubit

After the qubit has been encoded in the joint state of 5-qubits we imagine that these 5 qubits are sitting in some quantum memory register. We assume that the register is not perfectly isolated from its environment and every now and again an interaction with the environment causes an error to be introduced into the state of one of the qubits. This process is represented by the gray box in the center of Fig. 10.6.

For example, let us introduce a combined bit flip and phase shift error in the fourth of the 5 entangled qubits. We do this using the function `BitFlipAndPhaseShiftError[`*state, i, m*`]` that takes a state and applies a bit flip and phase shift error to the *i*th of *m* qubits. Thus the state of the 5 entangled qubits, which are being used to encode one logical qubit, becomes buggy.

In[]:=

```
buggyState =
      BitFlipAndPhaseShiftError[encodedQubit, 4, 5]
```

Out[]=
```
(-0.246294 + 0.200108 I) ket[0, 0, 0, 0, 1] +
(-0.11611 + 0.103992 I)  ket[0, 0, 0, 1, 0] +
( 0.11611 - 0.103992 I)  ket[0, 0, 1, 0, 0] +
(-0.246294 + 0.200108 I) ket[0, 0, 1, 1, 1] +
( 0.246294 - 0.200108 I) ket[0, 1, 0, 0, 0] +
(-0.11611 + 0.103992 I)  ket[0, 1, 0, 1, 1] +
(-0.11611 + 0.103992 I)  ket[0, 1, 1, 0, 1] +
(-0.246294 + 0.200108 I) ket[0, 1, 1, 1, 0] +
( 0.11611 - 0.103992 I)  ket[1, 0, 0, 0, 1] +
(-0.246294 + 0.200108 I) ket[1, 0, 0, 1, 0] +
(-0.246294 + 0.200108 I) ket[1, 0, 1, 0, 0] +
(-0.11611 + 0.103992 I)  ket[1, 0, 1, 1, 1] +
( 0.11611 - 0.103992 I)  ket[1, 1, 0, 0, 0] +
( 0.246294 - 0.200108 I) ket[1, 1, 0, 1, 1] +
(-0.246294 + 0.200108 I) ket[1, 1, 1, 0, 1] +
( 0.11611 - 0.103992 I)  ket[1, 1, 1, 1, 0]
```

Error Syndrome and Corrective Actions

The buggy state is then "decoded" by pushing the state through a circuit that is the mirror image of the encoding circuit. This is represented by the right-hand portion of the circuit shown in Fig. 10.6. After this step the resulting "decoded" state of all 5 qubits is:

In[]:=
```
decodedQubit = DecodeQubit[buggyState]
```
Out[]=
```
(-0.328409 + 0.294135 I) ket[0, 1, 1, 0, 0] +
(-0.696624 + 0.565992 I) ket[1, 1, 1, 0, 0]
```

We can obtain information about this decoded, but buggy, state by measuring the state of 4 of the 5 bits in the entangled state of 5-qubits. We can think of this as measuring the bit received on 4 of the lines a, b, c, d, or e. This information gives us the "error syndrome" of the code. This is the diagnostic test by which we determine which error afflicted the 5-qubits used to store the single logical qubit.

For example, by measuring the bits received on lines b, c, d, and e, we can identify the "error syndrome." For each possible error syndrome there is a unique rotation that can be applied to the state on the remaining (unmeasured) line that will restore the state on that line to the exact state of the original logical qubit. The function `MeasureErrorSyndrome[`*state, lines*`]` takes a given state and an instruction as to which lines in the output circuit to measure and returns the error syndrome found on those lines and the projected state that results after such measurements.

```
In[]:=
  {errorSyndrome, finalState} =
        MeasureErrorSyndrome[decodedQubit, {2,3,4,5}]
Out[]=
  {{1, 1, 0, 0},
    (-0.328409 + 0.294135 I) ket[0, 1, 1, 0, 0] +
    (-0.696624 + 0.565992 I) ket[1, 1, 1, 0, 0]}
```

Thus in this case the error syndrome was found to be {1,1,0,0}. We have calculated, ahead of time, what unitary operation must be applied to the unmeasured state (the state on line 1 in this case), to restore the state on line 1 to its original state. Table 10.4 shows the complete error syndrome table for the simplified Laflamme circuit. The notation P5 means a phase shift error on the 5th qubit, B2 means a bit flip error on the second qubit, BP4 means both a bit flip and phase shift error on the fourth qubit, and so on.

Table 10.4 Error syndromes and the required corrective actions for the 16 possible types of errors described in the left column.

Type of Error	Syndrome (on lines 2–5)	Corrective Action
P5	{0,0,0,0}	$\begin{pmatrix} 1 & 0 \\ 0 & -1 \end{pmatrix}$
P1	{0,1,1,0}	
P4	{1,1,1,0}	
BP5	{0,0,0,1}	$\begin{pmatrix} -i & 0 \\ 0 & i \end{pmatrix}$
BP4	{1,1,0,0}	
B2	{0,0,1,0}	$\begin{pmatrix} 0 & 1 \\ -1 & 0 \end{pmatrix}$
B4	{1,0,0,1}	
B5	{1,0,1,0}	
No Error	{1,0,1,1}	
B3	{1,1,1,1}	
P3	{0,0,1,1}	$\begin{pmatrix} 0 & 1 \\ 1 & 0 \end{pmatrix}$
B1	{1,0,0,0}	
P2	{1,1,0,1}	
BP2	{0,1,0,0}	$\begin{pmatrix} 0 & -i \\ -i & 0 \end{pmatrix}$
BP3	{0,1,1,1}	
BP1	{0,1,0,1}	$\begin{pmatrix} i & 0 \\ 0 & i \end{pmatrix}$

The table entries are encoded in the function `RotateState`. In our example, the error syndrome was {1,1,0,0} so the correct rotation to apply to the state on line 1 of the output is $\begin{pmatrix} -i & 0 \\ 0 & i \end{pmatrix}$.

In[]:=
```
RotateState[finalState, errorSyndrome]
```
Out[]=
```
(0.294135 + 0.328409 I) ket[0, 1, 1, 0, 0] +
(-0.565992 - 0.696624 I) ket[1, 1, 1, 0, 0]
```

Compare this to the original state of the qubit:

In[]:=
```
state1
```
Out[]=
```
(0.294135 + 0.328409 I) ket[0] +
(-0.565992 - 0.696624 I) ket[1]
```

They are the same, indicating that the error has been completely eliminated by the error-correction process. Notice that the error correction was accomplished without ever measuring the state of the logical qubit that we were trying to protect.

10.9 How Many Errors Can Be Tolerated?

The circuit shown in Fig.10.6 can correct a single phase shift error, a single bit flip error, or a single combined error in a group of 5 entangled qubits used to encode a single logical qubit. In other words, the state of a single qubit can be reliably restored by partitioning it amongst the entangled state of 5-qubits. As realistic quantum computations would require many qubits, it is worth asking how error-tolerant various encoding schemes could be. Specifically, if l logical qubits are encoded in the entangled state of n physical qubits, what is the greatest number of phase shift, bit flip, and combined errors, t_{max} that the encoded qubits can tolerate so that the state of the original qubit is still guaranteed to be retrievable? Arthur Ekert and Chiara Macchiavello of the Clarendon Laboratory at Oxford University have developed, for a certain class of codes, a rough quantum analogue of the Hamming bound, a famous inequality arising in classical coding theory[Ekert96]. This suggests a crude relationship between the parameters l, n, and t_{max} mentioned previously:

$$2^{l}\sum_{i=0}^{t} 3^{i}\binom{n}{i} \leq 2^{n}.$$

Unfortunately, this condition is neither *necessary* nor *sufficient*, but it seems to provide a pretty good bound in practice. Specifically, although the condition does not *guarantee* that we can always find a code having a set of parameter values for `l`, n, and t_{max} that satisfy the Ekert and Macchiavello "quantum Hamming bound", it does, nevertheless, provide a good rule of thumb for characterizing the kinds of quantum codes that might be feasible. For example, using the given bound, this suggests the greatest number of errors that can be tolerated is given by finding, for a given `l` and n, the largest t, t_{max} (still an integer) such that the inequality remains satisfied. We plot the resulting surface in Fig. 10.7.

Thus, according to the (rough) quantum Hamming bound, we can correct up to 6 errors amongst the entangled qubits if we encode 1 logical qubit amongst the correlations among 30 physical qubits. However, we stress that it may not be possible, in fact, to find such a code.

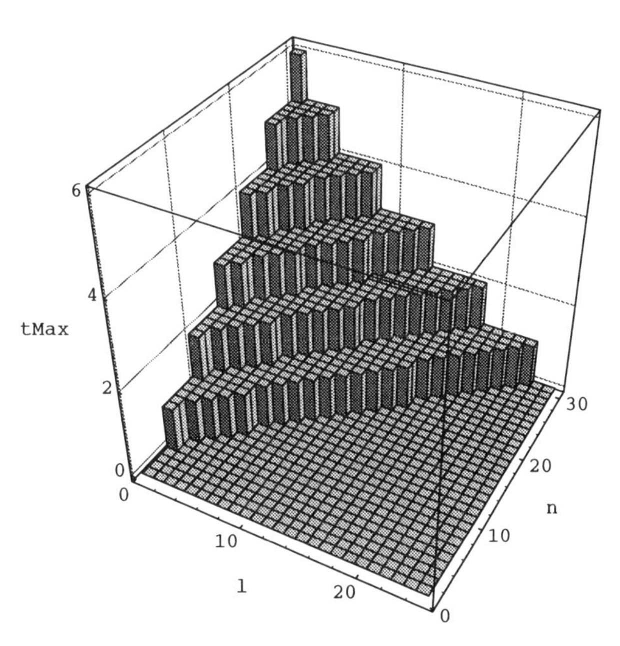

Fig. 10.7 The maximum number of errors that can be corrected using n physical qubits to encode the state of `l` logical qubits.

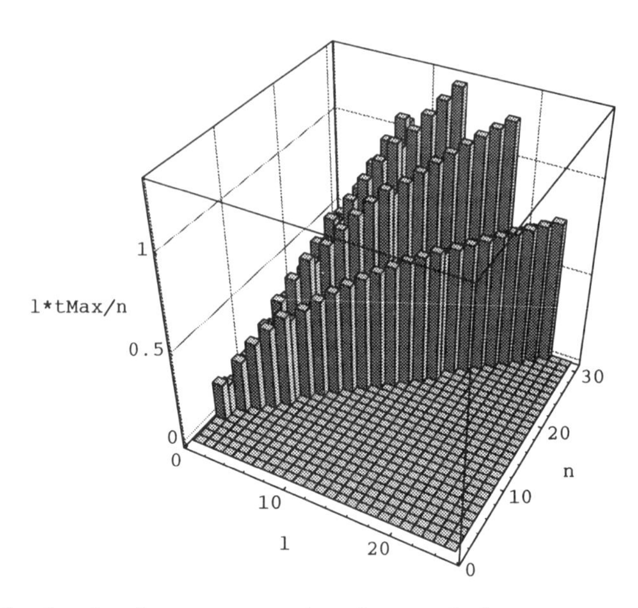

Fig. 10.8 Plot showing the parameter values that encode the greatest number of logical qubits, tolerant to the greatest number of errors, using the fewest physical qubits.

A related question concerns the *efficiency* of the various encoding schemes. Roughly speaking, a good encoding scheme would encode the greatest possible number of logical qubits that can tolerate the greatest possible number of errors, using the smallest possible number of physical qubits. Hence a measure of the efficiency of the coding scheme is to plot the quantity $\mathtt{l}t_{max}/n$ versus $\mathtt{l}$ and n.

Figure 10.8 suggests that of all the possible coding schemes, there are some that make better use of the quantum resources than others. For the range of values of $\mathtt{l}$ and n that we plotted, the best combination of parameters for a quantum code, by the $\mathtt{l}t_{max}/n$ measure, is to encode 13 logical qubits ($\mathtt{l}=13$) in 30 physical qubits ($n = 30$). Such a code ought to be resilient to up to 3 phase shift and/or bit flip errors amongst the 30 entangled qubits. Thus the theoretical bounds on quantum coding schemes provide some guidance as to the best coding scheme to use.

10.10 Computing Forever Without Error

The field of quantum error correction is advancing at an astonishing pace. In the past year several proposals have emerged for various encoding schemes in addition to designs for feasible implementations[Plenio96]. Recently, Peter Shor has discovered so-called "fault-tolerant" quantum error-correction schemes. Fault-tolerant codes can correct errors in entangled qubits used to represent a logical qubit even if they occur *during* the error-correction steps [Shor96], [DiVincenzo96]. Moreover, Raymond Laflamme, and Wojciech Zurek have shown that we can perform logical operations on the encoded qubits themselves without having to decode the state before it is used[Zurek96]. As understanding of quantum error correction schemes has deepened, the symmetries among the various protocols have been clarified using the mathematical tool of group theory[Calderbank96]. However, the most important discovery on quantum codes appeared quite recently when several researchers described so-called "concatenated" quantum codes. These are codes in which each code qubit is itself further encoded[Knill96], [Knill96a]. Thus a single qubit is encoded in a hierarchical tree of entangled qubits. Remarkably, if the individual probability of an error can be pushed below a certain nonzero threshold, then the use of concatenated codes allows correctable quantum computations of unlimited duration! In other words, we can quantum compute forever.

These results, and many others besides, point to the fact that the errors that inevitably arise during quantum computation due to decoherence may not be as disastrous as was, at one time, believed. By a judicious choice of naturally decoherence-resilient materials and a fault-tolerant, concatenated code, error-correction scheme, the possibility of a demonstration of a specialized quantum computer for a nontrivial quantum computation appears to be quite a reasonable proposition. In the next chapter we take a look at some of the candidate designs for such a device[Preskill97].

CHAPTER 11

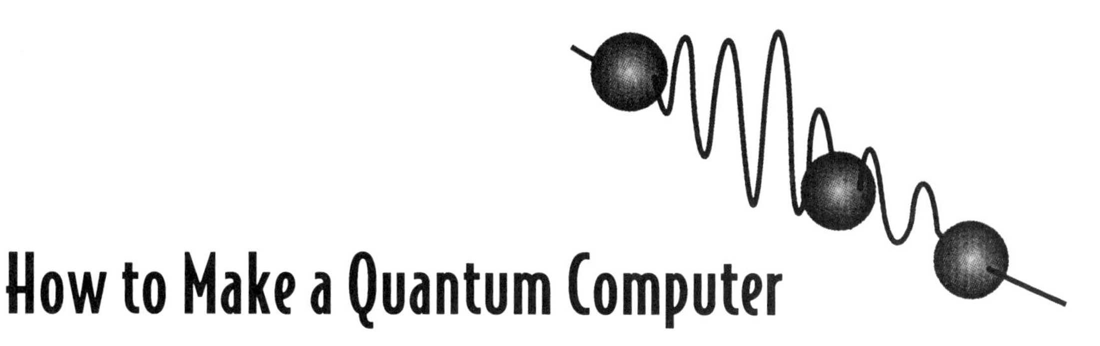

How to Make a Quantum Computer

"We believe that the present system provides a realistic implementation of a quantum computer which can be built with present or planned technology."
— J. I. Cirac and P. Zoller

In the previous chapters we have built up a case for the theoretical feasibility of quantum computers, even giving a real-life example of a specialized quantum computer, the quantum cryptography machine. But now we are finally in a position to talk about the technical feasibility of a general-purpose quantum computer, how it might be built, and its limitations. Many candidate schemes have already been proposed and more are appearing with surprising regularity. We will describe the four most promising schemes. We begin by considering a one-dimensional molecular computer that acts as a digital computer and then extend it to include superposition so that it behaves as a true quantum computer. Then we look at trapped ions, cavity QED, and NMR-based quantum computers.

11.1 Heteropolymer-Based Quantum Computers

A group from the Institut für Theoretische Physik at the University of Stuttgart, W. G. Teich, K. Obermayer, and G. Mahler[Teich88] have designed a molecular quantum switch that we use as the basis for our discussion. Their model was built upon by Seth Lloyd[Lloyd93] who extended the model to include superpositions. Both models can be extended to use a heteropolymer as the substrate for a multiprocessor system that acts as the hardware for the computer. So-called "quantum dots" would work in this architecture as well. A quantum dot is an electron confined to a small region where its energy can be quantized in a way that isolates it from other dots. The "corral" that contains the electrons uses "fences" of electromagnetic fields either from other atomic electric fields or from standing waves of photons.

Basically, the idea behind the heteropolymer computer is to use a linear array of atoms as memory cells. Each atom can be either in an excited or a ground state which gives the basis for a binary arithmetic. Neighboring atoms interact with each other in a way that enables a conditional logic to be supported. For example, an atom in an excited state that has both neighbors in the ground state will behave differently than an atom whose neighbors are in different excited states. The specifics of conditional logic for molecular gates are presented in detail in the following.

The software for the heteropolymer computer consists of a sequence of laser pulses of particular frequencies that induce transitions in certain atoms of the polymer.

The Stuttgart model is a molecular digital computer that relies on transitions among energy levels in atoms to switch states. Each atom has a three-energy level profile as shown in Fig. 11.1.

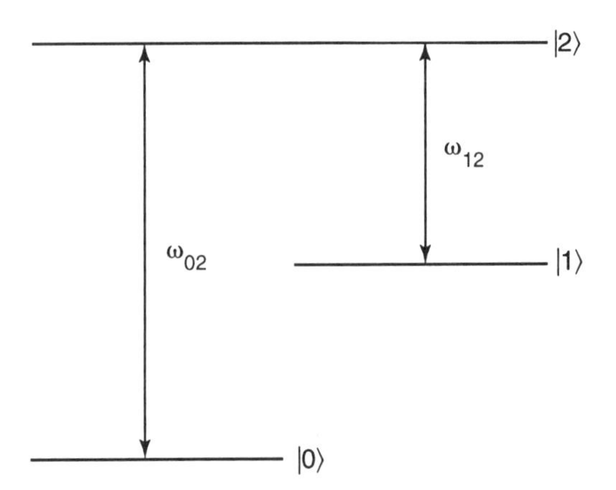

Fig. 11.1 A quantum switch utilizing the ground state, a metastable state, and an excited state of an atom. The frequencies, ω_{02} and ω_{12}, denote the photon energy needed at input to make an up transition or the energy released by a down transition.

State 0 ($|0\rangle$) is the ground state, state 1 ($|1\rangle$) is a *metastable* excited state, and state 2 ($|2\rangle$) is a rapidly decaying excited state to either $|0\rangle$ or $|1\rangle$. Metastable means the transition between $|1\rangle$ and $|0\rangle$ is very slow compared to the other transitions to $|0\rangle$; so for practical purposes we can ignore these slow transitions. In a computer this means that the processing done with the atom in the metastable state takes place before the transition $|1\rangle \to |0\rangle$ is likely to take place. As with all quantum phenomena, the decay of any excited state will be a random process. In this case, the decay will be according to an exponential distribution. Thus there is always some chance for the transition to take place, even if we have not instigated it as part of our computation.

Transitions between the atomic states are accomplished with laser pulses of the appropriate frequency; for example, the transition $|0\rangle \leftrightarrow |2\rangle$ proceeds in either direction when a laser pulse with photons of frequency $\omega_{02} = (E_2 - E_0)/\mathrm{h}$. In the case of the transition $|0\rangle \to |2\rangle$, the laser photon is absorbed and in the transition $|2\rangle \to |0\rangle$, the laser photon acts to stimulate the transition which results in the emission of a photon of the same energy as the laser pulse. The stimulated transition is the reason that a laser pulse of energy $\omega_{01} = (E_1 - E_0)/\mathrm{h}$ is not used to directly get the atom to $|1\rangle$.

Each laser pulse contains many photons because transitions are probabilistic; that is, not every photon will cause a transition. When the atom is excited to $|2\rangle$ it can decay to either $|0\rangle$ or $|1\rangle$. If it decays back to the ground state it will be immediately re-excited by the laser to $|2\rangle$. In other words, while the laser with frequency ω_{02} is turned on, the only (relatively) stable state is $|1\rangle$. In fact, since $|1\rangle$ is metastable the atom will remain in this state for some time after the laser is turned off.

So far this is not a very interesting device. Although it does have two stable states, if we were to have many of them together all we would be able to do would be to drive them all into one state or the other. To be useful as a computation device what we really need is the ability to address a particular atom and change its state without affecting the other atoms. We could use different atoms each with a different excitation energy, but we would need to have many different atoms with specific properties and there is no guarantee that any device could be made with these atoms that may have very different chemical properties. Even if we could accomplish this feat all we would have would be a collection of one-bit processors completely uncoupled to each other.

A much more interesting device is one that has some interaction among neighboring atomic one-bit processors which would then be a useful addressable switching device. What we need then is to have a *conditional dynamics* whereby transitions are made not just on the state a particular atom is in, but also depending on the states of neighboring atoms. Fortunately, Nature has provided us with just the interaction we need to accomplish this.

Logic Gates

If an atom does not have a spherically symmetric charge distribution, then it will have what is called a *dipole moment*. The resulting dipole field strength drops off as $1/r^3$ rather than the normal $1/r$ for an electric field interaction where r is the distance between the atoms. It turns out that when two atomic dipole fields are brought near enough to each other they will shift the atomic energy levels of each other by a measurable amount. Since the dipole field strength depends on an asymmetric charge distribution, a particular state of one atom will result in a particular field distribution and consequently a different energy shift for nearby atoms. For the purposes of simplicity we assume that because the dipole field drops off so rapidly with distance that only nearest-neighbors will have an effect on the energy levels. In fact, we can guarantee this is the case by using a wide enough bandwidth on the laser (so that only atoms with an energy level in a certain range will be affected) and having slightly unequal spacing between the atoms. Our energy diagram for two atoms A and B, each a three-level system as shown in Fig. 1.1, but now with dipole interactions looks like Fig. 11.2. Note that we now have to specify the state of atom B in order to know which frequency will be needed to affect a transition in atom A.

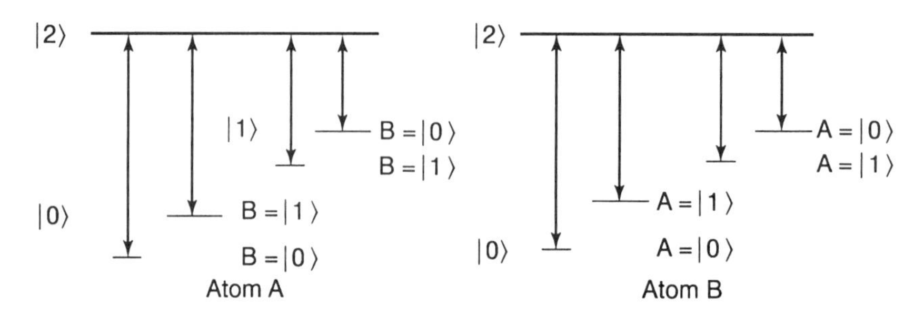

Fig. 11.2 Renormalized energy levels in a 2-state three-level atomic system used to implement conditional dynamics. Energy levels are not to scale.

One of the simplest operations that can be performed with a computer is to copy information (i.e., classical bits) from one location to another. For example, suppose we wish to copy the data in atom B to atom A. If B is in state 0, $|b=0\rangle$, and A is in state 1, $|a=1\rangle$, then we can change A to $|a=0\rangle$ by shining light with frequency ω_{A21}^{B0} which will cause A to transition to highly unstable state $|a=2\rangle$ which then rapidly decays to state $|a=0\rangle$. If on the other hand A is in $|a=0\rangle$, then we do not have to do anything and the laser pulse ω_{A21}^{B0} will have no effect on the state of A. By the same token, if B is in $|b=1\rangle$, then if we apply a laser pulse of frequency ω_{A20}^{B1} and if A is in $|a=0\rangle$, it will make a transition to $|a=2\rangle$ and then decay to $|a=1\rangle$. If A is already in $|a=1\rangle$, then the laser pulse ω_{A20}^{B1} will have no effect on A. Thus by applying two simultaneous pulses of frequency ω_{A21}^{B0} and ω_{A20}^{B1} we can copy the contents of B into A no matter what the state of B or A. This copying is summarized in Table 11.1. Note that this copying of a state is not the same as teleporting a state nor does it violate the quantum cloning theorem. When we say we have "copied the state" all that we have done is to give one atom the same bit value (corresponding to a particular energy level) as another atom. Quantum mechanics allows us to make an exact copy of a bit but it does not allow us to make an exact copy of a qubit in an unknown (superposed) state. Nevertheless, new techniques are appearing that allow an *approximate* copy of a qubit to be made[Buzek96, Buzek97].

On a practical level, the length of time of the laser pulse determines how reliably the information has been copied. The longer the pulse, the more likely the transition has been made. However, the length of the pulse must not be so long that the information to be copied is lost through decay of the state being copied. In this sense the switch bears a superficial similarity to so-called "flash" memory which must be refreshed periodically or else its information will be lost. The difference, of course, is that flash memory decay is

Table 11.1 Copying a bit.

Initial A	Initial B	Final A	Final B	Effective Pulse
$\|a=0\rangle$	$\|b=0\rangle$	$\|a=0\rangle$	$\|b=0\rangle$	none
$\|a=0\rangle$	$\|b=1\rangle$	$\|a=1\rangle$	$\|b=1\rangle$	ω_{A20}^{B1}
$\|a=1\rangle$	$\|b=0\rangle$	$\|a=0\rangle$	$\|b=0\rangle$	ω_{A21}^{B0}
$\|a=1\rangle$	$\|b=1\rangle$	$\|a=1\rangle$	$\|b=1\rangle$	none

not occurring at the level of individual atoms or molecules. As we see a little later, the length of this pulse can be used to put the system into a superposition for true quantum computation and not just molecular level digital computation.

From Bits to Qubits

Up to now we have discussed a molecular computer capable of classical computation. How can we use this architecture to build a quantum computer? Recall from the discussion earlier that it was important to shine the laser long enough to induce the transition from one state to another. What happens physically is that there is an induced electric dipole moment between $|0\rangle$ and $|1\rangle$. The effectiveness of the laser pulse in inducing the transition depends on the angle between the dipole moment and the polarization vector of the laser photon's electric field and the magnitude of the photon's electric field over the time of the pulse. Mathematically,

$$\frac{1}{\hbar}\int \vec{\mu}\cdot\hat{e}\,E(t)\,dt=\phi,$$

where $\vec{\mu}$ is the induced dipole moment, $\hat{e}$ is the polarization vector of the photon, $E(t)$ is the time-varying magnitude of the photon electric field, and ϕ is some angle. What this integral represents is the angle over which the dipole moment precesses during the time of the pulse.

To create a superposition we do not want to be in a particular state with high certainty; we want to be in more than one state, each with a significant probability. We can achieve this by using a "$\pi/2$"-pulse ($\phi = \pi/2$) instead of a "π"-pulse ($\phi = \pi$). This puts the selected atom, now a qubit instead of a bit, into a state where, if measured, it will be in $|0\rangle$ half the time and $|1\rangle$ half the time. If we start with a triple of atoms, denoted ABC, each in the ground state, then by a sequence of pulses we can put the triple into a superposition with Einstein-Podolsky-Rosen correlations as shown in Table 11.2. In general a sequence of pulses would effect a triple by an arbitrary phase angle and would give us a superposed state looking like[Lloyd93a]

$$\frac{1}{\sqrt{2}}\left|A\overset{\circ}{\bullet}B\overset{\circ}{\bullet}C\overset{\circ}{\bullet}\right\rangle+\frac{e^{i\phi}}{\sqrt{2}}\left|A\overset{\bullet}{\circ}B\overset{\bullet}{\circ}C\overset{\bullet}{\circ}\right\rangle=\frac{1}{\sqrt{2}}|000\rangle+\frac{e^{i\phi}}{\sqrt{2}}|111\rangle,$$

Table 11.2 Creating superpositions with pulse sequences.

Pulse	Freq.	State Transformation Caused By Pulse
$\pi/2$	ω_{0A0}	$\lvert 0\rangle_A \to \frac{1}{\sqrt{2}}\left(\lvert 0\rangle_A - \lvert 1\rangle_A\right)$
π	ω_{1B0}	$\frac{1}{\sqrt{2}}\left(\lvert 0\rangle_A - \lvert 1\rangle_A\right) \otimes \lvert 0\rangle_B \to \frac{1}{\sqrt{2}}\left(\lvert 00\rangle_{AB} - \lvert 11\rangle_{AB}\right)$
π	ω_{1C0}	$\frac{1}{\sqrt{2}}\left(\lvert 00\rangle_{AB} - \lvert 11\rangle_{AB}\right) \otimes \lvert 0\rangle_C \to \frac{1}{\sqrt{2}}\left(\lvert 000\rangle_{ABC} - \lvert 111\rangle_{ABC}\right)$

where the dark circle symbolizes an energy level occupied by an electron and an open circle indicates an available energy level, ϕ is the phase that depends on the particular pulses that have been applied and the $1/\sqrt{2}$ is for normalization purposes.

Loading Information

For the purpose of simplicity and ease of presentation we now consider each atom in the ABC cell of a long heteropolymer to be a 2-level system consisting of a ground state and a metastable excited state as before. Thus a pulse with frequency ω_{0b1} will change the state of the cell from $\lvert 0b1\rangle$ to $\lvert 0\bar{b}1\rangle$, where b is the state of atom B and $\bar{b}$ is the new state of atom B (its binary complement); that is, atom B's state is conditionally changed when atom A is $\lvert 0\rangle$ and when atom C is $\lvert 1\rangle$. Note that all cells in the heteropolymer with atom A in $\lvert 0\rangle$ and atom C in $\lvert 1\rangle$ will undergo this transformation achieving a parallelism in updating information. If we were to adjust the energy levels slightly, say, by moving the atoms in each cell to different distances from each other, we could selectively transform a particular cell and thus in effect address a particular bit.

Moving Bits

We can also exploit the asymmetry of a heteropolymer by noting that the A atom on the lead end of the polymer has no neighbor on one side and so will have a unique excitation energy with respect to the other atoms in the chain. This feature can be used to load information from one end. Suppose we start with the cells all in their ground state, which we can do either by appropriate laser pulsing or just by waiting for the excited states to decay. If we want to move a state (bit) along the string of cells we can apply a sequence of laser pulses so that the string of cells behaves like a group of Fredkin gates linked together by wires. For example, a bit can be moved (as opposed to copied as we showed earlier) from A to B with a sequence of five pulses as shown in the following.

$$\{(\omega_{0A1}+\omega_{1A1}),\omega_{1B1},(\omega_{0A1}+\omega_{1A1})\}|x01\rangle \rightarrow |0x1\rangle.$$

Again the frequencies of the laser pulses ω_x represent operators corresponding to a laser pulse of frequency ω_x. The subscript refers to the conditional dynamics, the leftmost subscript denoting the state of the leftmost atom of a triplet (one of $A_nB_nC_n$, $B_nC_nA_{n+1}$, or $C_{n-1}A_nB_n$), the middle subscript denoting the state of the middle atom and the rightmost subscript denoting the state of the rightmost atom. The + denotes simultaneous pulses. Pulses are applied from right to left as operators on the state (thus $\omega_{0A1}+\omega_{1A1}$ is the first pulse and so on). Table 11.3 shows all the possible configurations and the effect of each pulse in detail.

The preceding implementation describes a molecular digital computer that reaches the one atom per bit level. Again, on a practical level some sort of substrate will be needed to support this processing just as DNA relies on a phosphate backbone to support its double helix structure. Thus, when all things are taken together in the full computer, there is more than one atom per bit.

Performing Computations

Once a scheme for loading data onto the computer has been devised, the next question is how to make the computer perform various computations on those data. In the heteropolymer-based model, each program corresponds to a particular sequence of laser pulses[Teich88]. The frequency of each pulse determines which atoms, in which states, are the ones manipulated, and the duration and shaping of each pulse controls the detailed nature of the operation that is applied.

Table 11.3 Effect of pulse sequence.

Initial A	Initial B	Effect of 1st Pulse	Effect of 2nd Pulse	Effect of 3rd Pulse	Final A	Final B
0	0	none	none	none	0	0
0	1	$A \rightarrow 1$	$B \rightarrow 0$	none	1	0
1	0	none	$B \rightarrow 1$	$A \rightarrow 0$	0	1
1	1	$A \rightarrow 0$	none	$A \rightarrow 1$	1	1

Table 11.4 Logical formulae obtained from a 2-particle system under various sequences of laser pulses. The value of the Boolean formula is encoded in the final state of particle A.

Function	Pulse Sequence (i.e., "program")
0	$[\omega^{B0}_{A21}, \omega^{B1}_{A21}]$
$a \cdot b$	ω^{B0}_{A21}
$a \cdot \bar{b}$	ω^{B1}_{A21}
a	no pulse necessary
$\bar{a} \cdot b$	$\omega^{B0}_{A20} + [\omega^{A1}_{B20}, \omega^{B0}_{A21}] + [\omega^{B0}_{A20} + \omega^{B1}_{A21}]$
b	$[\omega^{B1}_{A20}, \omega^{B0}_{A21}]$
$(\bar{a} \cdot b) + (a \cdot \bar{b})$	not possible
$a + b$	ω^{B1}_{A20}
$\bar{a} \cdot \bar{b}$	$\omega^{B1}_{A20} + [\omega^{A1}_{B20}, \omega^{A0}_{B21}] + [\omega^{B0}_{A20}, \omega^{B1}_{A21}]$
$(\bar{a} \cdot \bar{b}) + (a \cdot b)$	not possible
$\bar{b}$	$[\omega^{B0}_{A20}, \omega^{B1}_{A21}]$
$a + \bar{b}$	ω^{B0}_{A21}
$\bar{a}$	$[\omega^{A1}_{B20}, \omega^{A0}_{B21}] + [\omega^{B0}_{A20}, \omega^{B1}_{A21}]$
$\bar{a} + b$	$\omega^{B1}_{A21} + [\omega^{A1}_{B21}, \omega^{A0}_{B21}] + [\omega^{B0}_{A20}, \omega^{B1}_{A21}]$
$\bar{a} + \bar{b}$	$\omega^{B0}_{A21} + [\omega^{A1}_{B21}, \omega^{A0}_{B21}] + [\omega^{B0}_{A20}, \omega^{B1}_{A21}]$
1	$[\omega^{B0}_{A20}, \omega^{B1}_{A20}]$

For example, with two atoms A and B we can implement various Boolean functions. The initial states of the atoms correspond to the initial inputs to the Boolean function. The final state of atom A corresponds to the output of the Boolean function. If atom A is in state "a" initially and atom B is in state "b" initially, Table 11.4 summarizes the Boolean functions that are achieved using the pulse sequences shown. Pulses in square brackets, [], can be applied simultaneously.

Reading the Answer

As we have discussed earlier, getting an answer from a quantum computer is not the triviality it is in the classical computer. In the 3-level system we have been discussing, the measurement process is accomplished by means of a process called *resonance fluorescence* [Teich90]. This is done by adding a fourth state $|3\rangle$ to the system that is strongly coupled to the ground state but not to $|1\rangle$; that is, it decays more rapidly to $|0\rangle$ than to $|1\rangle$. This is shown in Figure 11.3.

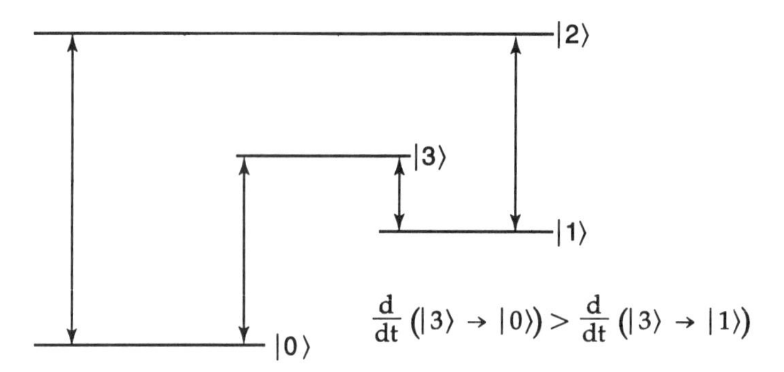

Fig.11.3 Measuring the state of the system is done via resonance fluorescence through a fourth state, $|3\rangle$ which decays much more rapidly to $|0\rangle$ than to $|1\rangle$; that is, $|3\rangle$ is more strongly coupled to $|0\rangle$ than to $|1\rangle$.

Thus by applying a laser pulse with frequency ω_{30} the system will only scatter photons (for later detection) of the same energy if it is in state $|0\rangle$; that is, it will first absorb the photon from the laser beam and then quickly re-emit a photon where its presence can be detected. If the system is in $|1\rangle$, then no absorption and hence no emission (scattering) and therefore no detection will occur. In practical terms, we know with high probability that if we detect a photon of energy $\hbar\omega_{30}$ in roughly the decay time of $|3\rangle \to |0\rangle$, then we know the atom was in $|0\rangle$. Similarly, if no photon is detected during this time we can be confident that the atom was in $|1\rangle$. Thus, by the presence or absence of a photon of a particular energy within a certain time interval, we can determine the value of any of the molecular bits.

Some people have suggested that rather than trying to build our own quantum computers we might be able to exploit naturally occurring quantum computers. This may sound extraordinary, but pretty much any molecule can be thought of as a quantum computer; it is just that most of them do not perform very interesting computations. However, Michael Biafore of the Massachusetts Institute of Technology has proposed a quantum cellular automaton whose Hamiltonian is realized in Nature in the form of the rare-earth compound praseodymium ethyl sulfate at a temperature of 1 Kelvin [Biafore94].

Likewise, another group of researchers has conjectured that certain one-dimensional magnetic Ising systems could function as a quantum computer[Berman94].

11.2 Ion Trap-Based Quantum Computers

In addition to the optically pulsed heteropolymer computer that we previously described, there are many other proposals for physical realizations of quantum computers. Some of these rely on photons that travel through beamsplitters, nonlinear media, and phase shifters to achieve the desired quantum computations[Chuang95], [Hagelstein95]. Others rely on nuclear spins. The latter have the advantage of much longer coherence times due to their smaller coupling with the environment[DiVincenzo95a].

A particularly promising scheme was proposed in 1995 by J. I. Cirac and P. Zoller from the Institut für Theoretische Physik at the Universität Innsbruck[Cirac95]. The Cirac-Zoller scheme uses a linear array of "trapped ions" as the basis for a quantum memory register. The trapping is arranged by way of electromagnetic fields, with the logical states of the qubits encoded in the energy states of the individual ions and in the vibrational modes between the ions (Fig. 11.4).

Like the atoms in the heteropolymer, each ion is considered as a 2-state system containing a ground state and an excited state, and thus can serve as a single qubit. The ions are arranged in a linear array such that each ion can be irradiated with light from its own laser. Such laser pulses have the effect of exciting specific transitions in specific ions allowing the array (i.e., the quantum memory register) to be placed in arbitrary superposed states.

There are two modes of excitation in the ion trap. One is the motion of the center of mass of the ions. In this mode each of the ions sloshes back and forth with the same frequency. This center-of-mass excitation has a different energy than the other modes so that it can

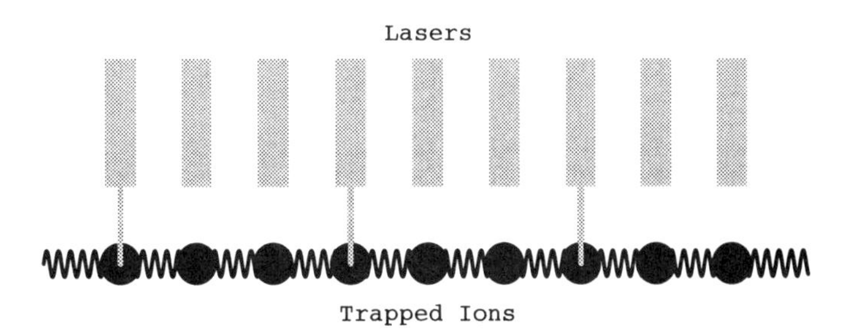

Fig. 11.4 Linear array of trapped ions.

be excited without disturbing any of the other states of the ion array. The coupling of the ion motion is via electrostatic repulsion which dominates any other force over a distance of a few optical wavelengths. Nonlocal entanglement among individual qubits is accomplished by transferring internal atomic coherence to and from the center-of-mass motion that is shared by all the ions. Thus the center-of-mass motion acts as a kind of "bus" to carry out manipulations on the individual qubits.

Very soon after the Cirac-Zoller proposal was announced, a group from the National Institute of Standards and Technology in Boulder, Colorado produced a working model of a quantum logic gate implemented as a trapped ion. The gate performs the controlled-NOT operation[Monroe95], [Schwarzschild96].

Recall that the controlled-NOT relies on the states of two bits for its output; that is, the output is the **NOT** of bit 2 if bit 1 is $|1\rangle$, whereas bit 2 is unchanged if bit 1 is $|0\rangle$. Rather cleverly, the Boulder group were able to encode *both* bits within a single Beryllium ion (Be^+) in an ion trap, which incidentally does even better than the "one atom per bit" limit. The 2-bit encoding is achieved by relying on the quantization of the energy levels of the outermost electron in a Be^+ ion and then putting this ion in a trap which, in turn, creates a quantization of vibrational energy levels of the atom as a whole (rather than that of the outermost electron of the atom; see Fig. 11.5).

The ion sloshes back and forth inside the trap, vibrating like a quantum harmonic oscillator. The trap is made from radio frequency waves rather than from atomic barriers.

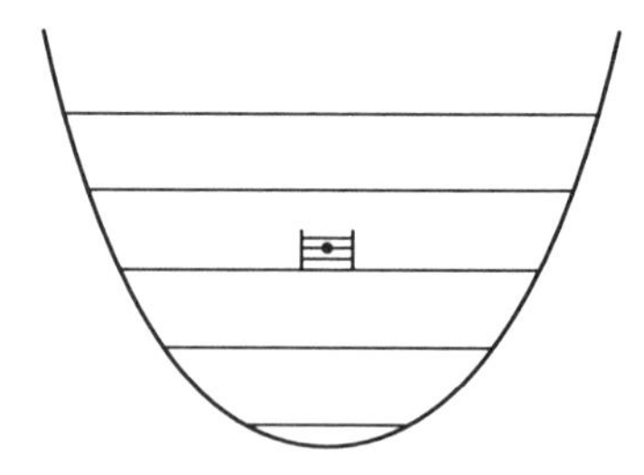

Fig. 11.5 Encoding of bits in an ion in a trap. The parabola represents the harmonic oscillator created by electromagnetic waves. The smaller energy levels at the third harmonic level represent the energy levels of the ion. The black dot is the energy level of the electron. By using the energy levels in the ion and the harmonic trap we have effectively used one atom to encode two bits.

Table 11.5 The operation of a controlled-NOT gate.

$\lvert cd\rangle$	controlled-NOT $\lvert cd\rangle$
$\lvert 00\rangle$	$\lvert 00\rangle$
$\lvert 01\rangle$	$\lvert 01\rangle$
$\lvert 10\rangle$	$\lvert 11\rangle$
$\lvert 11\rangle$	$\lvert 10\rangle$

When the Boulder group measured the performance of their ion-trap controlled-NOT gate, they discovered that the trap performed correctly about 90% of the time and had a decoherence time of about a millisecond. The device works in a similar fashion to the heteropolymer in that individually targeted laser beams can zap specific ions with $\pi/2$ and π pulses to set up specific superpositions.

Recall that the controlled-NOT has the effect shown in Table 11.5. The left column is the ket containing the control bit (first bit) and the data bit (second bit). The right column is the result of applying the `controlled-NOT` operation on the input ket.

Ions Traps for Factoring

The success of the controlled-NOT experiment sparked a lot of interest in the potential for an ion trap quantum computer to factor integers using Shor's algorithm. Cirac and Zoller had already done a preliminary analysis of how to compute Fourier transforms, a key step in Shor's algorithm, with an ion trap that used just eight barium ions.

Several researchers have built on these results. In particular, a group at Los Alamos National Laboratory has done a study of the relative merits of using different types of ions in the ion trap. Their results show that Shor's algorithm is more than a mere theoretical possibility and might someday be implemented on a real quantum computer. The decoherence time is of the order of tens of milliseconds which is comparable to the time needed to perform the transforms needed to factor very small integers. Unfortunately, the ion trap model, as it stands, is not a complete solution because the results are somewhat sensitive to interaction times, laser detunings, pulse phases, and ion positions. The researchers are hopeful, however, that a "proof of concept" quantum computer might be built within the not too distant future. Still we should not underestimate the practical difficulties of operating real ion trap-based quantum computers.

Figures 11.6 and 11.7 plot the bounds on the number of bits that can be factored without loss of coherence using Shor's algorithm using trapped ions[Hughes96]. In Fig. 11.6 the log of the number of laser pulses is plotted versus the number of trapped ions needed for the computation. The results for three different atoms are plotted; mercury, calcium, and barium. The solid curve with the dots shows the maximum number of bits that can be factored. These results are shown in Table 11.6 along with the physical parameters used in generating the plot. Without going into any details the number of pulses needed to do the factoring is:

$$N_{\max} = \frac{2Z\tau}{L^{1.84} A^{1/2} F^{3/2} \lambda^{3/2}},$$

where L is the number of trapped ions, Z is the degree of ionization of the ions ($Z = 1$), τ is the decoherence time, A is the atomic number of the respective ion, λ is the wavelength of the laser, and F is a measure of the focusing capability of the laser. Note that the values in the table all have used a value $F = 1.5$.

Although these numbers are not terribly encouraging, we should also point out that there are atoms with much longer-lived transitions, such as ytterbium (Yb). Yb has a transition with a lifetime of 1533 *days*. From Fig. 11.7 the maximum number of bits that can be

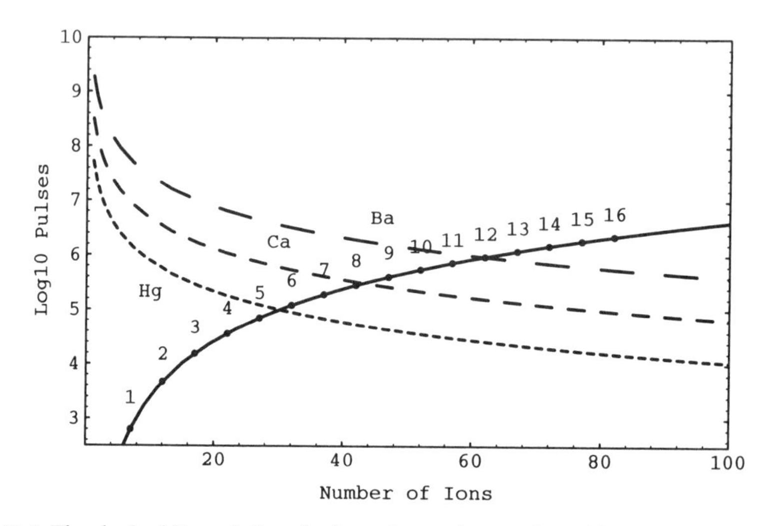

Fig. 11.6 The dashed lines define the bounds on the number of laser pulses that can be performed on various sizes of ion traps before coherence is lost for three different types of ions (mercury Hg, calcium Ca, and barium Ba). The solid line shows the number of laser pulses and ions needed in order to perform Shor's algorithm to factor a number requiring from 1 to 16 bits. The intercepts of the curves give an indication of the largest integers that can be factored using the kinds of ions shown.

Table 11.6 The size of the largest number that can be factored using an ion-trap quantum computer for various types of ions.

Atom	Z	A	$\tau(s)$	$\lambda(nm)$	N_{max}
Hg	1	198	0.1	282	5
Ca	1	40	1.14	729	8
Ba	1	137	47	1760	11
Yb	1	173	$1.32\text{x}10^6$	467	1926

factored is 385 which would take 30 billion laser pulses. Given the relationship between pulses and trapped ions:

$$N = 544\left(\frac{L-2}{5}\right)^3 + 78\left(\frac{L-2}{5}\right)^2 + 10\left(\frac{L-2}{5}\right),$$

we can compute that the number of trapped ions required to factor a 385 bit number is 1926.

Although theoretical results show us that there is nothing in principle to prevent us from building quantum computers, much work remains to be done to overcome the formidable technical problems before we see a realization of these fascinating machines.

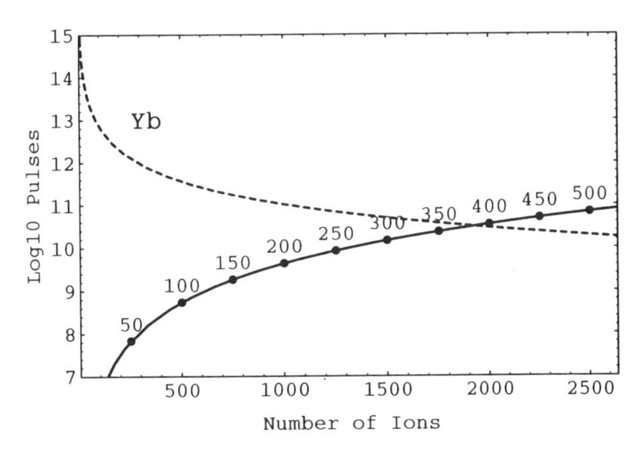

Fig. 11.7 The dashed lines define the bounds on the number of laser pulses that can be performed on various sizes of ion traps before coherence is lost for ytterbium (Yb). The solid line shows the number of laser pulse and ions needed in order to perform Shor's algorithm to factor a number requiring from 50 to 500 bits. The intercept of the curve gives an indication of the largest integer that can be factored using an ion trap quantum computer based on ytterbium.

11.3 Cavity QED-Based Quantum Computers

A separate group at the California Institute of Technology, led by physicist Jeff Kimble, has implemented a 2-qubit quantum logic gate using a very different methodology. They implement their "flying qubits" brand of quantum logic in what they refer to as a "quantum-phase gate." The quantum phase gate encodes information in the polarization states of interacting photons which induces a conditional dynamics in an analogous way to the models described earlier in this chapter.

If a set of gates is universal, then you can achieve any legitimate quantum computation in a quantum circuit built out of just those gates. At first, only 3-qubit gates were thought to be universal for quantum computation. Later, three independent research groups proved that many 2-qubit gates are universal also[DiVincenzo95a], [Barenco95], [Sleator95]. In particular, the 2-qubit XOR gate, augmented with certain 1-qubit gates, gives a universal set. This is good news for experimentalists, because it is easier to make qubits interact two at a time than it is to make them interact three at a time. Thus, if someone can devise a physical mechanism for achieving an XOR gate, the path is open, at least in principle, to building just about any quantum circuit.

At the heart of the XOR gate is the desire to do a conditional bit flip.

Yet another approach for implementing 2-qubit quantum gates is being pursued by Jeff Kimble's quantum optics group at the California Institute of Technology[Turchette95]. In the Caltech scheme, the control and target bits of the XOR gate are implemented as two photons of different optical frequencies and polarizations passing through a low-loss optical cavity. The interaction between these photons is aided by the presence of a cesium atom drifting through the cavity.

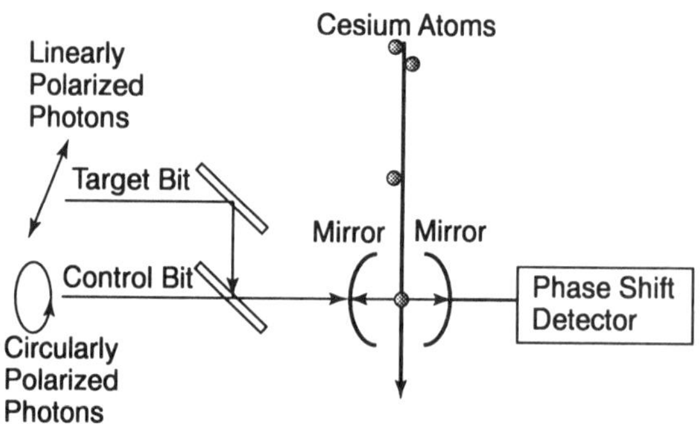

Fig. 11.8 Sketch of the experimental setup for a quantum logic gate implemented using cavity QED.

Now the goal is to achieve a 2-qubit quantum gate that allows for the flipping of one bit conditioned on the value of another. Moreover, to be a truly "quantum" gate, the physical implementation must also be able to create, preserve, and manipulate superpositions of qubits and entanglements between qubits. All of these goals can be accomplished using a cavity QED-based approach.

Let us imagine that the 2-qubit gate consists of a "control" bit and a "target" bit. We want to flip the target bit depending on the value of the control bit.

In the cavity QED approach, the control bit is implemented as a beam of photons all having either a "+" circular polarization or all having a "–" circular polarization. The intensity of the control beam is lowered so that, on average, there is only a single control photon in the cavity at any one time.

By contrast, the "target" bit is implemented as a beam of photons all having the same *linear* polarization. Mathematically, a linearly polarized photon can be described as an equal superposition of a left and a right circularly polarized state. Like the control beam, the intensity of the target beam is lowered so that there is at most one target photon in the cavity at one time.

If the cavity were empty, the interaction between the control and target photons bouncing around inside it would be rather weak. So to aid the interaction, a cesium atom is placed in (or rather drifts through) the cavity. The spacing between the mirrors in the cavity can be tuned to resonate with a particular transition between two energy levels of the cesium atom and the target and control photons.

The Caltech researchers then measured the rotation of the "target" bit's polarization as a function of the "control" bit's intensity, when the control bit beam consists of all + photons or all – photons. Empirically, there is a strong correlation between the phase shift of the target photon and control beam intensity only when the control photon's polarization is +.

You can think of the role of the cesium atom as being analogous to that of a birefringent crystal, as discussed in Chapter 8. Here the linearly polarized "target" photon can be thought of as a superposition of a left and right circularly polarized photon. When the target photon interacts with the cesium atom, the – component of its state is unchanged and the + component is phase shifted by an amount that depends on the excitation state of the cesium atom which, in turn, depends on whether the circular polarization of the control bit is + or –, with + states exciting the atom and – states leaving it in the ground state. Thus a conditional transformation of the state of the target qubit depending on the control qubit can be obtained.

Earlier cavity QED experiments by a group of researchers at the Ecole Normale Superiéure, in Paris, demonstrated similar conditional phase shifts in microwave cavities[Davidovich93]. It remains to be seen whether the cavity QED techniques can be extended into complicated quantum circuits.

11.4 NMR-Based Quantum Computers

The problems attendant in assembling, operating, and isolating ion traps, high finesse optical cavities, and heteropolymers have led some researchers to propose a radically different architecture for a quantum computer[Cory96], [Gershenfeld96]. The new idea is to adapt Nuclear Magnetic Resonance (NMR) techniques, which have been used for several years in chemical analysis and medical imaging, to accomplish the basic operations of a quantum computer.

In the ion trap, optical cavity, and heteropolymer-based quantum computers, we would need a memory register consisting of a few hundred to a few thousand qubits to factor a number beyond the reach of current classical computers[Hughes96]. These qubits would be implemented, physically, as quantized vibrational modes, quantized energy levels, and polarized photons. In comparison, an NMR-based quantum computer would consist of a test-tube-sized sample of some liquid, with each molecule of this liquid acting as an independent quantum memory register. In this model, each qubit would be implemented, physically, as the spin state of one of the nuclei in the atoms comprising each molecule of the sample. Thus the number of qubits in the memory register of an NMR-based quantum computer is equal to the number of atoms (and hence nuclei) per molecule of the sample. However, the total number of qubits involved in the computation is much larger than this, as each molecule of the sample acts as a separate quantum computer. A cubic centimeter of a typical liquid contains roughly Avogadro's number of molecules, that is, 6.0225×10^{23} per mole. So an NMR-based quantum computer has massive redundancy.

This huge shift in scale, from a single quantum computer with a few-qubit memory to billions upon billions of quantum computers each with a few qubit memory, means that we need to use a different technique for extracting answers from NMR-based quantum computers than from other types of quantum computers. For example, in quantum computers having a single memory register, we extract an answer by measuring some observable of the register, obtaining an eigenvalue of this observable in the process.

By contrast, to measure the state of the memory register of an NMR-based quantum computer, we would measure the ensemble average of some observable property of all the nuclear spins in the sample. This is a rather different approach to quantum computation that is, in many respects, closer to DNA-based computing [Adleman94].

The appeal of an NMR-based approach is that it seems to be feasible to build a rudimentary quantum computer using current technology. Certainly, the kinds of quantum manipulations needed to perform quantum computations are done on a daily basis in hospitals and laboratories around the world.

What is particularly exciting is the realization that an NMR-based quantum computer can solve NP-complete problems in polynomial time. Unfortunately, however, the size of the required computer (i.e., the quantity of liquid sample), grows exponentially in the size of the problem. Thus, although an NMR-based quantum computer might, someday, solve specific fixed-size problems that are beyond the scope of any classical supercomputer, the NMR-based approach, as it is currently conceived, is still not a panacea.

How NMR Works

So how does an NMR-based quantum computer work? Atomic nuclei often behave like tiny magnets. This magnetism arises because each proton and neutron in the nucleus has an intrinsic "spin" of $\frac{1}{2}$. These spins pair up in an antiparallel fashion, resulting in a net nuclear spin whose magnitude is determined by the number of unpaired protons and neutrons. In particular, if an atom has a nucleus consisting of even numbers of both protons and neutrons, it will have a zero net spin. If it consists of odd numbers of both protons and neutrons it will have a integer valued spin, 1, 2, 3, ..., and so on. Any other combination of protons and neutrons gives rise to a half-integer nuclear spin, $\frac{1}{2}, \frac{3}{2}, \frac{5}{2}, K$, and so on.

Now that is not quite the whole story because many chemical elements occur in the form of isotopes. These have the same number of protons as their sister nuclei but different numbers of neutrons. Consequently, if the nuclei of a particular isotope of some element has a zero net spin, there is often some other isotope of the same element whose nuclei have a nonzero spin. You can therefore synthesize just about any kind of molecule out of atoms all of whose nuclei have some nonzero value for spin. This means that the designer of an NMR-based quantum computer can tailor the isotopic composition of a molecule so that it can serve as a large enough quantum memory register.

Not only do the spins of the nuclei have quantized magnitudes, but they also adopt quantized orientations with respect to an applied external magnetic field of around 9 to 15 Tesla. If a nucleus has a spin I then it can point in just $2I + 1$ possible orientations relative to the applied field. These different orientations correspond to nuclear states of different energies.

What is remarkable is that these energy differences are exquisitely sensitive to their environment, although they exert hardly any influence on it in return. By sensing the energy differences in the nuclear spin states, and comparing the results against a pre-determined catalogue of "absorption fingerprints" of known chemical species, it is possible to infer the chemical composition of a sample. This "sensing" is done by scanning the sample with radio-frequency radiation. If the applied radiation has a frequency given by Planck's formula, $\Delta E = h\nu$, where h is Planck's constant, equal to $6.6256 \times 10^{-34}\,\mathrm{Js}$, ν is the frequency of the radio waves, and ΔE is the energy difference between two nuclear spin states, a nucleus in a lower spin state can absorb the radiation and be promoted to one of the higher states. But if the radiation is the wrong frequency the radiation will not be absorbed. Thus you can obtain an "NMR spectrum" of the sample by systematically scanning through a sequence of radio frequencies and monitoring the absorption of the radiation as a function of frequency.

NMR in the Service of Computation

Next we consider how to use the spin states of magnetically active nuclei as a physical basis for qubits, and hence a quantum memory register. First we need to understand how distinct qubits can be addressed and manipulated in a bulk quantity of some sample.

The exquisite sensitivity of each type of nucleus to its environment arises from two sources. First, the motion of the electrons in the electron clouds surrounding an atom gives rise to a magnetic field. A nucleus at the center of one atom feels the effect of the electronic orbitals of neighboring atoms. The shape of these oribitals, and hence the magnetic field they generate, varies with the type of atom. These fields combine with the externally applied magnetic field to make the field experienced by the nucleus slightly different depending on the chemical composition of the molecules in the sample.

This means that the location of the resonance peaks (i.e., the anomalously large absorption peaks seen at a specific frequency) of a particular type of nucleus can shift in a way that reflects the chemical composition of the sample. This effect is know as the "chemical shift."

In addition to chemical shifts, a nucleus can also couple with the magnetic field of neighboring nuclei, leading to an even finer structure in the NMR spectrum. The latter effect is called spin-spin coupling. What was one resonance peak initially can become several resonance peaks of slightly different frequencies due to spin-spin interactions. Fig. 11.9 illustrates the effect of spin-spin coupling on the energy levels, and hence resonance frequencies, of a 2-spin quantum memory register. The dashed lines illustrate the transitions allowed by the laws of quantum mechanics. These are transitions between states that differ by at most one bit (i.e., one spin state).

The left-hand side shows the energy levels ignoring spin-spin couplings. The right-hand side shows the effect that spin-spin coupling has on the energy levels. The levels split, and the four transitions now take place at four noticeably different frequencies. Hence each transition can be distinguished and brought about by bathing the sample in radio pulses of appropriate frequency.

The Fourier transform of the NMR spectrum of such a 2-spin system would consist of four distinct peaks, one for each allowed transition, as shown in Fig. 11.10. From left to right in the figure, these correspond to the transitions $|01\rangle \rightarrow |11\rangle$, $|00\rangle \rightarrow |10\rangle$, $|10\rangle \rightarrow |11\rangle$, and $|00\rangle \rightarrow |01\rangle$.

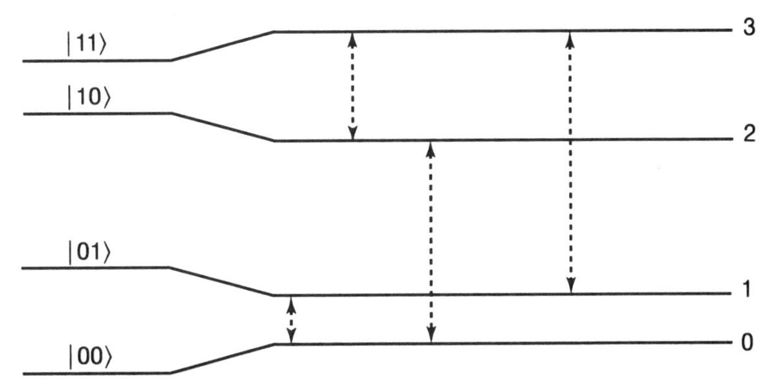

Fig. 11.9 Energy level diagram for a quantum memory register built from the spin states of two nuclei.

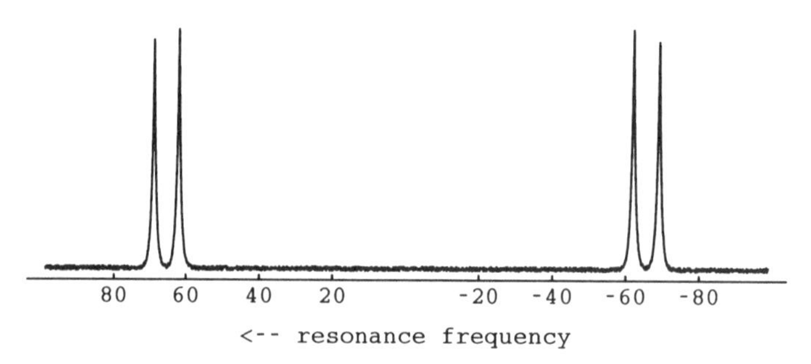

Fig. 11.10 Sketch of the Fourier transform of an NMR spectrum for a two-nuclei quantum memory register. The two left peaks are the split energies from the first nucleus and the two right peaks are the split energies from the second nucleus, due to spin-spin coupling.

Thus by way of chemical shifts and spin-spin coupling, the NMR spectrum of a particular type of nucleus can look slightly different depending on the details of its environment. In NMR spectroscopy these subtle differences in the spectra are exploited to identify the molecular structure of the sample. In NMR-based quantum computing, they allow us to select finely tuned radio frequency pulses that will allow desired unitary transformations to be applied to all of the nuclei at a particular location in a molecule of the sample. As each molecule is essentially acting as a separate quantum memory register, we can therefore apply arbitrary unitary transformations to all qubits at a specific location in each molecule by controlling the intensity and width of the radio frequency pulses. Hence, in principle, we can perform quantum computations.

In particular, we can effect the kinds of conditional logic operations on nuclear spin states that we discussed previously for atomic states. For example, to achieve the XOR operation, we need to be able to flip one bit in a pair of bits, when the other bit is in the "up" spin state. This would correspond to the transition $|01\rangle \to |11\rangle$ if the output is measured on the first bit or the transition $|10\rangle \to |11\rangle$ if the output is measured on the second bit. Fig. 11.11 shows the inverted NMR spectrum of a quantum memory register consisting of two nuclear spins before and after the application of a sequence of radio-frequency pulses that implement the XOR operation. The top figure corresponds to the initial state, having density matrix $|10\rangle\langle 10|$. The bottom figure corresponds to the final state, having density matrix $|11\rangle\langle 11|$, which is given by $\text{XOR} \cdot |10\rangle\langle 10| = |11\rangle\langle 11|$. Notice that the location of the peak has shifted. It is this change in the NMR spectrum that indicates the corresponding change in the state of the quantum memory register.

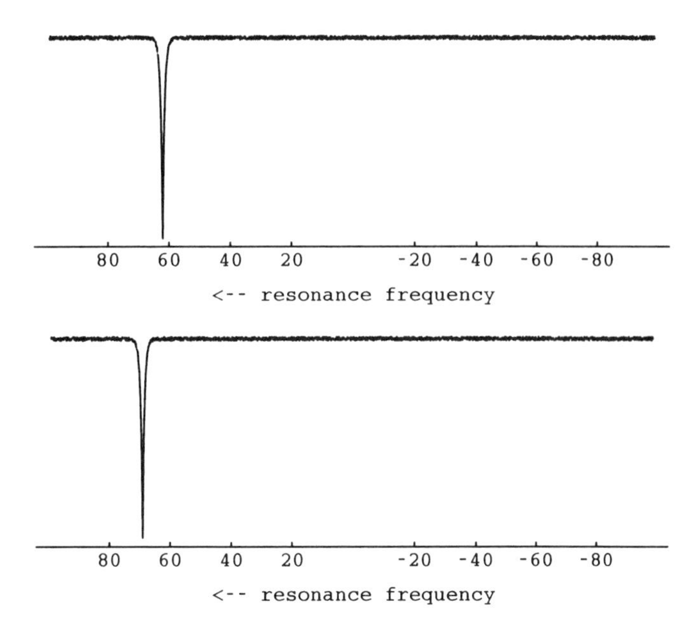

Fig. 11.11 Inverted NMR spectrum of a quantum memory register consisting of two nuclear spins. The top figure is the initial NMR spectrum corresponding to the state $|10\rangle$. The bottom figure is the NMR spectrum after the XOR operation has been applied, corresponding to the state $|11\rangle$.

In a similar fashion, it is possible to use a molecule having three magnetically active nuclei as the basis for a 3-qubit Toffoli gate. Toffoli gates are universal classical gates; so if we can build a Toffoli gate, we can, in principle, assemble a circuit that can perform *any* desired classical computation. A similar result for universal quantum gates should also be feasible[Deutsch89a, Barenco95].

If it is possible to make a universal NMR-based quantum computer, it is natural to wonder how large a quantum computation can be done in practice? To answer this, consider a NMR-based quantum computer consisting of M molecules each with n nonzero nuclear spins. Such a molecule could serve as a quantum memory register capable of storing up to 2^n separate inputs. The *total* number of spins in the computer is therefore $N = M \times n$. Let $r = M/2^n$ be the number of molecules per state. To have a reasonable chance at finding a solution we would want there to be at least one molecule per state; that is, we need $r > 1$. A test tube-sized NMR-based quantum computer would contain roughly $N \approx 10^{24}$ molecules. So, $r = M/2^n = N/(n 2^n)$. But we require r to be greater than 1 so this implies that $N/(n 2^n) > 1$. Thus n, the number of nonzero nuclear spins per molecule and hence the maximum size of the quantum memory register, must be at most 73. This puts a crude limit on the size of quantum memory register we could conceive of for a test-tube-sized quantity of sample. Using larger samples will improve this figure, of course, but the improvement grows only logarithmically with the mass of the sample. Nevertheless, this means that a quantum computer that can operate on roughly 73 qubits seems feasible given current NMR technology.

We have looked at four very different embodiments of a quantum computer. Some of the key differences are summarized in Table 11.7.

Whichever scheme proves to be the most feasible, the advances made along the way will no doubt have useful spinoffs in at least basic physics research. For example, even being able to create and manipulate a few entangled qubits allows some interesting tests to be made on certain predictions of quantum theory. Thus, ironically, quantum computing may eventually become a tool of experimental physics. Only a small number of qubits are needed to be able to simulate quantum systems much more efficiently than can be done classically. At around the 10-qubit level, a quantum computer is able to implement certain quantum coding schemes. At around the 100-qubit level a quantum computer could be used as a repeater in a noisy quantum cryptographic channel[DiVincenzo95]. And, as we saw at the few thousand level, a quantum computer can factor large integers.

This book has been a comprehensive and whirlwind trip through quantum computing, from its theoretical underpinnings to its potential realizations. Quantum computers are not yet with us, but the rate of progress over recent years has been quite staggering and we

Table 11.7 Summary of differences among the four leading embodiments of a quantum computer.

Quantum Computer	Qubit Representation	Memory Readout
Heteropolymer	Atomic Energy Levels	Individual Eigenstate
Ion Trap	Atomic Energy Levels/Phonons	Individual Eigenstate
Cavity QED	Photon Polarizations	Individual Eigenstate
NMR	Nuclear Spins	Ensemble Average

are optimistic that specialized quantum computers might soon be a reality. The schemes described in this chapter are just a beginning. Over the next few decades we fully expect progress in quantum computing to ignite a new revolution in our understanding of the science of computing and the quantum technology needed for radically new computer architectures. The devices that will be built will not only exceed the capabilities of their classical counterparts but they will serve as a testbed to illuminate much of the basic physics underlying quantum phenomena.

APPENDIX

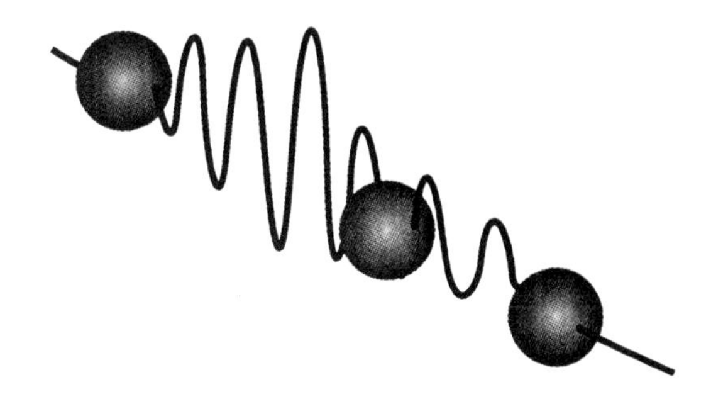

Using the Code Supplements

Explorations in Quantum Computing is published by TELOS, The Electronic Library of Science, a Springer Verlag imprint located in Santa Clara, CA. For information contact TELOS via the Internet at info@telospub.com.

If you have any problems with the CD-ROM software, or if you want to give us feedback, please contact Colin Williams at:

colin@solstice.jpl.nasa.gov
tel: (818) 306 6512
cpw@cs.stanford.edu
tel: (415) 728 2118

Instructions for Using the Mathematica Notebooks

This CD-ROM includes Mathematica Version 2.2 and Mathematica Version 3.0 Notebooks to accompany the book, *Explorations in Quantum Computing* by Colin P. Williams and Scott H. Clearwater.

Mathematica is a software system that allows you to perform mathematics on a computer. The latest version of the software (Version 3.0) was released in 1997. The Mathematica system allows you to mix together numerical, symbolic and graphical computations within a single programming environment.

Mathematica Notebooks are specially formatted files that contain text, graphics, animations, and executable Mathematica code. You can use the MathReader program on the CD-ROM to view the Mathematica Notebooks on your computer, even if you are not able to run the Mathematica system itself.

However, to get the most out of the CD-ROM files, we recommend that you open the Notebooks from within the full Mathematica software system. This is available on most college campuses. Alternatively, a personal copy of Mathematica can be purchased from Wolfram Research Inc. (http://www.wolfram.com). A student version is available at a discount price and can often be purchased through college campus bookstores/computer stores. With the full Mathematica system you will be able to both read the Notebooks and execute the commands contained in them, yourself. This is very helpful for developing a feel for the ideas behind quantum computing.

MathReader, a program made available for free by Wolfram Research, Inc., allows you to read the Notebooks, and to print them, but not to execute any of the commands. Six different versions of MathReader are contained on this CD-ROM in order to allow you to read the Notebooks even if you do not have access to the Mathematica system itself.

Notebooks created under Mathematica v2.2 have the filename extension *.ma. Those created under v3.0 have the extension *.nb. MathReader version 3 can view both *.ma and *.nb notebooks but, MathReader version 2 can view only the *.ma notebooks. We included both versions because not everyone has migrated to using Mathematica v3.0 yet (although they should - it's great!). Moreover, at the time of pressing the CD-ROM there was still no v3.0 MathReader for UNIX. If you are completely new to Mathematica we recommend using MathReader version 3 and the *.nb Notebooks only. The on-line help systems and installers for v3 MathReaders are far better than their v2 counterparts.

If you are already a Mathematica user, but you haven't switched to Mathematica v3.0 yet, you are at no disadvantage as the *.ma and *.nb Notebooks on the CD-ROM contain the same content even though they look different stylistically.

To open a Notebook, install either the appropriate version of MathReader or Mathematica for your computer platform and then double-click on the Notebook's icon. Alternatively, if you already have access to Mathematica, start Mathematica as usual, then go to the File menu and select Open. Next, locate the Notebook file within your personal directory structure and select it.

Once you have opened a Notebook, you will be able to scroll through it using a slider on the right hand edge of the page. Each Notebook contains a mini-tutorial on a specific topic related to quantum computing. You may follow through the tutorial, executing commands as you go, and then try some variations on your own.

To execute a command in a Notebook, position the cursor over the command, click the mouse button, and then hit the SHIFT key and the RETURN key simultaneously. A single RETURN alone will not work (this was reserved to allow users to type inputs that spill over several lines).

Note that the first time you attempt to evaluate a particular Mathematica command you may be asked whether you want to "evaluate initialization cells"?. If confronted with such a question, always hit the "yes" button in the dialog box that will appear. By choosing "yes" you are instructing Mathematica to load in all the definitions for the functions defined in that Notebook. This is crucial in order to get the software to run. If you ever find that you cannot get the software to run the most likely problem is that you have not initialized the Notebook. In version 3.0 you can force initialization by going to the Kernel/Evaluation/Evaluate Initialization menu. In v2.2 you can force initialization by going to the Action/Evaluate Initialization menu.

If you want to try out a lot of new commands, we recommend that you open a new Notebook (File menu, select "New"). Commands are shared between Notebooks. So your "New" (empty) Notebook will be able to run all of the commands that appear in any other Notebook that is open at the same time. You can also save each Notebook as a Package if you want to use the commands within it, repeatedly, for new experiments of your own. To do this under Mathematica v3.0 use the File/Save As Special.../Package Format menu option.

If you are new to Mathematica, we recommend that you read the introductory-to-intermediate level text "Mathematica: A Practical Approach (Second Edition) by Nancy Blachman & Colin P. Williams, Prentice-Hall (1997). This book is based on the introductory Mathematica course taught at Stanford University. It will teach you everything you need to know to understand the code in the quantum computing Notebooks.

Version 2.2 Notes

The 2.2 Notebooks (*.ma) work without any modification in version 2.2.

To view the animations in Anims2/Schroed2D.ma:

1. Move that Notebook to your hard drive first.
2. Open the file.
3. Double click on the cell bracket of the cell that contains the animation you want to see e.g. "Experiment 2: Double Slit".
4. Double click on the cell bracket of the cell called "Animation" (which reveals one frame of the animation)
5. Double click on the cell bracket immediately below this graphic (which reveals all the frames in the animation)
6. Double click on any of these frames then runs the animation.

Version 3.0 Notes

The 3.0 Notebooks (*.nb) work without any modification in v3.0.

To view the animations in Anims3/Schroed2D.ma ...

1. Move that Notebook to your hard drive first.
2. Attempt to open the (v2.2) file in Mathematica v3.0. When you attempt to open the file you will be asked if you want to convert it to a v3.0 Notebook before viewing it. Select "Convert" with the default setting for all other options. When asked, save the file as Schroed2D.nb on your hard drive. Then open Schroed2D.nb.
3. Double click on the cell bracket of the cell that contains the animation you want to see e.g. "Experiment 2: Double Slit".
4. Double click on the cell bracket of the cell called "Animation" (which reveals one frame of the animation).
5. Double click on the cell bracket immediately below this graphic (which reveals all the frames in the animation).
6. Double click on any of these frames then runs the animation.

Contents of the CD-ROM

README

- an introduction.

LEGAL

- the copyright information and legal disclaimer pertaining to this material.

MATHREADERS

Folder containing the applications for viewing and printing Mathematica v2.2 files and v3.0 files. There are 6 different versions included. For version 2.2 there are X-Windows, Windows, Macintosh. For version 3.0, Windows95 & WindowsNT, Macintosh for 680x0, and Macintosh Power PCs.

MATHEMATICA NOTEBOOKS V2.2

Folder containing the following files:

Animations
- Folder containing Schroed2D.ma

BraKet.ma
- Defines basic operations on bras and kets, direct product, nuts and bolts QM

ErrorCorrection.ma
- Simulation of Quantum Error Correction

FeynmanQC.ma
- Simulation of Feynman's Quantum Computer

Interference.ma
- Graphical Illustration of Interference Effect

OTPExampleOnly.ma
- Example of a Provably Secure Cryptosystem

QCdatabase.txt
- Database of References in Quantum Computing

QuantumBugs.ma
- Code for Monte Carlo Analysis of Error Propagation

QuantumCrypt.ma
- Simulation of Quantum Cryptography

RSAExampleOnly.ma
- Example of RSA Public-Key Cryptography

SearchEngine.ma
- A Search Engine for our Quantum Computing Database

ShorFactoring.ma
- Simulation of Shor's Algorithm

Schroed2D.ma
- Animation of Particle Impinging a Double Slit (Interference)

Teleportation.ma
- Simulation of Quantum Teleportation

Timing Factor Integer.ma
- Illustration that Factoring is a Hard Problem Classically

MATHEMATICA NOTEBOOKS V 3.0

Folder containing the following files:

Animations
- Folder containing Schroed2D.ma

BraKet.nb
- Defines basic operations on bras and kets, direct product, nuts and bolts QM

ErrorCorrection.ma
- Simulation of Quantum Error Correction

FeynmanQC.nb
- Simulation of Feynman's Quantum Computer

Interference.nb
- Graphical Illustration of Interference Effect

OTPExampleOnly.nb
- Example of a Provably Secure Cryptosystem

QCdatabase.txt
- Database of References in Quantum Computing

QuantumBugs.nb
- Code for Monte Carlo Analysis of Error Propagation

QuantumCrypt.nb
- Simulation of Quantum Cryptography

RSAExampleOnly.nb
- Example of RSA Public-Key Cryptography

SearchEngine.nb
- A Search Engine for our Quantum Computing Database

ShorFactoring.nb
- Simulation of Shor's Algorithm

Schroed2D.ma
- Animation of Particle Impinging a Double Slit (Interference)

Teleportation.nb
- Simulation of Quantum Teleportation

TimingFactorInteger.nb
- Illustration that Factoring is a Hard Problem Classically

A Simulator for Feynman's Quantum Computer

Colin P. Williams

```
Off[General::spell1]
```

■ Copyright Notice

■ Overview

This Notebook contains code for simulating Feynman's quantum computer (see "Explorations in Quantum Computing", Chapter 4). It contains tools for building mathematical representations of quantum gates, tools for embedding quantum gates in quantum circuits and a simulator for the Feynman quantum computer that implements a given quantum circuit. Be aware that classical simulations of quantum computers are very costly, computationally, so only simple circuits can be simulated within a reasonable time. Nevertheless, the simulator is sufficient to illustrate several important features of quantum computation.

You can build any classical circuit out of the following three types of reversible logic gates:

```
NOTGate
CNGate
CCNGate
```

These are the NOT gate, the CONTROLLED-NOT gate and the CONTROLLED-CONTROLLED-NOT gate respectively. A particularly "quantum" gate is the one input/one output square root of NOT gate:

```
SqrtNOTGate
```

The SqrtNOTGate can be used to devise a circuit for computing the NOT function that is inherently "quantum" in the sense that it relies upon the Superposition Principle of quantum mechanics.

Once you have designed your circuit, you can simulate a Feynman quantum computer that implements this circuit using either of the commands:

```
SchrodingerEvolution
EvolveQC
```

SchrodingerEvolution evolves the quantum computer for a fixed period of time, EvolveQC evolves the quantum computer until the computation is completed. In either case, you can visualize the resulting evolution using:

```
PlotEvolution
```

See the Notebook ErrorsInQCs.ma to investigate the effects of errors in the operation of a quantum computer. See the Notebook ErrorCorrection.ma to investigate techniques for quantum error correction.

■ What Computation are we going to Simulate?

We are going to describe a quantum circuit that computes the NOT function via the application of two "square root of NOT" gates connected back to back. There is no classical gate that can achieve the SqrtNOT operation i.e. there is no classical gate such that two consecutive applications of this (hypothetical, classical) gate yield the NOT operation.

■ Gates & their Truth Tables

```
NOT = {{0, 1},
       {1, 0}};

CN = {{1,0,0,0},
      {0,1,0,0},
      {0,0,0,1},
      {0,0,1,0}};

CCN = {{1,0,0,0,0,0,0,0},
       {0,1,0,0,0,0,0,0},
       {0,0,1,0,0,0,0,0},
       {0,0,0,1,0,0,0,0},
       {0,0,0,0,1,0,0,0},
       {0,0,0,0,0,1,0,0},
       {0,0,0,0,0,0,0,1},
       {0,0,0,0,0,0,1,0}};
```

■ Creation and Annihilation Operators

```
aOP = {{0,1},    (* annihilation operator on a single bit *)
       {0,0}};

annihilationOP[i_, m_]:=
    Apply[Direct,
          ReplacePart[Table[IdentityMatrix[2], {m}], aOP, i]
         ]

cOP = {{0,0},     (* creation operator that acts on a single bit *)
       {1,0}};

creationOP[i_, m_]:=
    Apply[Direct,
          ReplacePart[Table[IdentityMatrix[2], {m}], cOP, i]
         ]
```

■ Rewriting Gates using Creation & Annihilation Operators

We can express the NOT, CONTROLLED-NOT and CONTROLLED-CONTROLLED-NOT gates as sums and products of creation and annihilation operators. Moreover, we can embed gates in a circuit that contains multiple input and output lines. For example, NOTGate[i,m] is a NOT gate acting on the i-th of m inputs. CNGate[i,j,m] is a CONTROLLED-NOT gate acting on the i-th and j-th of m inputs etc.

```
NOTGate[i_, m_]:=
    creationOP[i,m] + annihilationOP[i,m]

CNGate[i_, j_, m_]:=
    (creationOP[i,m] .
     annihilationOP[i,m] . (annihilationOP[j,m] + creationOP[j,m]) +
     annihilationOP[i,m] . creationOP[i,m]
    )


CCNGate[i_, j_, k_, m_]:=
    (IdentityMatrix[2^m] +
      creationOP[i,m] .
      annihilationOP[i,m] .
      creationOP[j,m] .
      annihilationOP[j,m] .
       (annihilationOP[k,m] + creationOP[k,m] - IdentityMatrix[2^m])
    )
```

■ Square Root of NOT Gate (SqrtNOT)

A SqrtNOT gate is a 1 input / 1 output quantum gate.

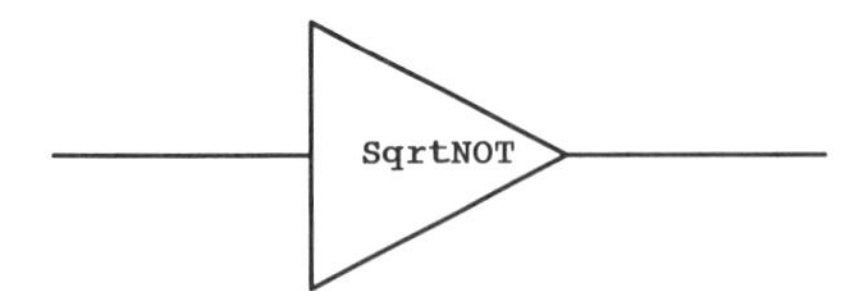

■ Code for sketching SqrtNOTGate

■

A SqrtNOT gate that acts on the i-th of m qubits can be defined in terms of creation and annihilation operators as follows (in *Mathematica* I is the square root of -1):

```
SqrtNOTGate[i_, m_]:=
    Module[{cOPim, aOPim},
        cOPim = creationOP[i,m];
        aOPim = annihilationOP[i,m];
        (1/2 (1 - I) (cOPim + aOPim) +
         1/2 (1+I) (aOPim . cOPim + cOPim . aOPim)
        )
    ]
```

It is interesting to examine the truth table for SqrtNOTGate. Unlike, any classical gate which always outputs bits (0s and 1s), the SqrtNOTGate can return superpositions of bits.

■ Try me!

```
TruthTable[SqrtNOTGate[1,1]]
```

$\text{ket}[0] \to (\frac{1}{2} + \frac{I}{2})\ \text{ket}[0] + (\frac{1}{2} - \frac{I}{2})\ \text{ket}[1]$

$\text{ket}[1] \to (\frac{1}{2} - \frac{I}{2})\ \text{ket}[0] + (\frac{1}{2} + \frac{I}{2})\ \text{ket}[1]$

Why do we call this gate a SqrtNOTGate? To answer this look at the truth table of two SqrtNOT gates connected back to back:

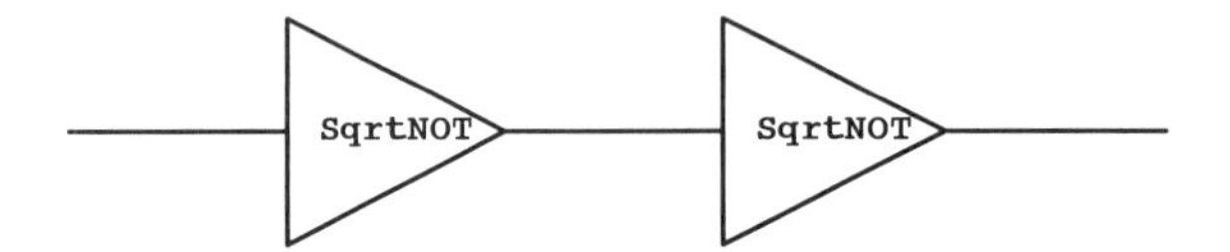

■ Try me!

```
TruthTable[SqrtNOTGate[1,1] . SqrtNOTGate[1,1]]
```

The net result of two SqrtNOTGates connected back to back, is a unitary operation (in fact the NOT operation).

■ Try me!

```
UnitaryQ[SqrtNOTGate[1,1] . SqrtNOTGate[1,1]]
```

```
UnitaryQ[NOT]
```

■ Representing the Computation as a Circuit

In general , the quantum memory register of a Feynman-like quantum computer will consist of a set of "cursor" qubits (which keep track of the progress of the computation) and a set of "program" qubits through which the "input" is fed into the computer and the "output" is extracted (when it becomes available). We want to apply specific operators, corresponding to the action of logic gates, on just the "program" qubits. Hence we must specify which qubits a given operator must act upon. In the case of the simple SqrtNOTGate squared circuit, we will use two SqrtNOTGates that both act on the 4th of 4 qubits. Thus the NOT circuit, built from two SqrtNOTGates connected back to back, is specified as follows:

```
sqrtNOTcircuit = {SqrtNOTGate[4,4], SqrtNOTGate[4,4]};
```

This contains an ordered list of quantum gates together with the input "lines" these gates work upon.

The embedded NOT gate (NOTGate[i,m]), the embedded CONTROLLED-NOT gate (CNGate[i,j,m]) and the embedded CONTROLLED-CONTROLLED-NOT gate (CCN[i,j,k,m]) are defined similarly. The arguments i, j, k label the lines of the gates and m signifies the total number of lines in the circuit. Thus, a controlled-NOT gate acting on the 2nd and 4th of 5 lines would be CNGate[2,4,5] etc.

You can compare the square root of NOT gate and its embedded form easily. A square root of NOT gate acting on the 1st of 1 qubit (i.e. an unembedded SqrtNOTGate) is given by sqrtNOT11 in the example below:

■ Try me!

```
SetOptions[$Output, PageWidth->Infinity];

sqrtNOT11 = SqrtNOTGate[1,1];   (* unembedded gate *)

MatrixForm[sqrtNOT11,
           TableSpacing->{0,4}, TableAlignments->{Center, Center}]
```

$$\begin{pmatrix} \frac{1}{2} + \frac{I}{2} & \frac{1}{2} - \frac{I}{2} \\ \frac{1}{2} - \frac{I}{2} & \frac{1}{2} + \frac{I}{2} \end{pmatrix}$$

The square root of NOT gate that acts on the 2nd of 3 qubits is given by sqrtNOT23 in the next example:

■ **Try me!**

```
SetOptions[$Output, PageWidth->Infinity];

sqrtNOT23 = SqrtNOTGate[2,3];

MatrixForm[sqrtNOT23,
           TableSpacing->{0,4}, TableAlignments->{Center, Center}]
```

$$\begin{pmatrix} \frac{1}{2}+\frac{I}{2} & 0 & \frac{1}{2}-\frac{I}{2} & 0 & 0 & 0 & 0 & 0 \\ 0 & \frac{1}{2}+\frac{I}{2} & 0 & \frac{1}{2}-\frac{I}{2} & 0 & 0 & 0 & 0 \\ \frac{1}{2}-\frac{I}{2} & 0 & \frac{1}{2}+\frac{I}{2} & 0 & 0 & 0 & 0 & 0 \\ 0 & \frac{1}{2}-\frac{I}{2} & 0 & \frac{1}{2}+\frac{I}{2} & 0 & 0 & 0 & 0 \\ 0 & 0 & 0 & 0 & \frac{1}{2}+\frac{I}{2} & 0 & \frac{1}{2}-\frac{I}{2} & 0 \\ 0 & 0 & 0 & 0 & 0 & \frac{1}{2}+\frac{I}{2} & 0 & \frac{1}{2}-\frac{I}{2} \\ 0 & 0 & 0 & 0 & \frac{1}{2}-\frac{I}{2} & 0 & \frac{1}{2}+\frac{I}{2} & 0 \\ 0 & 0 & 0 & 0 & 0 & \frac{1}{2}-\frac{I}{2} & 0 & \frac{1}{2}+\frac{I}{2} \end{pmatrix}$$

Other embedded gates may be derived in a similar fashion.

Notice that the embedded gates are still unitary:

■ **Try me!**

```
UnitaryQ[sqrtNOT23]
```

■ Determining the Size of the Memory Register

For the NOT circuit, built from two SqrtNOTGates connected back to back, there are 2 gates (k=2), so we need need k+1=3 cursor qubits (to track the progress of the computation) and 1 input/output qubit (m=1) which will serve a dual purpose; to enter the input to the circuit and to record the output. Hence we need 4 qubits in all; 3 cursor qubits plus one program qubit, making a total of m+k+1=4 qubits for the entire quantum memory register.

■ Computing the Hamiltonian Operator

In Feynman's quantum computer, the Hamiltonian is time independent and consists of a sum of terms describing the advance and retreat of the computation. The net effect of the Hamiltonian is to place the memory register of the Feynman quantum computer in a superposition of states representing the same computation at various stages of completion. The command for generating the Hamiltonian of a Feynman quantum computer is "Hamiltonian".

```
?Hamiltonian

Hamiltonian[m, k, circuit] returns the time independent  Hamiltonian matrix
  corresponding to the given circuit.  The circuit consists of k quantum gates and
  a total of  m lines (i.e. m inputs and m outputs).
```

■ **Try me!**

```
SetOptions[$Output, PageWidth->Infinity];
H = Hamiltonian[1, 2, sqrtNOTcircuit];
MatrixForm[H, TableSpacing->{0,4}, TableAlignments->{Center,Center}]
```

■ **Code**

■ Computing the Unitary Evolution Operator

The unitary evolution operator is derived from the solution to Schrodinger's equation for the evolution of the memory register i.e. | psi(t) > = U(t) | psi(0) > where U(t) = e^(-i H t / hBar) where H is the (time-independent) Hamiltonian, i is square root of -1, hBar is Planck's constant over 2 Pi and t is the time. Remember, U(t) and H are really square matrices so the necessary exponential is a matrix exponential. We set hBar=1 for simplicity. In terms of the code, we call U(t), for a particular circuit at a particular time, EvolutionOP[m,k,circuit,t].

```
?EvolutionOP

EvolutionOP[m, k, circuit, t] is the time dependent evolution  operator for the
  given circuit. The circuit consists of k  quantum gates and a total of m lines (
  i.e. m inputs and m  outputs).
```

U is given by the t-th matrix power of expH i.e. U = MatrixPower[expH, t] where expH is the matrix exponential of H. You may need to scroll sideways to see the structure of expH.

■ **Code**

■ **Try me!**

```
SetOptions[$Output, PageWidth->Infinity];
expH = MatrixExp[ -I Hamiltonian[1, 2, sqrtNOTcircuit]];
MatrixForm[expH, TableSpacing->{0,0}]
```

■ Running the Quantum Computer for a Fixed Length of Time

To simulate running the quantum computer, that implements a given circuit, for a specific length of time, use the function SchrodingerEvolution. This function takes 5 arguments: the initial state of the memory register, the number of inputs/outputs in the circuit, the number of gates in the circuit, the circuit matrix expressed as a product of embedded gates and the duration of the simulation. The result is the state of the memory register at the end of the simulation period.

```
?SchrodingerEvolution

SchrodingerEvolution[initKet, m, k, circuit, t]  evolves the given circuit for time t from the
  initial configuration  initKet (a ket vector e.g. ket[1,0,0,0]). You need to specify that
  there are k gates in the circuit (so there are k+1 cursor bits) and  that there
  are m program bits (i.e. the number of bits used as input  data not counting the cursor bits).
```

■ Code

■ Try me!

```
sqrtNOTcircuit = {SqrtNOTGate[4,4], SqrtNOTGate[4,4]};
SchrodingerEvolution[ket[1,0,0,0], 1, 2, sqrtNOTcircuit, 0.5]
```

```
-0.119878 ket[0, 0, 1, 1] + (0.229681 - 0.229681 I) ket[0, 1, 0, 0] +
  (-0.229681 - 0.229681 I) ket[0, 1, 0, 1] + 0.880122 ket[1, 0, 0, 0]
```

Is this a properly normalized ket?

■ Try me!

```
NormalizedKetQ[%]
```

```
True
```

Do you get the same answer is you run the simulator again, on the same input, for the same length of time?

■ Try me!

```
sqrtNOTcircuit = {SqrtNOTGate[4,4], SqrtNOTGate[4,4]};
SchrodingerEvolution[ket[1,0,0,0], 1, 2, sqrtNOTcircuit, 0.5]
```

Were you surprised by the outcome?

■ Explanation in here ...

■ Running the Quantum Computer Until the Computation is Done

In a Feynman-like quantum computer, the position of the cursor keeps track of the logical progress of the computation. If the computation can be accomplished in k+1 (logic gate) operations, the cursor will consist of a chain of k+1 atoms, only one of which can ever be in the |1> state. The cursor keeps track of how many, logical operations have been applied to the program bits thus far. Thus if you measure the cursor position and find it at the third site, say, then you know that the memory register will, at that moment, contain the result of applying the first three gate operations to the input state. This does not mean that only three such operations have been applied. In the Feynman computer the computation proceeds forwards and backwards simultaneously. As time progresses, the probability of finding the cursor at the (k+1)-th site rises and falls. If you are lucky, and happen to measure the cursor position when the probability of the cursor being at the (k+1)-th site is high, then you have a good chance of finding it there.

Operationally, you periodically measure the cursor position. This collapses the superposition of states that represent the cursor position but leaves the superposition of states in the program bits unscathed. If the cursor is not at the (k+1)-th site then you allow the computer to evolve again from the new, (partially) collapsed state. However, as soon as the cursor is found at the (k+1)-th site, the computation is halted and the complete state of the memory register (cursor bits and program bits) is measured. Whenever the cursor is at the (k+1)-th site, a measurement of the state of the program bits at that moment is guaranteed to return a valid answer to the computation the quantum computer was working on. So in the Feynman model of a quantum computer, there is no doubt at to the correctness of the answer, merely the time at which the answer is available.

To run the quantum computer until completion, you must periodically check the cursor position to see if all the gate operations have been applied. As soon as you find the cursor in its extreme position, you can be sure that, at that moment, a correct answer is obtainable from reading the program qubits. Note that we say "a" correct answer and not "the" correct answer because, if a problem admits more than one acceptable solution, then the final state of the Feynman's quantum computer will contain a superposition of all the valid answers. Upon measurement only one of these answers will be obtained however.

To read just the cursor, one simply restricts measurements of the memory register to be made on just those qubits used to keep track of the cursor. Thus the program qubits (that contain the answer) are NOT measured in this process. Nevertheless, the outcome of the measurement of the cursor position causes the relative state of the program qubits to be projected into a subspace that is consistent with the position of the cursor that is found. For example, if the cursor position indicates that, say, the first N gate operations have been applied, then the program qubits are projected into a superposition corresponding to the state created by applying just the first N gate operations in the circuit. This idea of measurements of one part of a memory register affecting the relative state of the other (unmeasured) part of the same register, is crucial to understanding the operation of quantum computers. In general you do not want to make any measurements on the program qubits (the qubits which contain the answer) until you can be sure that an answer is available. The command for reading the cursor position is ReadCursorBits:

```
?ReadCursorBits
```

ReadCursorBits[numCursorBits, superposition] reads the state of the cursor bits of a Feynman-like quantum computer. If the computation can be accomplished in k+1 (logic gate) operations, the cursor will consist of a chain of k+1 atoms, only one of which can ever be in the |1> state. The cursor keeps track of how many, logical operations have been applied to the program bits thus far. The state of the program bits of the computer are unaffected by measuring just the cursor bits. If the cursor is ever found at the (k+1)-th site, then, if you measured the program bits at that moment, they would be guaranteed to contain a valid answer to the computation the quantum computer was working on.

For our sqrtNOTcircuit, we have 2 gates (i.e. k=2) and therefore k+1=3 cursor positions. The initial state of the cursor is |100> and the input to the circuit is set to |0>. Thus the overall memory register is initially in the state |1000> (which we represent as ket[1,0,0,0]). To read the cursor position after a time t=1.1 has elapsed use ReadCursorBits[3, evoln1] where evoln1 is the state of the entire memory register after the computer has evolved for time t=1.1.

- **Code**

- **Try me!**

```
sqrtNOTcircuit = {SqrtNOTGate[4,4], SqrtNOTGate[4,4]};
state1  = SchrodingerEvolution[ket[1,0,0,0],1,2,sqrtNOTcircuit,1.1];

{cursor1, projected1} = ReadCursorBits[3, state1]
```

The output consists of a list, {cursor-position, projected-state}, showing the cursor position and the state of the entire memory register after the cursor position has been measured. For example, the output:

```
{{0, 1, 0}, (0.5 - 0.5 I) ket[0, 1, 0, 0] + (-0.5 - 0.5 I) ket[0, 1, 0, 1]}
```

would indicate that the cursor is in the second position ({0,1,0}), and the state of the entire memory register is then (0.5 - 0.5 i) |0100> + (-0.5 - 0.5 i) |0101> (where i is the square root of -1). You can compare the state of the register after measuring the cursor position (projected1) to its state before measuring the cursor position (state1) by simply asking for the values of these variables.

- **Try me!**

Compare the following two states:

```
projected1
```

```
state1
```

If you measure the cursor position of two identically prepared quantum computers at identical times would you find both cursors at the same place?

- **Try me!**

The initial set up is identical, is the result of reading the cursor?

```
sqrtNOTcircuit = {SqrtNOTGate[4,4], SqrtNOTGate[4,4]};
state2 = SchrodingerEvolution[ket[1,0,0,0],1,2,sqrtNOTcircuit,1.1];

{cursor2, projected2} = ReadCursorBits[3, state2]
```

Are the cursors found at the same position?

```
cursor1 === cursor2
```

Are the states of the memory registers the same after measuring the cursors?

```
projected1 === projected2
```

Were the states of the memory registers the same before the cursor positions were measured?

```
state1 === state2
```

▪ Explanation in here ...

In general, you will obtain different results for the cursor position in the two cases, although by chance you might have obtained the same answer both times! If so, try running the last experiment again a few times.

If two identically prepared quantum computers are allowed to evolve for identical times, then, as the Schrodinger equation is deterministic, the two memory registers will evolve into identical superpositions (ignoring errors of course). Hence state1 will always be identical to state2. However, when you measure the cursor positions, you cause each of the superpositions to "collapse" in a random way independent of one another. In one case you may find, say, 2 gate operations have been applied, and in the other you may find say 1 gate operation has been applied. Thus cursor1 is not, in general, the same as cursor2. Consequently, the relative states of the program qubits of each computer will then also be different after the measurements of the cursor. Hence projected1 is different from projected2, in general.

▪ Reading the Memory Register (Cursor Qubits & Program Qubits)

Once the cursor is found at a position indicating that all of the gate operations have been applied, i.e. at its k+1-th site, then you can extract an answer from the computer by reading all the qubits in the memory register.

```
?ReadMemoryRegister
```

```
ReadMemoryRegister[superposition] reads the state of each bit  in the memory
  register. As the i-th and j-th bit measurement  operators commute (for any i and
  j), it does not matter in what  order you measure the bits.
```

If a computation has only one answer, you will always obtain the single correct answer. If a computation admits many possible answers, you are equally likely to obtain any one of them upon measuring the final state of the register.

▪ Code

■ Evolving the Quantum Computer Until Complete

The above steps are bundled together in the command EvolveQC. EvolveQC allows you to evolve the quantum computer, checking whether an answer is available and, if so, measuring the memory register to extract it. The output from EvolveQC is a complete history of the evolution of the quantum computer. This history consists of a sequence of 4-element snapshots. Each snapshot shows the time at which the cursor was measured, the state of the register before the cursor was measured, the result of the measurment and the state into which the register is projected because of the measurement. You can use the option TimeBetweenObservations to control the time between observations of the cursor and whether the intervals should be regular or random. The default interval is 1 time unit.

```
?EvolveQC
```

```
The function EvolveQC[initState, circuit] evolves a Feynman-like  quantum computer, specified
  as a circuit of interconnected  quantum logic gates, from some initial state until the
  computation is complete. The output is a list of snapshots of the  state of the QC at
  successive cursor-measurement times. Each  snapshot consists of a 4 element list whose
  elements are the  time at which the cursor is measured, the state of the register
  immediately before the cursor is measured, the result of the  measurement and
  the state of the register immediately after  the cursor position is measured. The
  latter is the projected  state of the register. EvolveQC can take the optional
  argument  TimeBetweenObservations which can be set to a number or a  probability
  distribution. The default time between observations  is 1 time unit.
```

■ Code

■ Try me!

```
EvolveQC[ket[1,0,0,0], sqrtNOTcircuit]
```

■ Plotting the Evolution

You can visualize the time evolution of a Feynman quantum computer using the function PlotEvolution. PlotEvolution shows the probabilities of obtaining each possible state for the register at the cursor observation times, until the computation is done.

To use PlotEvolution you must first generate a particular evolution using EvolveQC.

```
?PlotEvolution
```

```
PlotEvolution[evolution] draws a graphic that illustrates the time evolution of the memory
  register of a Feynman quantum computer. It takes a single argument, the output from EvolveQC,
  and plots the probability (i.e. |amplitude|^2) of obtaining each eigenstate of the memory
  register at the times when the cursor is observed. By default PlotEvolution only plots the
  probabilities that prevail immediately before the cursor is measured.  You can give
  the option AfterObservingCursor->True to see the effect of measuring the cursor
  on the relative probabilities of finding the memory register in each of its possible states.
```

PlotEvolution takes a single input, the output of the function EvolveQC, and returns a graphic that shows the probability (i.e. | amplitude |^2) of each eigenstate of the memory register at the times when the cursor is observed. For compactness of notation we label the eigenstates of the (in this case 4-bit) memory register, |i>, in base 10 notation. For example, |5> corresponds to the eigenstate of the memory register that is really |0101> and |15> corresponds to the eigenstate of the memory register that is really |1111>. The vertical axis shows that probability of obtaining that eigenstate if the memory register were to be measured at the given time. Notice that there is a zero probability of ever obtaining certain eigenstates showing that certain configurations of the memory register are forbidden.

■ **Try me!**

We seed the pseudo-random number generator just to ensure reproducible results.

```
SeedRandom[1234];
evoln = EvolveQC[ket[1,0,0,0], sqrtNOTcircuit];
PlotEvolution[evoln]
```

PlotEvolution can take two options that controls the information that is output. By default, PlotEvolution plots the probabilities of finding the computer in the various eigenstates at times t1, t2, t3 etc before the cursor position is observed. By setting the option AfterObservingCursor->True you can plot the probability of finding the computer in the various eigenstates both before and after the cursor position is observed. Hence you can visualize the effect of the cursor measurement operations by setting the option AfterObservingCursor->True.

■ **Try me!**

```
SeedRandom[1234];
evoln = EvolveQC[ket[1,0,0,0], sqrtNOTcircuit];
PlotEvolution[evoln, AfterObservingCursor->True]
```

The time intervals between measurements need not be regular. Try setting the time between observations, in EvolveQC, to a probability distribution using the option TimeBetweenObservations.

■ **Try me!**

```
SeedRandom[1234];
evoln = EvolveQC[ket[1,0,0,0], sqrtNOTcircuit,
                 TimeBetweenObservations->NormalDistribution[1,0.7]];
PlotEvolution[evoln, AfterObservingCursor->True]
```

■ **Code**

■ Extracting an Answer

The answer is extracted by reading just the program qubits of the final state of the memory register. To figure out which qubits are program qubits, we need to know the number of lines in the circuit (i.e. m) and the number of gates (i.e. k). The last m qubits in the register are the program qubits and the first k+1 qubits in the register are the cursor qubits. For the sqrtNOTcircuit, m=1 and k=2.

■ **Try me!**

The state ket[1,0,0,0] represents an input of 0 to the square root of NOT squared circuit (the 4th bit is a 0, the first 3 bits are cursor bits with the cursor initialized at its start position). So we expect to see the answer 1 because the SqtNOT(SqrtNOT(0)) = NOT(0) = 1.

```
       input = ket[0];
      cursor = ket[1,0,0];
initialState = Direct[cursor, input];
           m = 1;
           k = 2;
       evoln = EvolveQC[initialState, sqrtNOTcircuit];
      output = ExtractAnswer[m,k,evoln]
```

■ **Try me!**

Conversely, if we input a 1 we expect to see an output of 0.

```
       input = ket[1];
      cursor = ket[1,0,0];
initialState = Direct[cursor, input];
           m = 1;
           k = 2;
       evoln = EvolveQC[initialState, sqrtNOTcircuit];
      output = ExtractAnswer[m,k,evoln]
```

■ **Try me!**

If we input a 0 and a 1 simultaneously the output ought to be equally likely to be 0 or 1. Try running this experiment several times.

```
       input = 1/Sqrt[2] (ket[0] + ket[1]);
      cursor = ket[1,0,0];
initialState = Direct[cursor, Expand[input]];
           m = 1;
           k = 2;
       evoln = EvolveQC[initialState, sqrtNOTcircuit];
      output = ExtractAnswer[m,k,evoln]
```

The final state of the register is a superposition that weights the probability of obtaining a 0 as high as that of obtaining a 1. For example, take a particular final state and imagine measuring 20 identical copies of this state (this is just a thought experiment, you cannot copy an arbitrary quantum state exactly).

■ **Try me!**

```
Table[ExtractAnswer[m,k,evoln], {20}]
```

```
inputBit = 1/Sqrt[2] (ket[0] + ket[1]);
cursor = ket[1,0,0];
initialState = Direct[cursor, inputBit]
    m = 1;
    k = 2;
evoln = EvolveQC[1/Sqrt[2] ket[1,0,0,0] + 1/Sqrt[2] ket[1,0,0,1],
                 sqrtNOTcircuit];
output = ExtractAnswer[m,k,evoln]
```

■ **Code**

■

```
On[General::spell1]
```

■ Bras and Kets defined in here ...

Bibliography

[Abbott95] Abbot, P.(ed.) "Tricks of the Trade: Random Numbers," Mathematica Journal, Vol. 5, Issue 1 (1995), pp. 20–21.

[Adami96] Adami, C. and Cerf, N. "Capacity of a Noisy Quantum Channel," Kellogg Radiation Laboratory preprint MAP-206, California Institute of Technology (1996).

[Adleman94] Adleman, L. "Molecular Computation of Solutions to Combinatorial Problems," *Science*, Vol. 266, 11 November (1994), pp. 1021–1024.

[Aspect82] Aspect, A., Dalibard, J., and Roger, G. "Experimental Test of Bell's Inequalities Using Time-Varying Analyzers," *Physical Review Letters*, Vol. 49 (1982), pp. 1804–1807.

[Barenco95] Barenco, A. "A Universal Two-Bit Gate for Quantum Computation," *Proceedings Royal Society London*, Vol. 449A (1995), pp. 679–683.

[Barenco95a] Barenco, A., Deutsch, D., and Ekert, A. "Conditional Quantum Dynamics and Logic Gates," *Physical Review Letters*, Vol. 74, May 15 (1995), pp. 4083–4086.

[Barenco96] Barenco, A., Berthiaume, A., Deutsch, D., Ekert, A., Jozsa, R. and Macchiavello, C. "Stabilisation of Quantum Computations by Symmetrization," Los Alamos preprint archive, http://xxx.lanl.gov/archive/quant-ph/9604028 (1996).

[Bell64] Bell, J. "On the Einstein-Podolsky-Rosen Paradox," *Physics*, Vol. 1 (1964). pp. 195–200.

[Benioff80] Benioff, P. "The Computer as a Physical System: A Microscopic Quantum Mechanical Hamiltonian Model of Computers as Represented by Turing Machines," *Journal of Statistical Physics*, Vol. 22 (1980), pp. 563–591.

[Benioff81] Benioff, P. "Quantum Mechanical Hamiltonian Models of Discrete Processes," *Journal of Mathematical Physics*, Vol. 22 (1981), pp. 495–507.

[Benioff82] Benioff, P. "Quantum Mechanical Models of Turing Machines That Dissipate No Energy," *Physical Review Letters*, Vol. 48 (1982), pp. 1581–1585.

[Benioff82a] Benioff, P. "Quantum Mechanical Hamiltonian Models of Discrete Processes that Erase Their Own Histories: Applications to Turing Machines," *International Journal of Theoretical Physics* Vol. 21 (1982), pp. 177–202.

[Benioff86] Benioff, P. "Quantum Mechanical Hamiltonian Models of Computers," *Annals of the New York Academy Science*, Vol. 480 (1986), pp. 475–486.

[Bennett73] Bennett, C. "Logical Reversibility of Computation," *IBM Journal of Research and Development*, Vol. 17 (1973), pp. 525–532.

[Bennett87] Bennett, C. "Demons, Engines and the Second Law," *Scientific American*, November (1987), pp.108–116.

[Bennett89] Bennett, C. and Brassard, G. "The Dawn of a New Era for Quantum Cryptography: The Experimental Prototype is Working!" *SIGACT News*, Vol. 20, Fall (1989), pp. 78–82.

[Bennett92] Bennett, C., Bessette, F., Brassard, G. Salvail, L., and Smolin, J. "Experimental Quantum Cryptography," *Journal of Cryptology*, Vol. 5 (1992), pp. 3–28.

[Bennett92a] Bennett, C., Brassard, G. and Ekert, A. "Quantum Cryptography," *Scientific American*, October (1992), pp. 50–57.

[Bennett92b] Bennett, C., and Wiesner, S. "Communication via One- and Two-Particle Operators on Einstein-Podolsky-Rosen States," *Physical Review Letters*, Vol. 69 (1992), pp. 2881–2884.

[Bennett93] Bennett, C., Brassard, G., Crepeau, C., Jozsa, R., Peres, A., and Wootters, W. "Teleporting an Unknown Quantum State via Dual Classical and Einstein-Podolsky-Rosen Channels," *Physical Review Letters*, Vol. 70 (1993), pp. 1895–1899.

[Bennett95] Bennett, C. "Quantum Information and Computation," *Physics Today*, Vol. 48, October (1995), pp. 24–30.

[Berdnikov96] Berdnikov, A., Turtia, S., and Companger, A. "A MathLink Program for High Quality Random Numbers," *Mathematica Journal*, Vol. 6, Issue 3 (1996), pp. 65–69.

[Berman94] Berman, G., Doolen, G., Holm, D., and Tsifrinovich, V. "Quantum Computer on a Class of One-Dimensional Ising Systems," *Physics Letters A*, Vol. 193 (1994). pp. 444–450.

[Bernstein93] Bernstein, E. and Vazirani, U. "Quantum Complexity Theory," *Proceedings of the 25th Annual ACM Symposium on the Theory of Computing* (1993), pp. 11–20.

[Berthiaume92] Berthiaume, A. and Brassard, G. "The Quantum Challenge to Complexity Theory," *Proceedings of the 7th IEEE Conference on Structure in Complexity Theory* (1992), pp. 132–137.

[Berthiaume94] Berthiaume, A. and Brassard G. "Oracle Quantum Computing," *Journal of Modern Optics*, Vol. 41, No. 12, December (1994), pp. 2521–2535.

[Berthiaume94a] Berthiaume, A., Deutsch, D., and Josza, R. "The Stabilisation of Quantum Computations," *Proceedings of the 4th Workshop on Physics and Computation*, PhysComp94, Dallas, TX, IEEE Computer Society Press, Los Alamitos, CA, November (1994), pp. 60–62.

[Biafore94] Biafore, M. "Can Quantum Computers Have Simple Hamiltonians," *Proceedings of the 4th Workshop on Physics and Computation*, PhysComp94, Dallas, TX, IEEE Computer Society Press, Los Alamitos, CA, November (1994), pp. 63–68.

[Brassard96] Brassard, G. "Teleportation as a Quantum Computation," *Proceedings of the 4th Workshop on Physics and Computation*, PhysComp96, extended abstract, Los Alamos preprint archive, http://xxx.lanl.gov/archive/quant-ph/9605035 (1996).

[Braunstein92] Braunstein, S., Mann, A., and Revzen, M. "Maximal Violation of Bell Inequalities for Mixed States," *Physical Review Letters*, Vol. 68 (1992), pp. 3259–3261.

[Braunstein96] Braunstein, S. "Perfect Quantum Error Correction Coding in 26 Laser Pulses," Los Alamos preprint archive, http://xxx.lanl.gov/archive/quant-ph/9604036 (1996).

[Buzek96] Buzek, V. and Hillery, M. "Quantum Copying: Beyond the No-cloning Theorem," Los Alamos preprint archive, http://xxx.lanl.gov/archive/quant-ph/9607018 (1996).

[Buzek97] Buzek, V., Braunstein, S., Hillery, M. and Bruβ, D. "Quantum Copying: A Network," Los Alamos preprint archive, http://xxx.lanl.gov/archive/quant-ph/9703046 (1997).

[Calderbank96] Calderbank, A., Rains, E., Shor, P., and Sloane, N. "Quantum Error Correction and Orthogonal Geometry," Los Alamos preprint archive, http://xxx.lanl.gov/archive/quant-ph/9605005 (1996).

[Cerf96a] Cerf, N. and Adami, C. "Quantum Mechanics of Measurement," Kellogg Radiation Laboratory preprint MAP-198, California Institute of Technology, May (1996). Also available at the Los Alamos preprint archive, http://xxx.lanl.gov/archive/quant-ph/9605002.

[Cerny93] Cerny, V. "Quantum Computers and Intractable (NP-complete) Computing Problems," *Physical Review A*, Vol. 48, July (1993), pp. 116–119.

[Chaitin77] Chaitin, G. "Algorithmic Information Theory," *IBM Journal of Research and Development*, Vol. 21, July (1977), pp. 350–359.

[Chuang95] Chuang, I., Yamamoto, Y. "A Simple Quantum Computer," Los Alamos preprint archive, http://xxx.lanl.gov/archive/quant-ph/9505011 (1995).

[Church41] Church, A. "The Calculi of Lambda-Conversion," *Annals of Mathematics Studies*, No. 6, Princeton University Press, (1941).

[Cirac95] Cirac, J. and Zoller, P. "Quantum Computations with Cold Trapped Ions," *Physical Review Letters*, Vol. 74 (1995), pp.4091–4094.

[Clauser69] Clauser, J., Horne, M., Shimony, A., and Holt, R. "Proposed Experiment to Test Local Hidden-Variable Theories," *Physical Review Letters*, Vol. 23 (1969), pp. 880–884.

[Cory96] Cory, D., Fahmy, A., and Havel, T., "Nuclear Magentic Resonance Spectoroscopy: An Experimentally Accessible Paradigm for Quantum Computing," *Proceedings of the 4th Workshop on Physics and Computation*, PhysComp96, Boston University, 22–24 November, New England Complex System Institute (1996). pp. 87–91.

[Crandall96] Crandall, R. "*Topics in Advanced Scientific Computation*," TELOS (Springer Verlag), New York, (1996), pp. 125–126.

[Davidovich93] Davidovich, L., Maali, A., Brune, M., Raimond, J., and Haroche, S. "Quantum Switches and Nonlocal Microwave Fields," *Physical Review Letters*, Vol. 71, No. 15, 11 October (1993), pp. 2360–2363.

[Deutsch82] Deutsch, D. "Is There a Fundamental Bound on the Rate at Which Information Can Be Processed?," *Physical Review Letters*, Vol. 42 (1982), pp. 286–288.

[Deutsch85] Deutsch, D. "Quantum Theory, the Church-Turing Principle, and the Universal Quantum Computer," Proceedings Royal Society London, Vol. A400 (1985), pp. 97–117.

[Deutsch89] Deutsch, D. "Quantum Communication Thwarts Eavesdroppers," *New Scientist*, December 9 (1989), pp.25–26.

[Deutsch89a] Deutsch, D. "Quantum Computational Networks," *Proceedings of the Royal Society of London*, A 425, (1989), pp.73–90.

[Deutsch92] Deutsch, D. and Jozsa, R. "Rapid Solution of Problems by Quantum Computation," *Proceedings Royal Society London*, Vol. 439A (1992), pp. 553–558.

[Deutsch92a] Deutsch, D. "Quantum Computation," *Physics World*, Vol. 5, June (1992), pp. 57–61.

[DiVincenzo95] DiVincenzo, D. "Quantum Computation," *Science*, Vol. 270, 13 October (1995), pp. 255–261.

[DiVincenzo95a] DiVincenzo, D. "Two-Bit Gates are Universal for Quantum Computation," *Physical Review A*, Vol. 51 (1995) pp.1015–1022.

[DiVincenzo96] DiVincenzo, D. and Shor, P. "Fault-Tolerant Error Correction with Efficient Quantum Codes," *Physical Review Letters*, Vol. 77 (1996), pp. 3260–3263.

[Drexler92] Drexler, K. E. "*Nanosystems: Molecular Machinery, Manufacturing and Computation*," John Wiley & Sons, New York (1992), pp. 342–371.

[Duan96] Duan, L. and Guo, G. "Reducing Decoherence in Quantum Computer Memory with all Quantum Bits Coupling to the Same Environment," Los Alamos preprint archive, http://xxx.lanl.gov/archive/quant-ph/9612003 (1996).

[Durr96] Durr, C. and Hoyer, P. "A Quantum Algorithm for Finding the Minimum," Los Alamos preprint archive, http://xxx.lanl.gov/archive/quant-ph/9607014 (1996).

[Einstein35] Einstein, A., Podolsky, B., and Rosen, N. "Can Quantum-Mechanical Description of Physical Reality be Considered Complete?" *Physical Review*, Vol. 47 (1935), pp. 777–780.

[Ekert91] Ekert, A. "Quantum Cryptography Based on Bell's Theorem," *Physical Review Letters*, Vol. 67 (1991), pp. 661–663.

[Ekert92] Ekert, A., Rarity, J., Tapster, P., and Palma, G. "Practical Quantum Cryptography Based on Two-Photon Interferometry," *Physical Review Letters*, Vol. 69, 31 August (1992), pp. 1293–1295.

[Ekert92a] Ekert, A. "Beating the Code Breakers," *Nature*, Vol. 358, July 2, (1992), pp. 14–15.

[Ekert93] Ekert, A. "Quantum Keys for Keeping Secrets," *New Scientist*, January 16, (1993), pp. 24–28.

[Ekert96] Ekert, A. and Macchiavello, C. "Quantum Error Correction for Communication," *Physical Review Letters*, Vol. 77 (1996), pp. 2585–2588.

[Ferrenberg92] Ferrenberg, A., Landau, D., and Wong, Y. "Monte Carlo Simulations: Hidden Errors from 'Good' Random Number Generators," *Physical Review Letters*, Vol. 69 (1992), pp. 3382–3384.

[Feynman60] Feynman, R. P. "There's Plenty of Room at the Bottom," *Engineering and Science*, Vol. 23, (1960) pp. 22–36.

[Feynman82] Feynman, R. "Simulating Physics with Computers," *International Journal of Theoretical Physics*, Vol. 21, Nos. 6/7 (1982), pp. 467–488.

[Feynman85] Feynman, R. "Quantum Mechanical Computers," *Optics News*, Vol. 11 (1985), pp. 11–20.

[Fredkin82] Fredkin, E. and Toffoli, T. "Conservative Logic," *International Journal of Theoretical Physics*, Vol. 21 (1982), pp. 219–253.

[Gardner79] Gardner, M. "Mathematical Games: The Random Number Omega Bids Fair to Hold the Mysteries of the Universe," *Scientific American*, November (1979), pp. 20–30.

[Gershenfeld96] Gershenfeld, N., Chuang, I., and Lloyd, S., "Bulk Quantum Computation," *Proceedings of the 4th Workshop on Physics and Computation*, PhysComp96, Boston University, 22–24 November, New England Complex System Institute (1996) p. 134.

[Gill77] Gill, J. "Computational Complexity of Probabilistic Turing Machines," SIAM Journal of Computing, Vol. 6, No. 4, December (1977), pp. 675–695.

[Giulini96] Giulini, D., Joos, E., Kiefer, C., Kupsch, J., Stamatescu, I.-O., Zeh, H. *"Decoherence and the Appearance of a Classical World in Quantum Theory,"* Springer Verlag (1996).

[Gödel31] Gödel, K. "Über formal unentscheidbare Sätze der Principia Mathematica und verwandter Systeme I," *Monatshefte für Mathematik und Physik*, Vol. 38, (1931), pp. 173–98.

[Grimmett92] Grimmett, G. and Stirzaker, D. "Probability and Random Processes, 2nd ed.," Oxford University Press, Oxford (1992), p. 175.

[Grover96] Grover, L. "A Fast Quantum Mechanical Algorithm for Database Search," *Proceedings of the 28th Annual ACM Symposium on the Theory of Computing* (1996), pp. 212–219.

[Grover96a] Grover, L. "A Fast Quantum Mechanical Algorithm for Estimating the Median," AT&T Bell Labs preprint (1996).

[Harel92] Harel, D. "Algorithmics: The Spirit of Computing," Second Edition, Addison Wesley Publishing Company, Inc. (1992).

[Hartle83] Hartle, J. and Hawking, S. "The Wave Function of the Universe," Physical Review D., Vol. 28 (1983) p.2960

[Hagelstein95] Hagelstein, P., Margolus, N., and Biafore, M. "Towards a Quantum Computer Project," MIT Research Laboratory of Electronics, unpublished report (1995).

[Hayes93] Hayes, B. "The Wheel of Fortune," *American Scientist*, Vol. 81 (1993), pp. 114–118.

[Hecht74] Hecht, E. and Zajac, A. "*Optics*," Addison-Wesley, Reading, MA (1974).

[Heisenberg27] Heisenberg, W. "Uber den anschaulichen Inhalt der quantentheoretischen Kinematik und Mechanik," Zeitschrift fur Physik, Vol. 43 (1927) pp. 172-198. English translation in "Quantum Theory and Measurement," Wheeler, J. A. and Zurek, W. H. (eds.), Princeton, New Jersey (1983) pp. 62–84.

[Hille92] Hille, B. "Ionic Channels of Excitable Membranes, 2nd ed.," Sinauer Associates, Sunderland, MA (1992), p. 389.

[Hopcroft84] Hopcroft, J. "Turing Machines," *Scientific American*, May (1984), pp. 86–98.

[Hughes95] Hughes, R. J., Alde, D. M., Dyer, P., Luther, G. G., Morgan, G. L., and Schauer, M. "Quantum Cryptography," *Contemporary Physics*, Vol. 36 (1995), pp. 149–163.

[Hughes96] Hughes, R., James, D., Knill, E., Laflamme, R., and Petschek, A. "Decoherence Bounds on Quantum Computation with Trapped Ions," *Physical Review Letters*, Vol. 77 (1996) pp. 3240–3243.

[Hutcheson96] Hutcheson, G. D., Hutcheson, J. D. "Technology and Economics in the Semiconductor Industry," *Scientific American*, January (1996), pp. 54–62.

[James90] James, F. "A Review of Pseudorandom Number Generators," *Computer Physics Communications*, Vol. 60 (1990), pp. 329–344.

[Joos85] Joos, E. and Zeh, H. "The Emergence of Classical Properties Through Interaction with the Environment," *Zeitschrift fur Physik B*, Vol. 59 (1985), pp. 223–243.

[Jozsa91] Jozsa, R. "Characterizing Classes of Functions Computable by Quantum Parallelism," *Proceedings Royal Society London*, Vol. A435 (1991), pp. 563–574.

[Jozsa94] Jozsa, R. and Schumacher, B. *Journal of Modern Optics*, Vol. 41 (1994), p. 2343

[Keyes70] Keyes, R. W. and Landauer, R. "Minimal Energy Dissipation in Logic," *IBM Journal of Research and Development*, March (1970), pp. 152–157.

[Keyes81] Keyes, R. W. "Fundamental Limits in Digital Information Processing," *Proceedings of the IEEE*, Vol. 69, February (1981) pp. 267–278.

[Keyes88] Keyes, R. W. "Miniaturization of Electronics and its Limits," *IBM Journal of Research and Development*, Vol. 32, January (1988), pp. 24–28.

[Keyes89] Keyes, R. W. "Physics of Digital Devices," *Reviews of Modern Physics*, Vol. 61, No. 2, April (1989) pp279–287.

[Keyes89a] Keyes, R. W. "Making Light Work of Logic," *Nature*, Vol. 340 (1989) p. 19.

[Keyes92] Keyes, R. W. "The Future of Solid-State Electronics," *Physics Today*, Vol. 45 August (1992), pp. 42–48.

[Keyes93] Keyes, R. W. "The Future of the Transistor," *Scientific American*, June (1993), pp. 70–78.

[Kitaev95] Kitaev, A. "Quantum Measurements and the Abelian Stabiliser Problem," Los Alamos preprint archive, http://xxx.lanl.gov/archive/quant-ph/9511026 (1995).

[Knill96] Knill, E. and Laflamme, R. "Concatenated Quantum Codes," Los Alamos preprint archive, http://xxx.lanl.gov/archive/quant-ph/9608012 (1996).

[Knill96a] Knill, E., Laflamme, R., and Zurek, W. "Accuracy Threshold for Quantum Computation," Los Alamos preprint archive, http://xxx.lanl.gov/archive/quant-ph/9610011 (1996).

[Laflamme96] Laflamme, R., Miquel, C., Paz, P., and Zurek, W. "Perfect Quantum Error Correcting Code," *Physical Review Letters*, Vol. 77 (1996), pp. 198–201.

[Landauer91] Landauer, R. "Information is Physical," *Physics Today*, Vol. 44, May (1991), pp. 23–29.

[Landauer95] Landauer, R. "Is Quantum Mechanics Useful?" *Philosophical Transactions Royal Society*, Vol. 353A (1995), pp. 367–376.

[Lenstra90] Lenstra, A., Manasse, M., and Pollard, J. "The Number Field Sieve," *Proceedings of the 22nd ACM Symposium on the the Theory of Computing* (1990), pp. 564–572.

[Lenstra93] Lenstra, A. and Lenstra, H. "The Development of the Number Field Sieve," *Lecture Notes in Mathematics 1554*, Springer-Verlag, New York (1993).

[Lloyd93] Lloyd, S. "A Potentially Realizable Quantum Computer," *Science*, Vol. 261, 17 September (1993), pp. 1569–1571.

[Lloyd93a] Lloyd, S. "A Potentially Realizable Quantum Computer," unpublished manuscript 1993 most of which found its way into [Lloyd93].

[Lloyd93b] Lloyd, S. "Quantum Mechanical Computers and Uncomputability," *Physical Review Letters*, Vol. 71, (1993), pp.943–946.

[Lloyd94] Lloyd, S. "Envisioning a Quantum Supercomputer," *Science*, Vol. 263, February, (1994), p.695.

[Lloyd94a] Lloyd, S. "Necessary and Sufficient Conditions for Quantum Computation," *Journal of Modern Optics*, Vol. 41, (1994), pp.2503–2520.

[Lloyd95] Lloyd, S. "Quantum-Mechanical Computers," *Scientific American*, October (1995), pp.140–145.

[Lloyd96] Lloyd, S. "Universal Quantum Simulators," *Science*, Vol. 273, 23 August (1996), pp. 1073–1078.

[Lloyd96a] Lloyd, S. "The Capacity of the Noisy Quantum Channel," Los Alamos preprint archive, http://xxx.lanl.gov/archive/quant-ph/960415, (1996)

[Maeder93] Maeder, R. "Turing Machines and Code Optimization," *The Mathematica Journal*, Vol. 3, Issue 3, Summer (1993), pp. 36–45.

[Malone95] Malone, "The Microprocessor: A Biography," TELOS (The Electronic Library of Science, Springer-Verlag), Santa Clara, (1995).

[Marand95] Marand, C. and Townsend, P. "Quantum Key Distribution Over Distances as Long as 30km," *Optics Letters*, Vol. 20, No. 16, 15 August (1995), pp. 1695–1697.

[Margolus96] Margolus, N. and Levitin, L. "The Maximum Speed of Dynamical Evolution," *Proceedings of the 4th Workshop on Physics and Computation*, PhysComp96, Boston, MA (1996), pp. 208–211.

[Marsaglia68] Marsaglia, G. "Random Numbers Fall Mainly in the Planes," *Proceedings of the National Academy of Sciences*, Vol. 61, September (1968), pp. 25–28.

[Marsaglia90] Marsaglia, G., Narasimhan, B., and Zaman, A. "A Review of Pseudorandom Number Generators," *Computer Physics Communications*,Vol. 60 (1990), pp. 345–349.

[Metropolis53] Metropolis, N., Rosenbluth, A., Rosenbluth, M., Teller, A., and Teller, E. "Equation of State Calculations by Fast Computing Machines," *Journal of Chemical Physics*, Vol. 21 (1953), pp. 1087–1092.

[Milburn89] Milburn, G. "Quantum Optical Fredkin Gate," *Physical Review Letters*,Vol. 62 (1989), pp. 2124–2127.

[Monroe95] Monroe, C., Meekhof, D., King, B., Itano, W., and Wineland, D. "Demonstration of a Fundamental Quantum Logic Gate," *Physical Review Letters*, Vol. 75, No. 25 (1995) pp. 4714–4717.

[Moore90] Moore, C. "Predictability and Undecibility in Dynamical Systems," *Physical Review Letters*, Vol. 64 (1990), pp. 2354–2357.

[Morrison61] Morrison, P. and Morrison, E. "*Charles Babbage and his Calculating Engines*," Dover, New York, (1961).

[Palfreman91] Palfreman, J. and Swade, D. "*The Dream Machine: Exploring the Computer Age*," BBC Books, London (1991), p. 20.

[Park88] Park, S. and Miller, K. "Random Number Generators: Good Ones are Hard to Find," *Communications of the ACM*, Vol. 31 (1988), pp. 1192–1201.

[Penrose89] Penrose, R. *"The Emperor's New Mind,"* Oxford University Press, Oxford (1989).

[Peres82] Peres, A. and Zurek, W. "Is Quantum Theory Universally Valid?," *American Journal of Physics*, Vol. 50, September (1982), pp. 807–810.

[Peres85] Peres, A. "Einstein, Godel, Bohr," *Foundations of Physics*, Vol. 15 (1985), pp. 201–205.

[Peres85a] Peres, A. "Reversible Logic and Quantum Computers," *Physical Review A32*, (1985), pp. 3266-3276.

[Peres96] Peres, A. "Error Symmetrization in Quantum Computers," Los Alamos preprint archive, http://xxx.lanl.gov/archive/quant-ph/9605009 (1996).

[Plenio96] Plenio, B., Vedral, V., and Knight, P. "Optimal Realistic Quantum Error Correcting Code," Los Alamos preprint archive, http://xxx.lanl.gov/archive/quant-ph/9603022 (1996).

[Post44] Post, E. "Recursively Enumerable Sets of Positive Integers and their Decision Problems," *Bulletin of the American Mathematical Society*, Vol. 50, (1944), pp. 284–316.

[Pratt87] Pratt, V. *"Thinking Machines: The Evolution of Artificial Intelligence,"* Basil Blackwell, Oxford, (1987).

[Preskill97] Preskill, J. "Reliable Quantum Computer," Los Alamos preprint archive http://xxx.lanl.gov/archive/quant-ph/970503/ (1997)

[Rarity94] Rarity, J., Ownes, P., and Tapster, P. "Quantum Random-Number Generation and Key Sharing," *Journal of Modern Optics*, Vol. 41 (1994), pp. 2435–2444.

[Rivest78] Rivest, R., Shamir, A., and Adelman, L. "A Method for Obtaining Digital Signatures and Public Key Cryptosystems," *Communications of the ACM*, Vol. 21 (1978), pp. 120–126.

[Robb93] Robb, T "Animating Schrödinger's Equation," *Wolfram Research MathSource CD-ROM*, http://www.wolfram.com, entry 0204-679 (1993) (also on the CD-ROM that came with this book).

[Rudolph96] Rudolph, T. "Quantum Computing Hamiltonian Cycles," Los Alamos preprint archive, http://xxx.lanl.gov/archive/quant-ph/9603001, (1996).

[Schrödinger26] Schrödinger, E. Annalen der Physik, Vol. 79 (1926), pp. 361–376.

[Schumacher95] Schumacher, B. *Physical Review A*, Vol. 51, (1995), p.2738

[Schwarzschild96] Schwarzschild, B. "Labs Demonstrate Logic Gates for Quantum Computation," *Physics Today*, March (1996), pp. 21–23.

[Shankar94] Shankar, R. "Principles of Quantum Mechanics," Second Edition, Plenum, New York (1994), p. 148.

[Shannon49] Shannon, C. and Weaver, W. "A Mathematical Theory of Communication," University of Illinois Press (1949).

[Shannon53] Shannon, C. "Computers and Automata," *Proceedings of the I. R. E.*, Vol. 41, October (1953), pp. 1235–1241.

[Shapiro90] Shapiro, S. (ed.) "Church's Thesis," *Encyclopedia of Artificial Intelligence*, John Wiley & Sons, New York (1990), pp. 99–100.

[Shor94] Shor, P. "Algorithms for Quantum Computation: Discrete Logarithms and Factoring," *Proceedings 35th Annual Symposium on Foundations of Computer Science* (1994), pp. 124–134.

[Shor95] Shor, P. "Scheme for Reducing Decoherence in Quantum Computer Memory," *Physical Review A*, Vol 52, October (1995), pp. R2493–R2496.

[Shor96] Shor, P. "Fault-Tolerant Quantum Computation," *Proceedings of the 37th Conference on Foundations of Computer Science*, Burlington, VT, October (1996), pp. 56–65.

[Silverman87] Silverman, R. *Mathematical Computing* Vol. 48 (1987), p. 329.

[Simon94] Simon, D. "On the Power of Quantum Computation," *Proceedings of the 35th Annual IEEE Symposium on Foundations of Computer Science* (1994), pp. 116–123.

[Sleator95] Sleator, T. and Weinfurter, H. "Realizable Universal Quantum Logic Gates," *Physical Review Letters*, Vol. 74, Number 20, 15 May (1995), pp. 4087–4090.

[Smullyan92] Smullyan, R. *"Gödel's Incompleteness Theorems,"* Oxford University Press, Oxford (1992).

[Swade93] Swade, D. "Redeeming Charles Babbage's Mechanical Computer," *Scientific American*, February (1993), pp. 86–91.

[Szilard29] Szilard, L. "On the Decrease of Entropy in a Thermodynammic System by the Intervention of Intelligent Beings," *Zeitschrift fur Physics*, Vol. 53 (1929), pp. 840–856.

[Teich88] Teich, W., Obermayer, K., and Mahler, G. "Structural Basis of Multistationary Quantum Systems. II. Effective Few-Particle Dynamics," *Physical Review B*, Vol. 37, No. 14 (1988) pp.8111–8120.

[Teich90] Teich, W. and Mahler, G. "Information Processing at the Molecular Level: Possible Realizations and Physical Constraints," *Complexity, Entropy, and the Physics of Information*, SFI Studies in the Sciences of Complexity, Vol. VIII, W. Zurek (ed.), Addison-Wesley, Reading, MA (1990) pp. 289–300.

[Tipler86] Tipler, F. "Cosmological Limits on Computation," *International Journal of Theoretical Physics*, Vol. 25 (1986), pp. 617–661.

[Townsend93] Townsend, P., Rarity, J., and Tapster, P. "Single Photon Interference in 10 km-Long Optical Fibre Interferometer," *Electronics Letters*, Vol. 29, 1 April (1993), pp. 634–635.

[Townsend93a] Townsend, P., Rarity, J., and Tapster, P. "Enhanced Single Photon Fringe Visibility in a 10 km-Long Prototype Quantum Cryptography Channel," *Electronics Letters*, Vol. 29, July (1993), pp. 1291–1293.

[Traub94] Traub, J. and Wozniakowski, H. "Breaking Intractability," *Scientific American*, January (1994), pp. 102–107.

[Tuckwell95] Tuckwell, H. "Elementary Applications of Probability Theory, 2nd ed.," Chapman & Hall, London (1995), pp. 87–88.

[Turchette95] Turchette, Q., Hood, C., Lange, W., Mabuchi, H., and Kimble, H. "Measurement of Conditional Phase Shifts for Quantum Logic," *Physical Review Letters*, Vol. 75, No. 25 (1995), pp. 4710–4713.

[Turing37] Turing, A. "On Computable Numbers with an Application to the Entscheidungsproblem," *Proceedings of the London Mathematical Society*, Vol. 42 (1937), pp. 230–265, erratum in 43 (1937), pp. 544–546.

[Unruh95] Unruh, W. "Maintaining Coherence in Quantum Computers," *Physical Review A*, Vol. 51, (1995), pp. 992–997.

[Vazirani94] Vazirani, U. Quotation from a newspaper article by Tom Siegfried, Science Editor of the Dallas Morning News, after a 1994 conference sponsored by the Santa Fe Institute, Los Alamos National Laboratory and the University of New Mexico.

[Welsh88] Welsh, D. "*Codes and Cryptography*," Oxford University Press, Oxford (1988), p. 183.

[Wiesner96] Wiesner, S. "Simulations of Many-Body Quantum Systems by a Quantum Computer," Los Alamos preprint archive, http://xxx.lanl.gov/archive/quant-ph/9603028 (1996).

[Williams94] Williams, C. and Hogg, T. "Exploiting the Deep Structure of Constraint Problems," *Artificial Intelligence Journal*, Vol. 70 (1994), pp. 73–117.

[Winograd72] Winograd, T. "*Understanding Natural Language*," Academic Press, New York (1972).

[Yao93] Yao, A. "Quantum Circuit Complexity," *Proceedings of the 34th IEEE Symposium on Foundations of Computer Science*, IEEE Computer Society Press, Los Alamitos, CA (1993), pp. 352–360.

[Zalka96] Zalka, C. "Efficient Simulation of Quantum Systems by Quantum Computers," Los Alamos National Laboratory, preprint archive, quant-ph/9603026 (1996).

[Zeh95] Zeh, H. "The Program of Decoherence: Ideas and Concepts," Los Alamos preprint archive, http://xxx.lanl.gov/archive/quant-ph/9506020, (1995).

[Zurek81] Zurek, W. "Pointer Basis of Quantum Apparatus: Into what Mixture does the Wave Packet Collapse," *Physical Review D*, Vol. 24, (1981) pp. 1516–1524.

[Zurek82] Zurek, W. "Environment-induced Superselection Rules," *Physical Review D*, Vol. 26 (1982), pp.1862–1880.

[Zurek84] Zurek, W. "Reversibility and Stability of Information Processing Systems," *Physical Review Letters*,Vol. 53 (1984), pp. 391–394.

[Zurek86] Zurek, W. H. "Reduction of the Wave Packet and Environment-Induced Superselection," *Annals of the New York Academy of Sciences*, Vol. 480 (1986) pp.89–97.

[Zurek91] Zurek, W. H. "Decoherence and the Transition from Quantum to Classical," *Physics Today*, October (1991), pp. 36–44.

[Zurek93] Zurek, W. "Preferred States, Predictability, Classicality and the Environment-Induced Decoherence," *Progress of Theoretical Physics*, Vol. 89 February 1993, pp.281–312.

[Zurek96] Zurek, W. and Laflamme, R. "Quantum Logical Operations on Encoded Qubits," *Physical Review Letters*, Vol. 77 (1996), pp. 4683–4686.

Index